DYNAMICS OF GAS–SURFACE INTERACTIONS

Advances in Gas-Phase Photochemistry and Kinetics

Each title in this series aims to provide critical reviews on a specific area of current interest, at the interface of molecular spectroscopy, photochemistry, and chemical kinetics.

> Molecular Photodissociation Dynamics
> Bimolecular Collisions
> Dynamics of Gas–Surface Interactions

How to obtain titles in this series
'Dynamics of Gas–Surface Interactions' is the third title in the series 'Advances in Gas-Phase Photochemistry and Kinetics'. Titles in this series may be obtained from:
    The Royal Society of Chemistry
    Turpin Transactions Ltd.
    Blackhorse Road
    Letchworth
    Herts. SG6 1HN

Telephone: Letchworth (0462) 672555
Fax: (0462) 480947
Telex: 825372

WITHDRAWN
FROM
UNIVERSITIES
AT
MEDWAY
LIBRARY

9409763 MBCSBN

Advances in
Gas-Phase Photochemistry and Kinetics

Dynamics of Gas–Surface Interactions

Edited by
C. T. Rettner
IBM Almaden Research Center
San Jose

M. N. R. Ashfold
School of Chemistry
University of Bristol

A Catalogue record for this book is available from the British Library

ISBN: 0-85186-853-3

Front Cover Illustration
Schematic representation of an H_2 molecule descending over the centre site of a surface (Courtesy, S. Holloway)

© The Royal Society of Chemistry 1991

*All Rights Reserved
No part of this book may be reproduced or transmitted in any form or by any means—graphic, electronic, including photocopying, recording, taping, or information storage and retrieval systems—without written permission from The Royal Society of Chemistry*

Published by The Royal Society of Chemistry,
Thomas Graham House, Science Park, Cambridge CB4 4WF

Typeset by Computape (Pickering) Ltd., North Yorkshire
and printed in Great Britain by Bookcraft (Bath) Ltd.

Preface

In the late summer of 1989, the two of us enjoyed a sunny reunion by the banks of the river Thames. We strolled with our families, chatting about the twists and turns in our lives since the late 1970s when we obtained our Ph.D.s under the supervision of Professor John Simons, then at Birmingham University. The basic idea for this volume originated around Staines, and the list of Chapter titles was all but complete by the time we had carried our small children the two miles to Penton Hook. From the beginning we believed that a volume on gas–surface interactions would be interesting and timely, but had some difficulty seeing how it could be justified within the series title, which specifies gas-phase phenomena. We decided to press forward regardless. Hence the contradiction of the title page. We nevertheless believe that the collection of Chapters assembled here should be of great interest not only to those researching the gas–surface interface but also to the larger gas phase chemical physics community.

In all other respects this volume follows closely along the lines of the two previous works. We feel that we have remained true to the goal of the Series by providing a forum for internationally acknowledged experts to present overviews of their respective fields. The individual Chapter topics and contents have been chosen to cover a wide range of issues relevant to studies of both dynamics and kinetics at the gas–surface interface. The emphasis throughout is towards surface chemistry, rather than, say, structure or spectroscopy. It is our intention that the full volume should be useful to both the surface science specialist and those seeking an introduction to this exciting field. It is intended that each of the Chapters should be understandable to graduates with a background in chemical physics, and we hope that the Volume will prove useful as a graduate level text.

The first Chapter reviews the basic process of energy transfer in gas–surface collisions. An understanding of such fundamental interactions is a prerequisite for a detailed appreciation of essentially all that follows. In order to predict the adsorption probability for an incident molecule, for example, we need to know how incidence kinetic energy can couple to surface excitations and/or to the internal degree of freedom in the molecule. Harris considers the dynamics of energy transfer in gas–surface collisions within the framework of a number of carefully chosen models, beginning with the case of energy transfer to a harmonic linear chain. Here the coupling to the chain usefully approximates the response of the lattice. It is shown that, for qualitative purposes, this can often be neglected, allowing the interaction to be modelled as a binary collision. Energy transfer to the rotational degree of freedom is considered in detail, in terms of a model anisotropic interaction potential. This largely classical treatment is complemented by an appraisal of the role of quantum effects and a consideration of the validity of classical expressions. One important additional feature in the quantum treatment is the possibility of purely elastic scattering. The specific treatment is restricted to a quantum lattice, which is again modelled as a linear chain, treating the particle's trajectory classically. Here the forced oscillator model is introduced. This chapter ends on a 'wondrously esoteric' note, related to the infra-red catastrophe.

Next we move on to the dynamics of dissociative chemisorption, which may be considered the simplest example of gas–surface chemistry, involving the breaking of molecular bonds in favour of new bonds to a surface. Since dissociated species can react to form new compounds, it is not surprising that this process is a key step in many complex reaction mechanisms, including most examples of heterogeneous catalysis. However, the molecular dynamics of this elementary reaction are deceptively complex. DePristo summarizes the various theoretical approaches to modelling this process, and reports on the considerable progress that has been made over the past few years. He begins by describing the general characteristics of gas–surface interaction potentials and proceeds to outline the construction of model potentials suitable for dynamical studies. He then reviews dynamical theories including approaches based on both quantum and classical mechanics. This is followed by a number of illustrations based on specific examples, notably the $H_2/Ni(100)$ and $N_2/W(110)$ systems.

Chapter 3 focuses specifically on the importance of quantum effects in gas–surface collisions and covers both energy transfer and dissociative chemisorption. As such there is some overlap with both of the first two chapters, but the bulk of the text is complementary, offering a distinctly different perspective. Holloway also opens with a section on potential energy surfaces. He then outlines his approach to solving the (quantum) dynamics before moving on to discuss specific case studies. Not surprisingly, these mostly concern H_2, for which quantum effects often dominate. Here we expect to see important effects due to zero-point energy, below barrier tunnelling, and coherent scattering, for example. It is less obvious that the scattering of NO requires a quantum treatment. Thus the detailed comparison of quantum and classical predictions for the $NO/Ag(111)$ system are

Preface

particularly interesting. One might be even more surprised to learn that the dissociation of methane and even nitrogen may proceed by a tunnelling mechanism. The dissociation of hydrogen on copper is considered in great detail, being treated as a model system to explore a range of dynamical effects.

This provides both motivation and useful background to the ensuing Chapter in which Hayden presents an overview of work on the hydrogen/copper system. He opens with a justification for our including an entire Chapter devoted to a single system. The interaction of hydrogen with copper has been the focus of many different studies of adsorption and desorption, as well as a host of theoretical treatments, spanning some 150 years. This review is also most timely, since much of the most exciting activity has occurred over the past two years. An important aspect of this Chapter is the relationship between adsorption and desorption, via the principle of detailed balance which is commonly invoked in discussion of gas–surface dynamics. It is fitting that one of the first detailed applications of this principle in this connection was for this very system, more than sixteen years ago. Thus angular and velocity distributions for desorbtion are discussed with reference to the variation of the dissociation probability with incidence energy and angle. Much of the recent work has been concerned with the effect of vibrational energy on dissociation, as was also apparent from Holloway's Chapter. Hayden explains how molecular beam techniques offer an important (if somewhat limited) degree of control in this respect; allowing the degree of vibrational excitation of the incident molecules to be varied separately from their kinetic energy. He concludes by reviewing the implications of the principle of detailed balance in the light of multi-dimensional potential energy surfaces and by commenting on the limitations of current methodology.

In Chapter 5 we turn to the kinetics of surface reactions. We will see that studies of kinetics and dynamics are mutually complementary, in much the same manner as for purely gas-phase processes. Dynamical knowledge cannot be applied to the prediction of the rates of gas–surface processes without an understanding of the kinetics relevant to this interface. Conversely, such an understanding can be used to deduce dynamical parameters from kinetic measurements. Weinberg presents an overview of the kinetics of gas–surface reactions, beginning with a fairly general discussion of the different broad categories of these processes. Direct reactions are compared with those in which molecules first trap on the surface before going on to react. Similarly, bimolecular reactions in which one reagent strikes directly from the gas phase are compared with those which occur between two adsorbed species. It is argued that the direct reactions should be relatively unimportant in both cases. The process by which these reagents become accommodated, trapping, is considered in considerable detail. The main text ends with a case study of a specific example of trapping-mediated dissociation.

We continue with kinetics in Chapter 6 which focuses on thermal desorption. This is arguably the most important kinetic process at the gas–surface interface. For example, the technique of temperature programmed desorption (TPD) is employed in a large fraction of all studies of gas–surface interactions. Kreuzer and Payne begin with a fairly general discussion of experimental methods and of TPD measurements. Next they consider desorption under quasi-equilibrium conditions, where it is assumed that the remaining surface coverage maintains an equilibrium distribution appropriate to the desorption temperature. This assumption breaks down when the time constant for surface diffusion is not appreciably faster than that for desorption. They proceed to examine more general situations using methods of non-equilibrium thermodynamics and a kinetic lattice gas model. The Chapter ends with a section on physisorption kinetics where new experimental results (from Menzel's group in Munich) are presented, displaying unprecedented resolution in thermal desorption spectrometry.

Chapter 7 concerns desorption by a distinctly different mechanism, the process of electron-stimulated desorption (ESD). This article by Ramsier and Yates, and the last Chapter, by Polanyi and Rieley, illustrate the importance of electronic transitions in promoting chemistry at the gas–surface interface. Ramsier and Yates provide an extensive review of the ESD process, beginning with consideration of the different mechanisms that have been proposed to account for this phenomenon. Next they provide a summary of experimental details, which includes a description of the measurement of the angular distributions of ions and electronically excited neutrals that are produced in ESD. The technique of ESD Ion Angular Distribution (ESDIAD) has its origins in a 1974 paper that Yates co-authored. Moreover, Yates and his co-workers have been responsible for many of the more important refinements. Here it is explained that such ESDIAD measurements can provide information on the orientation of adsorbate–surface bonds and on the dynamics of surface-bound molecules. The rich variety of behaviour is illustrated by descriptions of studies of a number of different systems.

Our Volume closes with a review of the relatively new field of surface photochemistry. This has much in common with Chapter 7, insofar that both Chapters are concerned with surface processes induced by electronic transitions. After a brief introduction and discussion of the historical perspective, Polanyi and Rieley summarize the progress in this field under four headings: photodissociation, photoreaction, photodesorption, and photoejection. In each case they consider, in turn, work on insulators, semiconductors, and metals. Their nomenclature and definitions will be of considerable value in helping to clarify the broad areas of this growing field. The discussion of the photoreaction process is particularly intriguing, especially to those familiar with gas-phase reaction dynamics. In principle, one may take advantage of the fact that adsorbates often form highly ordered structures, suggesting the possibility of arranging for photodissociation products to collide with preferred orientations and impact parameters. While this level of control has not been realized to date, great strides have

Preface

been made towards this end. This concluding Chapter serves to reinforce our opening contention that a Volume devoted to gas–surface interactions must also be of interest to the pure gas-phase specialist.

Charles T. Rettner, San Jose
Michael N. R. Ashfold, Bristol
February 1991.

Contributors

A. E. DePristo, *Iowa State University, Ames, Iowa, U.S.A.*
J. Harris, *Institut für Festkorperforschung der KFA, Julich, Germany*
B. E. Hayden, *University of Southampton*
S. Holloway, *University of Liverpool*
H. J. Kreuzer, *Dalhousie University, Halifax, Nova Scotia, Canada*
S. H. Payne, *Dalhousie University, Halifax, Nova Scotia, Canada*
J. C. Polanyi, *University of Toronto, Toronto, Ontario, Canada*
R. D. Ramsier, *University of Pittsburgh, Pittsburgh, Pennsylvania, U.S.A.*
H. Rieley, *University of Liverpool*
W. H. Weinberg, *University of California, Santa Barbara, California, U.S.A.*
J. T. Yates, Jr., *University of Pittsburgh, Pittsburgh, Pennsylvania, U.S.A.*

Contents

Chapter 1 Mechanical Energy Transfer in Particle–Surface Collisions
J. HARRIS

1 Introduction	1
2 Classical Scattering from a Linear Chain	6
3 The Baule Formula and Binary Collision Models	16
4 Translational–Rotational Conversion	22
5 Scattering from a Quantum Lattice	35
References	45

Chapter 2 Dynamics of Dissociative Chemisorption
A. E. DEPRISTO

1 Introduction	47
2 Potential Energy Surfaces (PES)	50
General Characteristics	50
Construction of a Model Potential Energy Surface	54

	3 Dynamical Theories	58
	4 Selected Illustrations	67
	References	85

Chapter 3 Quantum Effects in Gas–Solid Interactions
S. HOLLOWAY

1 Introduction	88
2 Potential Energy Surfaces	91
Formal Derivation	91
Adiabatic Representation	92
Diabatic Representation	92
Quantum *vs.* Classical Mechanics	94
3 Solving the Quantum Dynamics	95
Propagation Methods	95
Choice of the Initial Parameters and Final State Projection	98
4 Inelastic Scattering	100
The Trapping of $H_2/Cu(100)$	100
T–R: the Scattering of NO/Ag	101
T–V: the Scattering of NO/Ag	112
5 Reactive Scattering	113
H_2 Scattering	113
Other Diatomics	128
6 Future Directions	131
References	133

Chapter 4 The Dynamics of Hydrogen Adsorption and Desorption on Copper Surfaces
B. E. HAYDEN

1 Introduction	137
2 Adsorbed Hydrogen on Copper	139
Dissociatively Adsorbed Hydrogen	139
The Apparent Activation Barrier to Dissociation	141

3 Angular Distributions in Desorption	143
Non-equilibrium Distributions	143
Permeation and Desorption	144
A One-dimensional Potential	145
4 Translational Energy Partitioning	147
Dissociation Promoted by Translational Energy	147
Excess Translational Energy following Recombination	150
Direct Measurements of Dissociative Sticking	152
5 Vibrational and Rotational Energy Partitioning	155
Hot Vibrational and Cold Rotational Distributions	155
Vibrationally Enhanced Dissociation	156
Vibrational and Translational Coupling	158
A Two-dimensional Potential Energy Surface	160
6 Conclusions	164
Detailed Balance and the Multi-dimensional Potential Energy Surface	164
An Experimental Prospective	167
References	168

Chapter 5   Kinetics of Surface Reactions
W. H. WEINBERG

1 Introduction	171
2 Direct *vs.* Trapping-mediated Surface Reactions	172
3 Transition State Theory of Surface Reaction Rates	175
Langmuir–Hinshelwood Reaction Mechanism	175
Eley–Rideal Reaction Mechanism	179
Langmuir–Hinshelwood *vs.* Eley–Rideal Reaction Rates	184
4 Trapping and Accommodation	192
Trapping	192
Accommodation *vs.* Trapping	196

5 Example of a Trapping-mediated Surface
   Reaction: Dissociative Chemisorption of
   Ethane on Ir(110)–(1 × 2)   205

6 Synopsis   215

References   216

Chapter 6   Thermal Desorption Kinetics
H. J. KREUZER and S. H. PAYNE

1 Introduction   220

2 Experimental Preliminaries   222

3 Desorption under Quasi-equilibrium Conditions   227

4 Non-equilibrium Thermodynamics of Surface Processes   237

5 Kinetic Lattice Gas Model   245

6 Physisorption Kinetics   249

7 Concluding Remarks   252

References   254

Chapter 7   Electron Stimulated Desorption and its Application to Chemical Systems
R. D. RAMSIER and J. T. YATES, JR.

1 Introduction   257

2 Models of Electron Stimulated Desorption (ESD) Processes   258
    The Menzel–Gomer–Redhead (MGR) Model   259
    The Antoniewicz Model   260
    The Knotek–Feibelman (KF) Model   262
    Quantum Mechanical Models   264
    Dissociative Attachment (DA) Mechanisms in ESD   269
    Auger Stimulated Desorption (ASD)   271

Other Mechanisms of ESD 274

3 Experimental Observation of ESD 276
   Measurement of Ion Yield, Threshold Energy,
     and Ion Identity 276
   Measurement of Angular Distributions of Ions
     and Electronically Excited Neutrals Produced
     in ESD 279

4 Electron Stimulated Desorption Ion Angular Distribution (ESDIAD) 282

5 Chemical Systems Studied by ESD Techniques 284
   CO/Ru(001) 284
   CO/Pt(111) 285
   CO/Se/Pt(111) 288
   CO/Pt(112) 289
   CO/Ni(110) 294
   CO/Cr(110) 296
   Cyclopentene/Ag(221) 300
   F/Si(100) 302
   H/Ni(111)(Alkali Metal Co-adsorbates) 303
   H_2O/Ni(110) 304
   NO/Pt(112) 308
   NH_3/Si(100) 309
   NH_3/Ni(110) 312
   PF_3/Ni(111) 314
   O_2/Ag(110) 316
   Pyridine/Ir(111) 320

6 Conclusions 321

References 323

Chapter 8 Photochemistry in the Adsorbed State
J. C. POLANYI and H. RIELEY

1 Introduction 329
   Historical Perspective 329

2 Photochemical Processes 332
   Photodissociation; PDIS 332
   Photoreaction; PRXN 333
   Photodesorption; PDES 333
   Photoejection; PEJ 334

3 Photodissociation; PDIS	334
Methodologies	334
Insulators	335
Semiconductors	339
Metals	342
4 Photoreaction; PRXN	346
Insulators	346
Semiconductors	351
Metals	352
5 Photodesorption; PDES	354
Insulators	354
Semiconductors and Metals	355
6 Photoejection; PEJ	358
References	359
Subject Index	364

CHAPTER 1

Mechanical Energy Transfer in Particle–Surface Collisions

J. HARRIS

1 Introduction

When a particle from the gas phase reflects from a solid surface it will in general exchange some energy, $\delta\epsilon$, with the surface. Traditionally, in surface chemistry, this energy exchange was described in terms of *accommodation* and characterized by an *accommodation coefficient* that measures the extent to which, as a result of collisions, the gas acquires the temperature of the surface. The concept of accommodation is natural when one is dealing with experiments conducted under quasi-thermal conditions with gas and surface characterized by temperatures T_g, T_s. With the advent of molecular beam technology, experiments are performed with specific incidence and emergence conditions for the gas particles (initial and final energy, angle, and azimuth, $\epsilon_{i,f}$, $\theta_{i,f}$, $\phi_{i,f}$) and a *mechanical* description of the energy exchange becomes desirable and in fact necessary. By studying particle–surface collisions as individual *scattering events* we can gain information about the behaviour and the interactions that cause this behaviour at the microscopic level. The technological advances that have made this possible, along with developments in the field up to *ca.* 1984, have been reviewed by Barker and Auerbach,[1] in which a comprehensive list of references can be found.

From the theoretical point of view, a major hurdle to be overcome in analysing the results of molecular beam experiments is lack of knowledge of the interaction. Some information is available from explicit calculations but this is very limited and at best partial. For most systems, we have to work with plausible models of the interaction and be content with *qualitative* comparisons with data. Although dynamical calculations can be performed at elevated surface temperature, T_s, this tends to obscure details and adds a considerable measure of uncertainty to any comparison. Most data, however, are taken at elevated temperature in order to prevent accumulation of particles on the surface and ensure that the atoms or molecules in the beam 'see' a clean target. At low T_s this can be achieved only by cycling measurement with periodic 'flashing' of the crystal to remove adsorbed particles which is expensive in resources (and patience). Under any circumstances defects and impurities will be present on the surface and it is far from easy to know when their concentration is 'low enough'. Impurities give rise to weak long range interactions and can have scattering cross-sections that are many times greater than the area of the unit cell where they happen to be adsorbed. At elevated T_s one has, in addition, to separate the directly scattered fraction from the fraction due to particles that trap initially and desorb some time later via a thermal fluctuation. A clean separation can be made via time-resolved chopped-beam methods where the time delay of the trapping-desorption fraction is demonstrated explicitly. Again, however, this is expensive in resources and it is more common to resolve the two fractions on the basis of assumed differences in their angular patterns, with the *direct-scattered* fraction retaining memory of the incidence conditions and the *trapping-desorption* fraction leading to a thermal distribution. As we will see, this is a somewhat simplistic view and may in some cases lead to serious misinterpretation of data. Behaviour at elevated T_s and in the presence of adsorbates is of the very greatest interest in connection with surface reactions and catalysis. Nevertheless, from the point of view of gaining a detailed understanding of the elementary processes that occur, high T_s is a complicating factor.

These qualifications must be borne in mind if one is to gain the correct view of the *status quo* in the field. We have arrived at the point where molecule–surface collisions must be treated as *scattering* rather than thermal processes, but, except in rare cases, do not have the kind of control of the scattering conditions that is possible in *e.g.* crossed beam molecule–molecule scattering experiments. Accordingly, the questions of current interest tend to be more of a qualitative than quantitative nature. To clarify these, it is perhaps useful to begin with a description of the scattering event envisaged; the collision of a particle having a specific energy ϵ_i with a well-characterized, adsorbate-free solid surface.

The interaction between a neutral particle, P, and a surface at large distance is the van der Waals interaction $V_{vw}(z_P) = -C_{vw}/z_P^3$, where z_P is the normal distance of P's centre-of-mass from a plane close to the topmost layer of ion-cores. The constant C_{vw} depends on the dielectric properties of the solid and the frequency dependent polarizability of P. If P is an ion the asymptotic interaction is the *image potential* $V_I(z_P) = -1/4z_P$. In either case the approaching particle will initially be accelerated along the normal direction and its effective angle of incidence with respect to the shorter range part of the interaction will decrease. This can be a quite important effect for particles with energies in the thermal range. The course of the collision then depends on the interaction in a very material way and this depends in turn on the chemical nature of P and any internal degrees of freedom P possesses. If P is chemically inert so that no re-arrangement of its electronic structure occurs throughout the collision, the relevant short-range interaction will be repulsive and P will reflect from a 'corrugated wall'. The manner in which the reflection occurs depends on a number of factors, the degree of the surface corrugation, the point with respect to the surface mesh where P makes its impact, the steepness of the wall, the amount of energy the wall absorbs as it recoils, and the coupling through the interaction of translational and any rotational or vibrational degrees of freedom P may possess. If P is an ion we have also to account for *neutralization* and the possibility that the particle reflects from the surface as a neutral. For high energies there may also be substantial *penetration* of the surface lattice and particles may sample several layers before re-emerging or many even become incorporated in the lattice. In general, P will rebound from the wall into the tail of the particle–surface interaction where the inelastic couplings are weak and the motion can again be treated approximately as a one-body problem with conservative potential $V_s(\mathbf{x}_P)$ whose limiting value far from the surface is $V_{vw}(z_P)$ or $V_I(z_P)$. If P's total energy, $E_P(\mathbf{x}_P) = \frac{1}{2}m_P\dot{\mathbf{x}}_P^2 + V_s(\mathbf{x}_P)$, where m_P is the mass, becomes negative, P is trapped in the potential well and can escape only if on the next bounce it gains energy from the surface. For low T_s this is unlikely to occur and we can assume P to be irreversibly stuck. If $E_P(\mathbf{x}_P)$ remains positive, then P may escape from the surface and emerge as an inelastically scattered particle. However, the kinetic energy may be due primarily to velocity components parallel to the surface in which P may at some point reflect and return to the surface as a positive-energy trapped particle. Its subsequent fate will depend on the outcome of further collisions with the 'hard' part of the surface which will occur at points some distance removed from the initial point of impact.

If classical mechanics is valid each particle incident on the surface will follow a unique trajectory so that the outcome of the collision is deterministic in the sense that a given set of incident parameters results in a unique and specific set of exit parameters (or sticking). In a beam experiment, the *impact parameter* giving the location within the surface unit cell where the impact occurs, is not controllable. A mono-energetic, well-collimated beam of particles striking the surface will sample a range of impact parameters and what is measured in the exit channel is an average over impact parameter, $P(\epsilon_f, \hat{\Omega}_f)$, which gives the probability that the particle P will emerge with energy ϵ_f in a direction about emission angles $\hat{\Omega}_f$. If the collision is in the quantum regime the initial state of P is plane-wave-like and no averaging over impact parameters is involved. The outcome of the initial impact is deterministic in the sense that the evolution of the wavefunction is unique and determined by the time-dependent Schrödinger equation, but quantum theory nevertheless delivers only a probability for scattering into a final state characterized by $\epsilon_f, \hat{\Omega}_f$ so we can define $P(\epsilon_f, \hat{\Omega}_f)$ with the same meaning as in the classical case. For $T_s = 0$, this probability distribution is (roughly) the particle–surface equivalent of the scattering cross-section familiar in molecule–molecule scattering.

Clearly, $P(\epsilon_f, \hat{\Omega}_f)$ will depend on the incidence conditions in a manner dictated by the interaction. For instance, if the surface is flat the interaction is essentially *one-dimensional* and the energy transfer occurs at the expense of the normal component, $\epsilon_i^\perp = \epsilon_i \cos^2 \theta_i$, of the incident energy with approximate conservation of the parallel component. This means that backscattered particles will emerge primarily in the super-specular direction $\theta_f > \theta_i$ and, furthermore, that the distributions with respect to final energy and angle are linked. This can easily be tested experimentally. Another consequence of a flat surface is the occurrence of substantial positive-energy trapping at wide angle incidence. Trapping requires only that the energy transfer, $\delta\epsilon$, exceeds the incident normal energy, $\epsilon_i^\perp$, which goes to zero as $\theta_i \to 90°$. This is less easy to check experimentally because the positive-energy trapped fraction is not directly measurable. These particles will transport along the surface undergoing further interactions. On an ideal flat surface with conservation of $\epsilon_i^\parallel$ they can return to the gas phase only by gaining energy as a result of a thermal fluctuation. However, there will always be some corrugation of the surface potential so that some positive-energy trapped particles will revert to the gas phase, where they will make a contribution to the reflected flux. Others will lose their energy to the lattice and become irreversibly stuck (at $T_s = 0$). As mentioned above, the situation is complicated at elevated T_s because the direct scattering may be greatly influenced by *energy gain processes* and *thermally induced roughness* which lead to incoherent scattering. Also, as noted above, the 'stuck' fraction will re-emerge on the time scale of the measurement making a further contribution to the measured distribution. The opposite limit to the flat surface obtains when the potential energy surface (PES) is so strongly corrugated that, essentially, P makes binary collisions with the atoms of the surface mesh. The surface will then behave like a periodic array of scattering centres. The scattering behaviour of the array will depend on the *range* of the interaction potential governing each binary collision. If this is of the same order as the lattice spacing, the surface may retain some degree of 'flatness'. If the range is short compared with the lattice spacing the surface will become 'porous' and substantial penetration of the first layer will occur. Since the range, as measured by the classical turning point for a binary collision, will decrease with incident energy we expect a general tendency to increased roughness as the energy of the probe increases so a given scattering system can behave quite differently at low and high energy.

A strong motivation for the development and deployment of molecular beam technology is to gain contact with surface reactions and heterogeneous catalysis. Here the probes of main interest are strongly bound molecules that may dissociate into fragments on initial collision, with the fragments then undergoing further reactions as adsorbed species. The 'steering' of such reactions to yield a desired product is of tremendous industrial and economic importance. In many cases, the initial break-up of an incident molecule on collision with a surface is believed to be a rate limiting factor in the overall reaction rates achieved and is obviously a process that can be studied profitably using molecular beams. Although the dissociation of a molecule at a surface must necessarily involve a vital re-arrangement of the electronic structure, such collisions need not necessarily be accompanied by a large release of chemical energy and may therefore, at low incident energy, be essentially adiabatic. This will depend on the energy originally stored in the bond or bonds that are broken and the chemisorption energy of the fragments released. As a rule these two energies will differ by only a small fraction of either and it is this difference that controls the violence of the reaction. In many cases, the dissociation is *activated* and low energy particles will not react but will scatter from the PES much as truly inert particles do. A great deal of information about the PES governing such reactions can then be gained by analysing this scattering. Even where a dissociation reaction is not activated and there is a reaction co-ordinate along which the energy decreases monotonically, low energy paths may form only a small region in a multi-parameter space into which relatively few incident particles find their way (steric hindering). The majority event may then again be back-scattering or, alternatively, sticking as a quasi-stable intermediate. Even where a particle eventually dissociates on the surface, the reaction dynamics may involve an initial stage where the behaviour is symptomatic of a chemically intact (or only weakly perturbed) probe. In such cases, a question of major interest and importance concerns the role of internal degrees of freedom, vibrations and rotations and the extent to which these influence the energy transfer and sticking probability. These questions are addressed from different points of view and with different emphases in the next three chapters. In this article, the stress is placed on phenomena connected directly to the *energy transfer* in the initial collision. The discussion will also be restricted to collisions that are governed, at least approximately, by a *single PES*. For low incident energies this means the collisions must be *electronically adiabatic*. The assumption is that the motion of the nuclei is sufficiently slow that the electron distribution adjusts instantaneously to changes in their positions so that the wavefunction of the electrons is always close to the *instantaneous ground-state*. The ground-state energy of the electrons is a function of all the co-ordinates of all the nuclei (including those of the substrate atoms) and forms the potential energy function or PES that governs the motion of the nuclei. If the probe atoms are much lighter than the surface atoms it is often legitimate to ignore the dynamics of the substrate and work with a *stiff-lattice* PES. The scattering is then purely elastic. This approximation is commonly employed in the analysis of the intensities of the Bragg beams when, *e.g.* He atoms scatter from a surface. To describe inelastic scattering and sticking, the substrate dynamics is vital. The collision sets the substrate atoms in motion and the interactions between these atoms ensure that some energy is carried away from the impact area and is dispersed into the bulk. This is the *mechanical energy transfer*, $\delta\epsilon$, with which we will be concerned. Non-adiabatic processes will always occur with some probability and give rise to energy transfer mechanisms that are inherently *non-mechanical* because they devolve from the *excitation* of the electronic system during the collision. Such mechanisms may even dominate, as, for example, when the collisions involve *electron transfer* from or to the probe particle. These cases require a different theoretical approach to the mechanical mechanisms we consider in this article. The interested reader is referred to some recent reviews by Gadzuk,[2] where further references on non-mechanical energy transfer processes and methods for describing them theoretically can be found.

The relatively sparse information available on particle–surface interactions leaves much ground open for interpretation. This is true even within the framework of a theory including only mechanical collisions. Any PES purporting to describe particle–surface collisions will entail a number of free parameters (whether this is acknowledged by the PES-'architect' or not!) and may, or may not be 'falsifiable' (meaning that there exists a set of data that cannot be 'fitted' by the theory). It is therefore particularly important to *understand* as much as one can in the simplest possible terms and to avoid attempting to 'fit data'. In this article, we will focus on a few of the more important phenomena associated with the energy transfer and will treat these in detail with the aid of simplified mechanical models. Data will be referred to sparingly and only to the extent that they shed light on the points that are singled out for treatment. No attempt is made to 'review the field'. We begin in Section 2 with a detailed discussion of the scattering of a particle from a *linear chain* of atoms bound together by springs. This model is useful in demonstrating how a surface lattice responds to the impact of a particle and isolating some of the factors that influence the energy transfer, in particular, the *mass ratio* and the *time scale* of the collision. We will see that for illustrative purposes, the precise response of the lattice has only a weak influence on the outcome of a collision and use this information in Section 3 to consider the qualitative effect of *surface temperature* and *impact parameter* on the scattering. In Section 4 we discuss the role of *internal degrees of freedom* when molecules scatter from surfaces and extend our linear chain model to illustrate the vital importance of molecular *rotations* and give a simple and transparent illustration of the origin of an *anticorrelation* that exists between rotational excitation and energy transfer to the lattice. An extension to two spatial dimensions then illustrates the influence of rotations on the *sticking* behaviour and the *angular distributions* of scattered particles. The discussion thus far is restricted to classical mechanics and, in Section 5 we consider, with the aid of the *forced oscillator model* the response of a quantum lattice and illustrate the transition from classical to extreme quantal behaviour with regard to the energy transfer and the sticking coefficient.

2 Classical Scattering from a Linear Chain

A simple model of the 'target surface' in a gas–surface scattering event is a linear chain of atoms, each joined to its neighbours by springs (a one-dimensional harmonic lattice). The springs describe increases in the potential energy when a given atom is displaced from its equilibrium position and it is assumed that the displacements are sufficiently small that an expansion of the potential energy about the equilibrium positions can be terminated in lowest (quadratic) order. The model is characterized by the atom mass, M, and the *force constant*, κ, that determines the force acting on atom i as a result of displacements of the atom's separation from its neighbours from *equilibrium separation*, a. Let the positions of the atoms at any given instant be x_i with $i = 1 \to N$, where N is the total number of atoms in the chain and suppose the chain is initially at rest with $x_i = x_i^0 \equiv (i-1)a$. The forces that act on the atoms of the harmonic lattice depend *linearly* on their displacements $u_i \equiv x_i - x_i^0$ from the equilibrium positions so the application of Newton's second law to each atom of the chain results in the equations of motion for the displacements:

$$M\ddot{u}_1 = \kappa(u_2 - u_1)$$
$$M\ddot{u}_N = \kappa(u_{N-1} - u_N)$$
$$M\ddot{u}_i = \kappa[(u_{i+1} - u_i) + (u_{i-1} - u_i)] = \kappa(u_{i+1} + u_{i-1} - 2u_i)$$
$$i \neq 1 \quad \text{or} \quad N \qquad (1)$$

The force terms are all zero provided the displacements are all the same (*i.e.* these need not be zero). This is because a rigid displacement of all the atoms in space does not involve the extension of any of the springs. Equations (1) have a number of solutions corresponding to standing waves, or *normal modes* of the chain. By direct substitution it is easy to verify that the displacement patterns of the normal modes have the form:

$$u_i(k) = \sqrt{\left(\frac{2}{N}\right)} \cos\left[ka\left(i - \frac{1}{2}\right) + \omega(k)t\right]$$

$$k = \frac{n\pi}{Na}$$

$$\omega(k) = 2\sqrt{\left(\frac{\kappa}{M}\right)} \left|\sin\frac{ka}{2}\right| \qquad (2)$$

where n is an integer and the prefactor ensures orthonormality of the mode amplitudes. The $u_i(k)$ are solutions of equation (2) for any (odd) value of N but we will be interested only in large N for which the discrete nature of the allowed k-values is unimportant. In the limit, $N \to \infty$, there are solutions for any value of k. However, a solution corresponding to $k > \pi/a$ is identical to a solution with $k \leq \pi/a$ so the linearly independent solutions fall in the interval $0 \leq k \leq \pi/a$ and correspond to a band of frequencies, $\omega(k)$, which span the range $0 \leq \omega(k) \leq 2\sqrt{(\kappa/M)}$. If the wavelength of the normal mode is much longer than the lattice spacing we can expand the sine function in equation (2) to obtain:

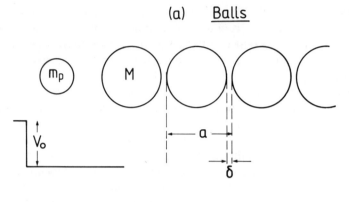
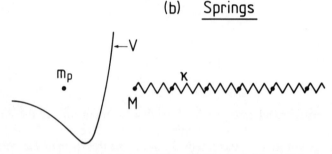

Figure 1 (a) *Particle scattering from a 'billiard ball target'.* (b) *Particle scattering from a 'harmonic linear chain' (see text)*

$$\omega(k) \approx \sqrt{\left(\frac{\kappa}{M}\right)} ka, \quad k \to 0 \tag{3}$$

so the *sound velocity* (group velocity of long wavelength waves) is:

$$v_s \equiv \left.\frac{d\omega(k)}{dk}\right|_{k \to 0} = a\sqrt{\left(\frac{\kappa}{M}\right)} \tag{4}$$

The top of the frequency band corresponds to an oscillation where each atom is exactly out of phase with its neighbours. The frequency, $\omega_0 \equiv \omega(\pi/a) = \sqrt{[2\kappa/(M/2)]}$, is that of a single pair of atoms [with reduced mass $M/2 = M^2/(M+M)$] whose 'spring constant' has twice the strength of each individual spring in the chain (because each member of the pair is bound to neighbours). As $k \to \pi/a$ the group velocity goes to zero which means that a wavepacket constructed from normal modes close to the top of the band and representing a localized oscillation will propagate extremely slowly along the chain. The normal modes of a three-dimensional lattice comprise *bulk* modes whose amplitudes span the entire lattice and *surface* modes, for which the amplitudes fall off away from the surface. Surface modes can exist also in one dimension if the end atoms of the chain have a different mass or different force constant from those of the bulk. Although the three-dimensional case is considerably more complicated than the linear chain the two are, from the conceptual point of view, similar and a theory valid for the simpler system can relatively easily be extended, often by inspection.

To simulate a particle–surface scattering system we imagine that the end atom of the chain, $i = 1$, is struck by an external particle, P, with which it interacts via a two-body potential (a Morse potential, for instance, as in Figure 1b). This model of atom–surface scattering was proposed and analysed by Zwanzig[3] and amounts to supplementing the equation involving u_1 in equations (1) with an additional force term and writing an additional equation for the motion of P. Specifically:

$$M\ddot{u}_1 = \kappa(u_2 - u_1) - \frac{\partial V(u - z)}{\partial u}$$

$$m_p \ddot{z} = -\frac{\partial V(u - z)}{\partial z} \tag{5}$$

where $z(<0)$ is the location of the particle and $V(u - z)$ is the Morse potential. The remainder of equations (1) are unchanged because P interacts only with the end particle. When P, incident at the chain with velocity v_i and energy $\epsilon_i = \frac{1}{2} m_p v_i^2$, comes within range of the interaction, it is first accelerated towards the chain by the attractive branch of the Morse potential and then rebounds from the back wall. (A Morse potential falls off exponentially and can be used only when the 'van der Waals tail' of the true interaction is not particularly important which may not be the case if the well is shallow. Under normal circumstances, a particle subject to the van der Waals interaction couples only very weakly to the lattice and the effect can be neglected compared with the large coupling that comes into play when the particle hits the wall.) Because the potential is a two-body force, the outermost atom of the chain is first drawn outwards towards P and then driven inwards. The spring joining atom 1 to atom 2 is therefore first extended somewhat and then compressed so that atom 2 is set in motion. The spring joining atom 2 to atom 3 is then activated *etc.* and further atoms of the chain are sequentially set in motion. Meanwhile, P has rebounded and after a certain time τ_c, the *collision time* or *first round-trip* time, will escape the influence of the two-body potential. Many of the chain atoms will be in motion at this time, and their energy, however it may subsequently get redistributed, will remain behind in the chain, which is now decoupled from P. Since the system as a whole conserves total energy, this means that P will rebound with energy $\epsilon_f < \epsilon_i$ and the difference $\delta\epsilon \equiv \epsilon_i - \epsilon_f$ is the mechanical energy transferred from P to the chain as a result of the collision.

While the origin of the energy transfer is easy to view pictorially, the actual calculation of $\delta\epsilon$ is not entirely trivial. In his elegant and concise paper, Zwanzig[3] showed how the problem, potentially involving the solution of simultaneous equations for all the atoms of the chain, can be reduced to an effective two-body problem involving the relative displacement of the scattering particle and the outermost chain atom with the influence of the remaining atoms treated via a *memory* term that serves as a non-conservative force. This technique has been taken over and developed further by Adelman and Doll[4] to the case of three spatial dimensions, where the 'outermost atom' becomes a *primary zone*. The resulting calculational method has been used extensively by Tully and co-workers,[5] and by others in connection with atom/molecule–surface scattering. Here we restrict ourselves to a qualitative, physically motivated discussion, for which the linear chain, as a simplified example, suffices. The reader interested in pursuing matters beyond this or in technical details is referred to the relevant literature.

A major factor influencing the energy transfer is the ratio between the collision, or round-trip time τ_c of P and the response time $\tau_0 \sim \omega_0^{-1}$ of the lattice, which is determined by the frequencies of the normal modes, characterized by ω_0. We can consider two limiting cases. If $\omega_0 \tau_c \ll 1$, the collision is *fast* and P escapes before the chain has had time to respond. At first sight, one might suppose this means that the energy transfer will then vanish but this is not the case because, however slow the response of the chain, the outer atom 'feels' the interaction with the same strength as the scattering particle. The fast-particle or *impulse* limit, therefore, corresponds to an energy transfer that is symptomatic of a *two-body* collision. Immediately following the collision the energy that has been transferred to the chain is stored primarily as kinetic energy of the outermost atom. The re-distribution of this energy amongst the atoms of the chain follows subsequently on time scale $\tau_0 \gg \tau_c$. In this limit, the energy transfer is independent of all details of the interaction and can be calculated *kinematically* by applying conservation of energy and linear momentum in a two-body collision. If the chain is originally at rest then after time t such that $\tau_0 \gg t > \tau_c$, the outermost chain atom will be moving to the right with velocity u, while P is moving to the left with velocity v_f, where:

$$\frac{1}{2} m_p v_i^2 = \frac{1}{2} m_p v_f^2 + \frac{1}{2} M u^2$$
$$m_p v_i = M u - m_p v_f \qquad (6)$$

Solving these equations we find that the energy transfer, equal to the final energy of the target particle which initially is stationary, is:

$$\delta\epsilon \equiv \frac{1}{2} M u^2 = \frac{4\mu}{(1+\mu)^2} \epsilon_i$$

$$\mu = \frac{m_p}{M} \qquad (7)$$

This is commonly referred to as the 'Baule' formula and gives the energy transfer in a linear binary collision exactly in terms of the mass ratio, μ, and the incident energy. For a collision with a chain, the result becomes exact only in the extreme impulse limit, though as we will see, it remains a rather good approximation for conditions that depart quite markedly from this limit. Although the 'Baule' formula is valid for two-body collisions with arbitrary values of μ, it ceases to be relevant for collision with a chain if $\mu > 1$ because in this case the initial collision does not stop the scattering particle [i.e. the solution of equation (6) gives $v_f < 0$], which therefore barrels on and re-collides with the outermost chain atom as this rebounds from atom 2. For a sufficiently massive particle, therefore, there is no first round-trip time and the impulse, or two-body limit does not obtain. P may re-emerge from the chain but can do so only under the joint action of several particles i.e. as a result of the compression of the springs and only on a time scale of order τ_0. For $\mu < 1$, the solution of equation (6) gives $v_f > 0$, corresponding to the rebound of the particle. In a true two-body collision, P will *always* rebound if $\mu < 1$. That is, *trapping* of P on the target particle *cannot occur*. For a chain, however, trapping will *always* occur at sufficiently small incident energy because the chain will always absorb some energy. This is an important difference between two-body, and particle–chain or particle–surface collisions.

In spite of its simplicity (or perhaps, *because* of its simplicity), the origin of the energy transfer given by the Baule formula may not be obvious, especially if one thinks of energy loss as due to 'friction' or energy 'dissipation'. As the manner of its derivation makes obvious, the Baule energy transfer is a purely mechanical effect and arises because within the frame of reference the centre-of-mass of the two particles involved in the collision is not stationary. In a true two-body collision the entire concept of 'energy transfer' can be eliminated by viewing the collision in the reference frame where the centre-of-mass is stationary. The scattering problem involves the relative co-ordinate of the two partners and is exactly equivalent to the scattering of a single particle having reduced mass $M_r = M_1 M_2/(M_1 + M_2)$ from the pair potential. Obviously, no 'trapping' (or 'reaction') can occur because any particle with positive energy will traverse the well region of the interaction, climb up the wall and reflect out again. When one partner is a chain this is no longer true because chain atoms other than the one bonded to the scattering particle can carry away energy. The manner in which this occurs depends to some extent on details of the collision, but we can illustrate this easily for an impulsive collision and at the same time resolve an issue regarding momentum conservation in particle–surface collisions.

Suppose P undergoes a fast collision and rebounds with speed v_f. Then its change of linear momentum is $P_0 \equiv m_p(v_i + v_f)$. If we now apply blindly the conservation of linear momentum to the collision we may suppose that the 'final state' involves a motion of the entire chain to the right with a finite (if small) velocity, $v_c = P_0/NM$, in order to satisfy momentum conservation. This is *not* what happens. It is surely true that linear momentum must be conserved and so the chain must make up the change of momentum that results because P has been reflected. However, the chain does not do this by

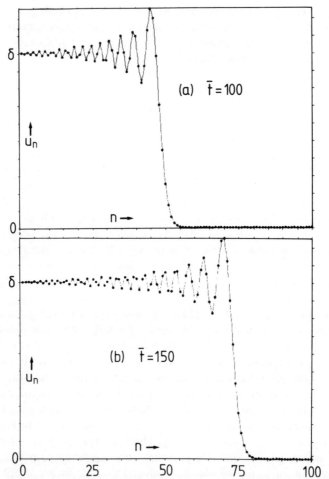

Figure 2 *Displacements of the atoms of a harmonic linear chain for two times, $\tilde{t} = \omega_0 t = 100$(a) and 150(b) following an initial impulse on the outermost atom at time $t = 0$. The figure illustrates that the energy deposited in the chain is carried away from the 'surface' by a wavefront with atoms behind the wavefront coming to rest at new equilibrium positions displaced from their original ones by δ. The 'final state' of the chain subject to an initial impulse is, therefore, a spatial displacement of each atom by δ*
(Reproduced with permission from *Phys. Scr.*, 1987, **36**, 156)

moving *as a whole* and, in fact, the 'final state' of a long chain subject to an impulsive collision does not involve a motion of the chain at all but merely an overall *spatial displacement* by an amount $\delta = 2P_0/M\omega_0$. To show this, consider an initial impulsive collision which results in the sudden acquisition, by the outermost atom of the chain, of momentum P_0. The subsequent motion of the chain can then be analysed in the absence of the particle, P, that caused this momentum change (which has, on a short time scale, reflected out of range of its interaction with the chain). We need, therefore, solve equations (1) subject to the boundary conditions that all displacements are initially zero and all velocities are initially zero except for $u_1 = P_0/M$. This is straightforward (if not entirely trivial) and the result for the atomic displacements at time t is:[6]

$$u_i(t) = \frac{2\delta}{\pi} \int_0^{\tilde{t}} dt' \frac{\sin t'}{t'} \cos\left\{(2i-1)\sin^{-1}\left(\frac{t'}{\tilde{t}}\right)\right\} \qquad (8)$$

where $\tilde{t} \equiv \omega_0 t$. This looks nasty, but has the simple limiting behaviour:

$$u_i(t) = \begin{cases} \delta & 2i \ll \tilde{t} \\ 0 & 2i \gg \tilde{t} \end{cases} \qquad (9)$$

A plot of the actual displacements for two times, $\tilde{t} = 100, 150$ is shown in Figure 2 and illustrates what actually happens to the chain. The initial disturbance propagates as a well defined *wavefront* beyond which the lattice atoms are undisturbed. Behind the wavefront the atoms oscillate but eventually settle down in new equilibrium positions separated by one lattice spacing, but they are displaced by δ from the initial positions. The momentum imparted to the front atoms of the chain is carried *by the wavefront*, which moves along the chain with the speed of sound. If the chain is of infinite length the displacement front propagates for ever and once it has traversed a given length of the chain the atoms behind it are stationary. Thus the 'final state' of the chain is indeed a spatial displacement and this does not at all contravene the conservation of linear momentum. The energy originally deposited in the chain is *dissipated* after a time of order τ_0 in the sense that it is transferred spatially from one atom to the next along the chain leaving primary zone atoms at rest, though displaced by δ. The actual value of δ is readily evaluated. For instance, for a CO molecule with an energy of ~ 1 eV striking a chain of copper atoms, and assuming for the sake of simplicity that the CO rebounds without change of energy, we find $\delta \sim 2$ a.u., which is a third of a lattice spacing! If the chain is truly isolated this displacement *never heals*. If the chain is part of a three-dimensional lattice, however, each atom of the chain can be viewed as constrained by side-springs to atoms in neighbouring chains. The initial impulse couples only to the chain that is directly struck and gives rise to compressions and expansions down this chain according to equation (8). The side-spring extensions will then pull the chain atoms back to their original positions, but since these extensions are much smaller than the longitudinal extensions the time scale for the healing process is longer than that governing the initial propagation down the chain. This picture is somewhat simplistic with respect to a real surface, which may not be well represented by a harmonic model with nearest neighbour chains. However, the picture of a strong initial localized disturbance that propagates rapidly into the lattice followed by a rather slower healing of the lattice behind the wavefront remains a reasonable first approximation to what happens in nature.

The above discussion describes how the chain responds to a collision that is sufficiently fast that the scattering particle traverses the region where it interacts with the outermost atom in a time that is short compared with the response time of the chain. This will be the case if the incident energy is sufficiently high. For a slower collision it is no longer possible to decouple the motion of the particle, P, from that of the chain. A common way of viewing this, particularly in the context of *trapping* is to imagine that the outermost atom of the chain remains at rest and the two-body collision occurs when P has arrived at the bottom of the well in its interaction potential. This view is a bit hard to argue when the scattering particle interacts only with the end atom via a two-body force so let us consider an even simpler model, (illustrated in Figure 1a), with a billiard ball target. The balls are separated by a small distance δ and are of equal mass so that a collision between one that is stationary and one that is moving results in a complete transfer of energy (*cf.* the Baule factor, $4\mu/(1+\mu)^2$, has its maximum value of unity for mass ratio $\mu = 1$). If the outermost ball is struck, the displacement pattern shoots down the chain with a 'velocity' determined by δ and the initial momentum imparted to the outer ball, *i.e.* the displacement pattern is the same as for the spring model except that the velocity of propagation is no longer independent of the initial impulse. We suppose that our projectile, P, strikes the outermost atom but before doing so is accelerated by a potential step of depth V_0, supposedly simulating the surface interaction. P then hits the outermost billiard ball with a kinetic energy enhanced by the well depth, $\epsilon_i \rightarrow \epsilon_i + V_0$. According to the Baule formula the energy transfer is then just:

$$\delta\epsilon = \frac{4\mu}{(1+\mu)^2} [\epsilon_i + V_0] \tag{10}$$

but is now non-zero in the limit that the incident energy goes to zero. This means that a low energy particle will arrive back at the potential step with insufficient energy to escape and so will bounce back and be 'trapped on the first round trip'. When it returns to strike the outermost billiard ball again, the energy originally imparted to this has been dissipated (in the sense that it has been removed forever from the interaction region), there is no source whereby the particle can gain energy and so this is permanently stuck between the outermost ball and the potential step. Thus the criterion for the *trapping* of the projectile is that the energy transfer exceed the incident energy before acceleration by the potential step, which will be the case for all incident energies less than the critical energy given by:

$$\epsilon_i^c = \frac{4\mu}{(1-\mu)^2} V_0 \tag{11}$$

which therefore represents a *trapping threshold* for P. Note that $\epsilon_i^c \rightarrow \infty$ as $\mu \rightarrow 1$ because a collision between like mass particles results in a complete

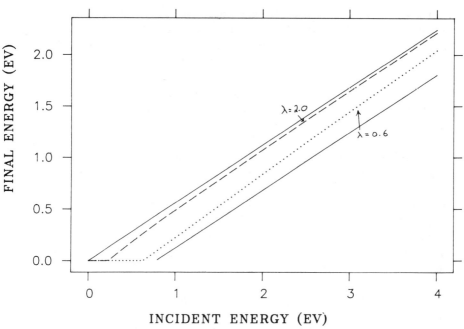

Figure 3 *Final energy of a projectile of mass 28 scattering from a linear chain of Pt atoms (see text). The projectile interacted directly with the outermost chain atom via a Morse potential with well depth 1.0 eV and exponent $\lambda = 0.6, 2.0$. The exponent controls the collision time τ_c of a low energy collision. The full lines give the final energies predicted by the two-body Baule formula, equation (7) (uppermost line) and the Baule formula with the incident energy corrected by the well depth, equation (10)*

transfer of energy with the 'probe' particle brought to rest. This therefore 'sticks' even in the absence of an attractive interaction.

Returning to our chain model, with P bound to the end chain atom by a two-body force, it is not at all clear what meaning we can assign to V_0 in equation (10). In three-dimensional scattering we might try to argue that V_0 represents that part of the interaction that results from surface atoms other than the one that gets 'hit', or we might argue that in the limit of a slow collision the recoil of the end atom is inhibited by the remaining atoms in the chain which carry the energy away quickly. No argument is particularly convincing, and let us not worry about it but think of the billiard ball model as a simple illustration of how a 'trapping threshold' actually comes about. Figure 3 shows two explicit calculations for the linear chain which illustrate the actual energy transfer that occurs in particle–chain collisions and its relation to the two Baule formulae equations (7) and (10). The masses and spring constant that were used in the calculation are those appropriate for a CO molecule striking a chain of platinum atoms, and the Morse potential acting between the CO and the outermost platinum atom had a well-depth of 1 eV and exponents $\lambda = 2.0$ and 0.6 a.u., respectively. For $\epsilon_i \ll V_0$, the 'traversal times' of P in the Morse potential are of order $\tau_0/3$ and $2\tau_0$, for $\lambda = 2.0, 0.6$, so the two cases refer to collisions which at low incident energy are relatively fast and slow, respectively. Figure 3 shows the final energy of the CO as a function of incident energy in the two cases, along with the final energies that would obtain if the energy transfer were given by the Baule two-body formula, equation (7), and the Baule formula with incident energy corrected by the full well depth, equation (10) (full lines). These bracket the actual energy transfers from above and below and, in fact, the faster collision (dashed line) displays an energy transfer that is very close to the 'two-body' limit, except at energies substantially smaller than the well depth where the curve dips down and displays a trapping threshold at about 0.25 eV. The 'slow collision' on the other hand (dotted line) displays a substantially stronger energy transfer and a trapping threshold at about 0.6 eV, only slightly less than that given by the well depth corrected Baule transfer. As the incident energy increases the traversal time across the interaction region decreases and the 'slow collision' changes to a 'fast' one. Accordingly, the energy transfer approaches the 'two-body' limit, though rather slowly. This example illustrates several important points. First, the energy transfer depends approximately *linearly* on the incident energy. Secondly, as previously noted, the two Baule results, equations (7) and (10), bracket the actual energy transfer and so allow a reasonable 'ball-park' estimate for $\delta\epsilon$. The *faster* the collision with respect to time-scale τ_0, the closer the energy transfer is to the two-body collision limit, equation (7). Whether the collision is fast or slow, however, the slope of the $\delta\epsilon$ vs. ϵ_i curve is determined primarily by the *mass ratio*, μ, through the Baule factor $4\mu/(1 + \mu)^2$. Thus, *exchange of energy is inefficient for light particles and efficient for heavy particles, which therefore tend to stick readily*. Thirdly, the *trapping threshold* depends sensitively on *details* of the interaction. In general, we may expect this to be smaller than ϵ^c given by equation (11) by an amount that depends on the traversal time in the surface well.

3 The Baule Formula and Binary Collision Models

Having considered the response of the lattice and established that for qualitative purposes the Baule 'two-body' formula gives a reasonable estimate of the energy transfer even when the collision is quite far from the impulse limit, we can now dispense with the lattice and use the binary collision model to give crude, illustrative arguments as regards the effect of surface temperature and impact parameter. Consider a two-body collision between P and the outermost surface atom, but assume that this is not now stationary but starts out with velocity u, supposedly as a consequence of its thermal motion. Then we can apply conservation of kinetic energy and linear momentum as in equation (6) with additional contributions from the surface particle. After a little algebra we find that the energy transfer from P to the surface atom is given by:

$$\delta\epsilon = \frac{4\mu(\epsilon_i - \tfrac{1}{2}Mu^2) + 2m_p v_i(1-\mu)u}{(1+\mu)^2} \tag{12}$$

The energy transfer depends on whether u is positive or negative, but not symmetrically so that when we average u over a Boltzmann distribution at temperature T_s, with $\langle u \rangle = 0$; $\tfrac{1}{2}M\langle u^2\rangle \sim k_B T_s$, we end up with:

$$\delta\epsilon(T_s) = \frac{4\mu}{(1+\mu)^2}(\epsilon_i - k_B T_s) \tag{13}$$

This 'thermal averaging' is a bit glib (see, for example, reference 7), but suffices to establish that the effect of surface temperature is to *reduce the energy transfer by an amount of the order* $k_B T_s$. Experimentally a linear dependence of $\delta\epsilon$ on T_s has been observed for a wide range of low-energy scattering systems (*e.g.* references 8 and 9). Equation (13) is not much use for quantitative purposes but does illustrate the important point made in the introduction as regards comparing data taken at elevated surface temperature with theoretical calculations which assume $T_s = 0$. Such comparisons are meaningful only if $\epsilon_i \gg k_B T_s$. Calculations can be performed at elevated T_s but at the cost of averaging over the phase of the thermal fluctuations as well as over impact parameter and whatever internal degrees of freedom the probe particle possesses. Even in relatively simple cases, the calculation of a reliable energy distribution at elevated T_s requires accumulation of averages over a massive number of trajectories. Given the lack of knowledge of the PES and the fact that at finite T_s differences between PESs having different features tend to get washed out, it is doubtful that the effort required to perform such calculations is commensurate with the likely return.

To illustrate the manner in which the energy transfer depends on impact parameter consider a two-body collision between mass m, having incident energy ϵ_i, and mass M, initially stationary, that results in m being back-scattered at angle θ (with the convention that $\theta = 0$ corresponds to direct back-scattering and $\theta \sim \pi$ to a glancing collision). Then we can apply conservation of linear momentum in two directions plus conservation of kinetic energy to relate the energy transfer to the angle of back-scattering (assuming $\mu \equiv m/M < 1$) and find:

$$\delta\epsilon = \frac{2\mu}{(1+\mu)^2}[1 + \cos\theta \sqrt{[1-\mu^2(1-\cos^2\theta)]} + \mu(1-\cos^2\theta)]\epsilon_i \tag{14}$$

For a given incident energy, $\delta\epsilon$ has maximum value equal to the result derived earlier for direct back-scattering [equation (7)]. At scattering angle $\theta = \pi/2$ the energy transfer is reduced from its maximum value by a factor $(1+\mu)/2$, while for near forward scattering, $\delta\epsilon$ goes to zero. These results are, of course, known intuitively to all billiard players, who, surely, would be immensely impressed to learn that they are strictly *kinematical* and hold rigorously for any binary collision whatever the interaction involved. From

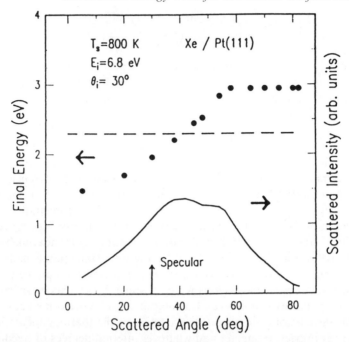

Figure 4 *Final energies of* Xe *atoms scattered from* Pt(111) *as a function of exit angle. The full line gives the intensity, the dots the observed energy transfer, and the dashed line the energy transfer averaged over exit angle. The figure, illustrates the reduction of the energy transfer with increasing back-scattering angle as suggested by the Baule formula equation (14)*
(Reproduced with permission from *Phys. Rev. B*, 1990, **41**, 6240)

the point of view of surface scattering, the important point is the *reduction* of the energy transfer when the scattering occurs at an angle rather than directly back into the incident direction. Averaging over the *impact parameter* with respect to a three-dimensional surface latitude will therefore of necessity *reduce* the energy transfer with respect to the value obtained in a one-dimensional model where the 'back-scattering angle' is always zero (an effect that is sometimes, incorrectly, attributed to the surface atoms behaving as though they are heavier than they actually are).

The influence of scattering angle on the energy transfer is demonstrated very nicely indeed in Figure 4, where recent data of Winters *et al*,[10] giving the final energy of Xe atoms scattered from a Pt(111) surface, are shown as a function of angle of emergence. The incident energy was 6.8 eV and the incident angle 30°. The dashed line shows the average value of the final energy which corresponds to an energy transfer of about 70% of the incident energy. The angular distribution of the emergent Xe, shown as the full line, is quite broad and centred at about 15° super-specular. This kind of behaviour is common and, in the absence of information as to the energy transfer, might have been taken as an indication that the surface is essentially flat, so that the parallel energy survives the collision and the energy transfer is at the

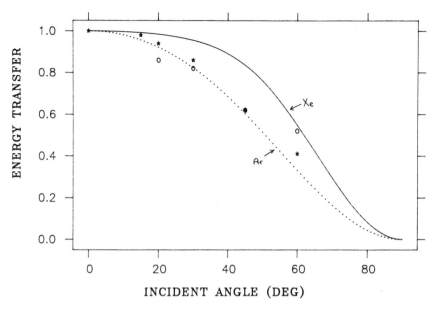

Figure 5 *Average energy transfer (corresponding to the dashed line in Figure 4) as a function of incident angle normalized to its value at $\theta_i = 0$. The filled and open symbols represent values measured for Xe and Ar atoms with incident energies 14.3 and 7.2 eV respectively scattering from a Pt(111) surface maintained at temperature $T_s = 800$ K (reference 10). The full (dashed) line gives the result of applying equation (14) for mass ratio appropriate for Xe–Pt(Ar–Pt) and with scattering angle equal to twice the incident angle (specular scattering)*

expense of the normal component. This would imply that the final translational energy and final angle are correlated, with atoms emerging close to grazing suffering the largest energy transfer. In fact, the observed trend in final energy (filled circles) is the opposite and the correct explanation for it is that the surface is not at all flat but is behaving like an array of individual scattering centres. The observed trend in final energy is then simply understood in terms of the fall-off of the energy transfer in *binary collisions* with scattering angle implied by equation (14). The data are influenced to some extent by the elevated surface temperature, which reduces the energy transfer and, more importantly, by multiple collisions which, presumably, are responsible for the saturation of the final energy towards higher scattering angles (Figure 4). The observed average energy transfer as a function of incident angle normalized to its value at normal incidence is shown in Figure 5. The filled and open symbols refer to Xe and Ar atoms with incident energies of 14.3 and 7.2 eV, respectively, scattering from Pt(111). If the scattering were due solely to specular 'two-body hits' with single surface atoms, the variation would be given by the lines, which are plots of equation (14) with the back-scattering angle $\theta = 2\theta_i$ and mass ratios $\mu = 0.2$ and 0.67

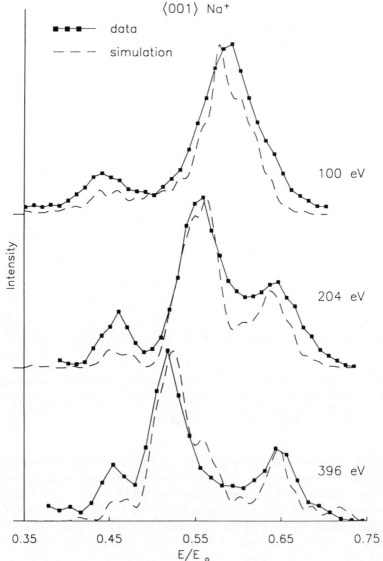

Figure 6 *Energy distributions of Na^+ ions back-scattered from a Cu(110) target for three incident energies, E_0 = 100, 204, and 396 eV. The incident and emergent angles were 45° (specular scattering) and the plane of scattering was along the (001) azimuth in the surface. The data (filled symbols) are compared with the distributions obtained via trajectory calculations using a sum of Hartree–Fock Na^+–Cu pair potentials plus an image potential that was allowed to saturate at 3.0 eV. The excellent agreement shows that this construction describes the interaction at the energies concerned rather well. The three features can be attributed to single- and double-scattering events from the first layer and single events from the second layer (see text)*
(Reproduced with permission from *Phys. Rev. B*, 1989, **39**, 13129)

for Ar–Pt and Xe–Pt respectively. While this very simple kinetic theory cannot be correct in detail because it does not account for the distribution of impact parameters that is responsible for the angular pattern or for multiple scattering, it does reproduce qualitatively the fall off of the energy transfer as the incident angle increases. Whether a full calculation using an accurate pair potential and accounting for multiple collisions would reproduce the data of Winters *et al.* quantitatively, in particular the angular patterns and the observed deviation of the energy transfer from linearity in the incident energy, is, at present, a matter of speculation.

A very detailed comparison of ion-scattering data for Na^+ and K^+ ions in the energy range 100–400 eV scattering from a Cu(110) surface with a theory based on the model of successive binary collisions has been carried out by Goodstein *et al.*[11] Using accurate Na^+–Cu and K^+–Cu pair potentials, they showed that the observed energy loss spectra could be accounted for quantitatively and were able to identify features in the spectra in terms of single and multiple scattering events. At these high energies there is significant lattice penetration and single scattering events from the *second* layer were found to give the strongest feature in the spectra. Some sample spectra are shown in Figure 6 for scattering along the (001) direction in the surface. The ordinate shows the intensity of the beam, proportional to the scattering probability, emerging in the specular direction for fixed incident angle of 45° (scattering angle $\theta = 90°$) as a function of ϵ_f/ϵ_i, where ϵ_f is the final energy, and for three values of the incident energy $\epsilon_i = 100, 204, 396$ eV. As can be seen, the calculations agree well with the data both with respect to peak position and relative intensity. This is a quite severe test of the pair potential, the precise functional form of which has a strong influence on the multiple scattering features. For the higher incident energies three peaks are readily identifiable while for the lowest incident energy two of these merge. The peak at the smallest final energy (and therefore the largest energy transfer) arises from direct single collisions from top layer atoms and falls at an energy close to that given by the kinematic formula equation (14) (mass ratio $\mu = 0.36$). The peak at highest final energy arises from a double scattering event where the ion is first deflected almost parallel to the surface by one surface atom and then re-deflected outwards by another. The much smaller energy transfer can readily be understood from the kinematic formula, equation (14), which shows that the energy transferred via two *forward* scattering events (*i.e.* through successive angles of about 135°) is substantially less than via a single back-scattering collision. The most interesting feature in the spectrum is the central peak which has by far the largest intensity. This is due to a single scattering event from the *second* layer and its large intensity arises from a *focusing* effect of the first layer atoms as the ion passes through. The energy transfer for this collision is larger than that of the double collision because it is a back-scattering event, and smaller than the top layer peak because deflection of the ions by the first layer increases the scattering angle for the binary collision. The binary collision model reproduces all effects, including the steering and focusing, very satisfactorily indicating that in the energy range considered the ion–surface interaction is very well approximated by a sum of pair potentials. At these high energies, the collision is so fast that the 'many-particle' interactions between the surface atoms play essentially no role and the manner in which the energy transferred is carried away from the interaction region does not affect in any material way the trajectory of the outgoing particle.

4 Translational–Rotational Conversion

When a molecule strikes a surface the interaction may depend in an essential way on the *internal* co-ordinates of the molecule as well as the location of its centre-of-mass with respect to the surface, and this dependence can influence the energy transfer significantly. The translational and internal energies of the molecule are coupled by the interaction so that a scattered molecule may on emergence display internal excitation at the expense of the incident translational energy. Measurement of the translational and internal energies for scattered molecules as a function of the incident conditions allows, in principle, conclusions to be drawn about the nature of the couplings in the collision region. Such measurements have become possible only recently with the advent of laser detection techniques and are resource intensive and difficult (for a recent review, see reference 12). The field is therefore relatively new and the available data sparse and limited to those molecules whose spectra lend themselves readily to laser detection. Much of the data are for the NO molecule and we restrict the discussion of this section to this and similar heteronuclear diatomics, *e.g.* CO, whose internal excitations comprise vibration of the bond co-ordinate and rotations of the bond axes. Of these, the *rotations* are the more important because the rotational anisotropy is very large so that *all* molecules undergo translation to rotational conversion to some extent (and most, as we will see, to a very large extent). Vibrational excitation, on the other hand, is not a particularly important feature of *mechanical* collisions because molecular 'springs' are typically much stiffer than the springs between the surface atoms, but can be extremely important for PESs that display a relatively easy route to dissociation. Since molecular dissociation is treated extensively in the next chapter we will not here discuss the role of vibrational excitation in this context.

Molecules like CO are strongly bound and have bond distances that are small compared with typical optimal separations between the atoms when these are chemisorbed individually on the surface (typically ~ the lattice spacing). There is therefore a *mis-match* between optimal C—O separations in the molecular and in the chemisorbed states. This is a simple way of understanding the origin of the large *energy barriers* to dissociation which exist in such cases. Even when a state of complete dissociation is thermodynamically stable on the surface, the energy required to access this state from the gas phase may be quite a sizeable fraction of the molecular binding energy. For incident energies well below this threshold the variation of the potential energy when the CO bond co-ordinate is stretched will depend only weakly on other co-ordinates, implying that vibrational excitation or deexcitation does not occur to any large extent during the collision. This has been confirmed in several theoretical studies, notably by Muhlhausen et al.,[13] who used a model PES to study the scattering of NO from Ag(111). Experiments on this system[14] showed vibrational excitation in the range of a few percent for incident energies up to 1 eV that could not be accounted for by the PES of Muhlhausen et al. An explanation in terms of an electronic mechanism[15,16] was proposed to account for these data.[14] However, excitation at the percent level can result from a single adiabatic PES if the translational→vibrational coupling is somewhat larger than in the Muhlhausen et al. model, so the experiment is not conclusive in this regard. Whatever the origin of the vibrational effect that is observed, the experiment confirms that this is relatively small.

This is by no means the case with regard to *rotational* excitation, which we now consider. By extending the one-dimensional model of Section 2 to include a rotational degree of freedom, we illustrate two important consequences of the coupling. These are: (a) an *anticorrelation* between energy transfer and rotational excitation *i.e.* molecules emerging with enhanced rotational energy transfer less energy to the lattice and (b) the occurrence of *multiple round trips* in the surface well as a result of *translational→rotational conversion*. First we use elementary chemical considerations to establish the basis for a model interaction.

The 'hard wall' which reflects inert atoms and molecules from a surface arises from the *closed shell* electronic structure of these probes. The electronic wavefunctions of the solid protruding into the selvedge must orthogonalize to the wavefunctions of the closed shells on the probe and this drives up the electronic energy of the system by an amount that depends roughly on the overlap between the undisturbed metal and probe orbitals. The mechanism is the same as that responsible for the repulsion between two helium atoms in the gas phase and is known as *Pauli repulsion* (because it is the Pauli principle which prevents the metal electrons from 'dropping down' into the filled levels of the probe). Pauli repulsion is responsible for the back-wall of the *physisorption* interaction of all truly inert probes and for the small binding energies (~ 6–100 meV) of these probes. In such cases the attractive branch of the interaction is the van der Waals energy, which is 'zeroth order' in the overlap of wavefunctions. A probe is genuinely 'inert', however, only if, in addition to a closed-shell electronic structure, it has only high-lying *unfilled* levels. For if there is such a level in the neighbourhood of the solid's Fermi energy, the metal electrons can occupy it and thereby achieve orthogonality to the lower lying probe levels. Accordingly the penetration of the surface is less punitive and the probe can profit more from attractive branches of the interaction and bind more strongly to the surface. The weakening of the Pauli repulsion due to the presence of low-lying orbitals is called *polarization*. To give a gas-phase illustration of the effect, the Be atom has a low-lying, unoccupied $2p$ level while He does not and this is the main reason why the energy curve of the Be_2 dimer has a well depth of

~0.1 eV that is larger than that of He_2 by two orders of magnitude. The nature and location of the lowest unoccupied orbital (the 'LUMO') of a probe is therefore a vital aspect of the surface chemistry. Molecular LUMOs are often 'antibonding' with respect to the molecular constituents (and are always antibonding for closed-shell, tightly bound, molecules like H_2, CH_4 that are highly unreactive in the gas phase), and their behaviour close to the surface determines to a large extent whether a molecule will dissociate or not. For the CO molecule, the LUMO is the $2\pi^*$ orbital and is relatively low lying. From the present point of view, the vital feature of the LUMO is that it is highly asymmetric with respect to the two constituents, being concentrated primarily on the C-end of the molecule. Just as the Pauli repulsion itself depends on the overlap of metal and molecular 'closed-shell' orbitals, so the weakening of the repulsion due to polarization depends on the overlap of the metal wavefunctions with the LUMO. Accordingly, we must expect the interaction of CO to be asymmetric according as to whether the molecule approaches the surface with its C-end or its O-end down. If the O-end is down, overlap with the LUMO is quite weak and we can expect only a marginal weakening of the Pauli repulsion and thus a 'well depth' not too far removed from typical physisorption values. If the C-end is down, there will be a large polarization and the well formed when the molecule approaches the surface will be much deeper than a physisorption well and will have its minimum much closer to the surface.

A simple model that displays this effect obtains on treating the molecule as a rigid rotor characterized by centre-of-mass co-ordinate z and orientation γ. In its simplest version, the model is an extension of the particle plus linear chain considered in Section 2, in which the potential binding the probe to the outermost chain atoms depends on both z and γ. The equations of motion for the chain atoms $i \neq 1$ are unchanged and given still by equation (1), while those for the probe and the outermost chain atom $i = 1$ are:

$$M\ddot{u}_1 = \kappa(u_2 - u_1) - \frac{\partial V(u - z, \gamma)}{\partial u}$$

$$m_p \ddot{z} = -\frac{\partial V(u - z, \gamma)}{\partial z} \quad (15)$$

where the potential now depends explicitly on γ. In addition, we must write an equation of motion for the angular co-ordinate:

$$I_p \ddot{\gamma} = -\frac{\partial V(u - z, \gamma)}{\partial \gamma} \quad (16)$$

where I_p is the molecule's moment of inertia. To complete the model we need to specify the pair-potential, $V(z, \gamma)$, which can conveniently be written as a modified Morse potential:

$$V(z, \gamma) = V_0 f(z - z_0)\{[1 + \lambda(1 - \cos\gamma)] f(z - z_0) - 2\}; \quad f(z) \equiv e^{-\beta z} \quad (17)$$

The rotational asymmetry of the potential is governed by the parameter λ. If this is zero a pure Morse potential with well depth V_0 is recovered and the potential is independent of the angular co-ordinate. If $\lambda \neq 0$, the well depth of the potential is:

$$V_{\min}(\gamma) = \frac{V_0}{1 + \lambda(1 - \cos\gamma)} \quad (18)$$

and the location of the minimum given by:

$$z_{\min}(\gamma) = z_0 + \frac{1}{\beta} \ln[1 + \lambda(1 - \cos\gamma)] \quad (19)$$

The maximum and minimum values of $V_{\min}$ are thus V_0 and $V_0/(1 + 2\lambda)$ and correspond to $\gamma = 0$ and π respectively. For the CO molecule, then, $\gamma = 0$ implies C-end down. As reasonable parameters characterizing the interaction with a surface we take the maximum well depth (equal to the binding energy) to be $V_0 = 1$ eV, $\beta = 1$ a.u. and $\gamma = 1.2$ which gives an O-down well depth of about 0.3 eV. An important feature of the pair potential, equation (17), is that the rotational corrugation is not just a constant factor but is applied only to the repulsive wall of the Morse. This ensures that the location of the well depth moves outwards as its value decreases and also that the rotational corrugation 'switches off' as the distance from the surface increases. Both effects represent reasonable physical behaviour (*cf.* the physical interaction at long range is of the van der Waals type and is only weakly dependent on the orientation).[17]

A rotor having translational energy but no rotational energy and experiencing the potential given in equation (17) will first accelerate towards the surface with little change of its rotational co-ordinate until the 'back-wall' of the interaction is encountered and the strong coupling between translation and rotation sets in. If the initial rotational angle $\gamma \neq 0$ or π, the rotor will then experience a torque and begin to rotate. Up to this point the gain in potential energy due to the well has been counterbalanced by an increase in the translational energy. The rotational kinetic energy gained will be at the expense of this and we can refer to *translational→rotational* conversion as occurring in the surface well. When the molecule rebounds into the outer regions of the potential where the coupling between rotational and translational motion is weak an amount of energy depending on the details of the trajectory will remain in the rotation. If the molecule escapes, it will be rotationally excited. However, if the rotational energy is large, sufficient energy may have been drawn out of the translational motion to prevent the particle escaping from the potential well *i.e.* the *trans→rot* conversion has *trapped* the particle on the first round trip. Note that this does not necessarily require that the chain absorb any energy at all: *i.e.* in principle trapping can

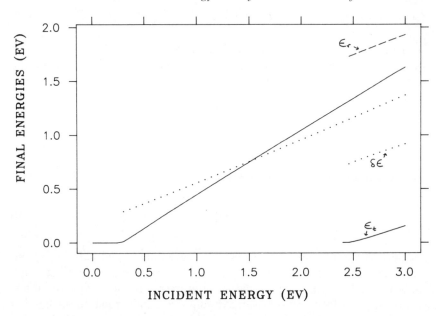

Figure 7 *Final energies of a CO molecule subject to a bond-orientation dependent potential following scattering from a linear chain of Pt atoms. The full line shows the final translational energy, the dashed line the rotational energy, and the dotted line the energy transfer. The molecule was oriented initially with its bond-axis perpendicular to the chain. This led to a 'first round-trip trapping threshold' that falls at an incident energy of 2.5 eV. For comparison, the final translational energy and the energy transfer for a CO molecule experiencing only the orientation-averaged potential are shown. In this case the trapping threshold is only 0.3 eV. The figure illustrates the massive transfer of initial translational energy into rotational energy that can occur for scattering systems of this type*

occur purely elastically, though in classical mechanics at least, it never will because some energy will always be transferred.

Figure 7 shows the final rotational and translational energies of the rotor of our model system, along with the energy transfer to the chain, as a function of incident energy. As before, the masses and spring constant are those appropriate for CO–Pt, the well parameters are as given above, and the CO was started out with $\gamma = \pi/2$ so that the bond axis is perpendicular to the chain (*i.e.* parallel to the 'surface'). For comparison the final translational energy and energy transfer with no rotational coupling and for an interaction averaged over γ are also shown. The huge difference between the two sets of results is due entirely to *trans→rot* conversion which is so effective that all particles with incident energy less than 2.5 eV are trapped on the first round trip. That is, the 'first round-trip trapping threshold' moves upwards in energy from ~ 0.3 eV in the absence of the rotational coupling to 2.5 eV in its presence (for the starting angle of $\gamma = \pi/2$). Above threshold, the

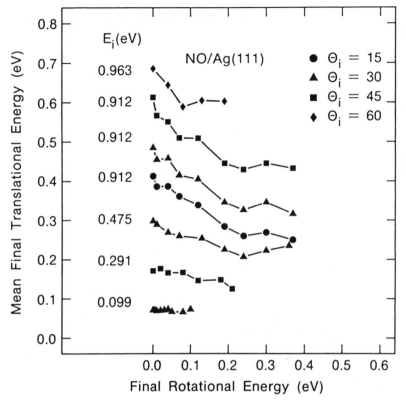

Figure 8 *Observed average translational energy of NO molecules scattering from an Ag(111) surface as a function of final rotational energy. The figure illustrates the anticorrelation between translational and rotational energy following the scattering, i.e. the higher the rotational energy the lower the translational energy in the final state. Since the slope of the curves is less than 45° (and the probability of vibrational excitation was small), the figure demonstrates also that the final rotational energy and the energy transfer to the surface are anti-correlated*
(Reproduced with permission from *Phys. Rev. Lett.*, 1986, **57**, 2053)

energy of the CO is overwhelmingly rotational (dashed line), so effective is the conversion of the energy during the collision. Note that the huge increase in the trapping threshold is *not* due to increased energy transfer. In fact, the energy transfer (dotted line) is *less* in the presence of the coupling than in its absence and we encounter an effect first established experimentally, that the final translational energy and energy transfer are *anticorrelated* with the final rotational energy. This is illustrated in Figure 8 where the measured final translational energies of NO molecules scattering from Ag(111)[18] are plotted against the measured emergent rotational energy. The reason for the anticorrelation is that the acceleration of the rotational co-ordinate feeds energy continuously out of the translational co-ordinate as the particle traverses the well and *cushions* the impact. The particle thus behaves as though its kinetic

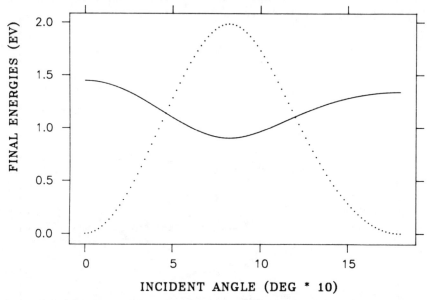

Figure 9 *Final rotational energy (dotted line) and energy transfer (full line) for a CO molecule scattering from a linear chain of Pt atoms as a function of initial orientation angle ($0°$ and $180°$ implies C- and O-end down, respectively). The incident energy was 3.0 eV. The figure illustrates the 'cushioning' effect of the rotational excitation that leads to an anticorrelation between final rotational energy and energy transfer*

energy is reduced from its actual value and, as Figure 7 illustrates, the behaviour of the energy transfer of our model system in the presence of rotations parallels that in their absence, with an upward shift of ~ 1 eV (*i.e.* it requires ~ 1 eV more translational energy in the presence of rotational coupling to give the same energy transfer as would be obtained in the absence of this coupling). The cushioning effect of the rotations is particularly clear in Figure 9, where the final rotational energy of emerging CO (dotted line) and the energy transfer (full line) is plotted as a function of the initial orientation of the bond axis for fixed incident energy of 3.0 eV. For initial bond angles $\gamma = 0$ or π the CO is exactly end on, no torque is exerted and the rotational coupling is exactly zero throughout the collision. The energy transfer then displays a local maximum and has a slightly larger value for C-end down than for O-end down because the potential well is deeper. (Note that the difference is not very dramatic because the collision at 3 eV is quite fast and the final energy close to the 'two-body' limiting value of 1.7 eV, as given by the Baule formula, equation (7).) If the molecule is only slightly tilted as it approaches the chain, the torque developed is weak and the particle makes a full round trip and escapes before any substantial acceleration of γ can occur. The emergent rotational energy is thus very small. For an initial orientation close to side-on, on the other hand, the torque is sufficient to build up a massive rotation as the CO traverses the well

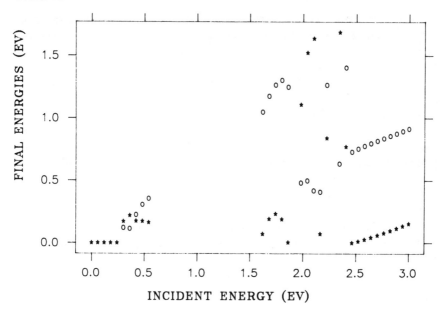

Figure 10 *Final translational (filled symbols) and rotational (open symbols) for CO molecules scattering from a linear chain of Pt atoms. The conditions are identical to those of Figure 7 but the trajectories are allowed to make two round trips in the surface well. The figure illustrates 'quasi-chaotic' 'second round-trip emergence' for incident energies below the first round-trip trapping threshold (see text)*

and a large portion of the final energy, up to almost 100% for $\gamma \sim \pi/2$, goes into the rotational co-ordinate. The correlation between final rotational energy and energy transfer in this example is, of course, more striking than in the experiment because we have isolated specific trajectories in one spatial dimension and have not averaged over an ensemble.

Referring back to Figure 7, and the large value of 2.5 eV for the trapping threshold, we consider what happens to the particles that are trapped on the first round trip due to the *trans→rot* conversion. Most of these molecules have a positive total energy and may escape on subsequent round trips if their rotational energy is reconverted into translational energy. In drawing up Figure 7, trajectories were stopped if the first round trip culminated in a return of the particle towards the surface (*i.e.* the truncation condition was that the centre-of-mass velocity undergo exactly two changes of sign). Figure 10 shows the result of carrying the calculation one round trip further and requiring four changes of sign before declaring a trajectory as trapped. Trajectories that escaped on the first round trip are unaffected by this but we see that below the first round-trip trapping threshold there is now substantial second round-trip emergence. Unlike the first round-trip emergence, however, the dependence of the final translational energy (filled symbols) and the energy transfer (open symbols) on initial conditions is very far from simple. Particles emerge after two round trips in narrow bands of incident energy with the behaviour within each band being regular but very different from band to band (the spacing of the trajectories in incident energy is too coarse to reveal the full details of the behaviour). This is typical of mechanical systems displaying chaotic behaviour (rather, for nomenclatural purists, 'extreme sensitivity to the starting conditions'). The 'chaos' is due in the present case to the coupling, and the sensitivity of the $rot \rightarrow trans$ (or further $trans \rightarrow rot$) that occurs on the second round trip to the precise phase with which the trajectory was started out. The longer the trajectory is allowed to run the more drastic this sensitivity will be. (True chaos, in the mathematical sense, sets in only if the trajectory remains within range of the coupling for ever. For a discussion of the relevant concepts see reference 19 and references therein.) The origin of this sensitivity is readily visualized if one imagines dropping a rubber egg on a hard table and trying to control via the initial orientation how it will move after, say, the second or third bounce. Whilst our model is highly idealized, it actually displays a reduced tendency to 'chaos' *vis-à-vis* a real surface, where the potential will be translationally as well as rotationally corrugated so that particles trapped by $trans \rightarrow rot$ conversion will also translate along the surface. The behaviour following the second bounce will then depend not only on the phase with respect to the rotation, but also the precise point within the unit cell where the second 'hit' occurs.

In the face of this demonstration of 'mechanical unpredictability' one may be tempted to conclude that CO–surface collisions are a hornet's nest we would do well to steer clear of. In fact, matters are not so bad and, when viewed in a *statistical* light via averaging over the ensemble of parameters that are not uniquely specified in an experiment, a degree of simplicity is restored, and we do not need to delve too deeply into 'chaos theory' to gain some qualitative insight into observed behaviour. We illustrate this via two examples. The first concerns the angular distributions of back-scattered particles. To calculate angular distributions we need to extend the dimensionality of the system and consider scattering from a two-dimensional array of Pt atoms spaced as on a Pt(111) surface. We need be concerned only with a few details of the model and refer to the appropriate reference for the remainder.[20] The interaction was constructed as a pairwise sum of CO–Pt potentials given by equation (17) but with the orientation angle γ referred to the radius vector joining the atom in question to the centre-of-mass of the CO. This introduces a strong coupling between the lateral and rotational corrugations of the interaction because any given orientation angle can be optimal for only one of the surface atoms of the array. One consequence of this is that the optimum tilt angle in the surface well depends on the lateral position with respect to the surface mesh. For an on-top site, the pair potential at that site dominates and the optimum orientation of the bond axis is along the normal direction with the C-end down. As the molecule moves away from the on-top site, its pair potential with this site remains dominant and the bond orientation follows the radius vector and so tilts with respect to the normal. As a bridge site is approached, the molecule rights itself again, this being the optimal 'mis-match' of the orientational dependences of two equal nearest neighbour pair potentials. Precisely this behaviour has been found in explicit calculations of the interaction of CO with an Al cluster.[21] Because of the strength of the interaction (the binding energy of CO on Pt is 1.4 eV) and of the coupling between normal and parallel translational energy and bond orientation, the fate of a given trajectory is difficult to predict for virtually all values of the incident parameters. However, the angular distributions have a simple interpretation nevertheless and an example is shown in Figure 11a. This was calculated by running many-particle trajectories for a range of impact parameters and initial CO bond orientations and binning the outgoing particles at 5° intervals. The incident energy was 1.0 eV and the angle of incidence 60° (the specular direction is marked with an arrow). The two distributions shown refer respectively to trajectories that emerged after just one round trip and after up to four round trips. The latter distribution was normalized to unity but includes nevertheless only about half of all trajectories run (the remainder 'scudded' along the cluster and reached its end before a decision was made as to whether re-emergence would occur). The distributions consist of a broad 'lobe' centred slightly towards the normal from specular superposed on a background that spans practically the entire half space. The background is prominent only for molecules that made more than one round trip, for which it forms the dominant component (as one can see by subtracting the two curves to obtain the distribution of molecules that emerged after 2–4 round trips). Distributions of this type (though invariably with a weaker background) are commonly observed in beam scattering and are invariably interpreted in terms of direct scattering (main lobe) and trapping-desorption (background). The latter fraction is assumed to trap on initial collision, accommodate on the surface, and desorb on a time scale much longer than the collision time via thermal fluctuations. This interpretation is quite widespread and, in fact, it is fairly common practice to regard the relative strengths of the backgrounds observed at different energies and angles as a measure of the behaviour of the sticking coefficient (on the assumption that at low surface temperature all the particles contributing to the background would stick). Figure 11a includes *only direct scattering* and shows that, while this interpretation *may* be correct in some cases, it is not the only possible interpretation of the background, particularly when we have reason to believe the PES may display 'mechanically chaotic behaviour'. All that is necessary for the particle to emerge in the 'background' rather than the main lobe is 'loss of memory' of the incident conditions, and this can be achieved after just a few round trips in the well. The important point that was not properly appreciated before the 'chaos era' is that the time scale on which mechanical systems 'thermalize' is not necessarily long compared with a typical round-trip time but depends on the rapidity with which classical trajectories started at neighbouring points in phase space diverge.

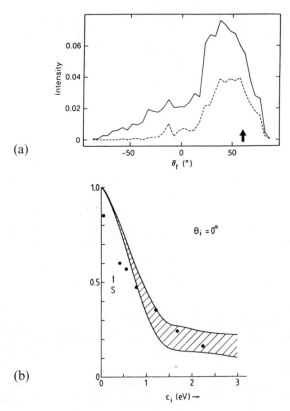

Figure 11 (a) *Angular distributions for CO molecules scattered from a two-dimensional array of Pt atoms. The dashed line refers to particles that emerged after a single round trip, the full line to particles that emerged after up to four round trips. The difference is basically a diffuse background illustrating that molecules that perform more than one round trip have 'lost memory' as to the incident conditions.* (b) *Measured sticking coefficient of CO on Pt(111) (filled symbols) for normal incidence and as a function of incident energy. The hatched area (see text) reflects the sticking behaviour of CO interacting via a model PES with a two-dimensional array of Pt atoms. The comparison indicates that the slow fall off of the measured sticking coefficient with incident energy can be attributed to 'trapping via translational→rotational conversion'*
(Reproduced with permission from *J.Chem.Phys.*, 1989, **91**, 6421)

A second point that sometimes leads to confusion is the implicit assumption that seems sometimes to be made that observation of a strong main lobe around the specular direction implies that the surface is in some sense 'flat'. Again, this *may* be correct, but *need* not be. Of the trajectories contributing to the 'first round-trip' distribution in Figure 11a, not a single one underwent anything like 'specular reflection'. The distribution is skewed in the 'forward direction' not because the trajectories reflect specularly but merely because in a single traverse of the well the parallel kinetic energy is not very efficiently transferred to the substrate.

The effects discussed above are surely exaggerated in our simple two-dimensional model, which restricts the trajectories to a line where the lateral corrugation of the potential is a maximum. However, they are equally surely present to some degree when molecules like CO are scattered from a real surface. In fact, the data of Verheij et al.[22] for CO scattering from Pt at 65 meV incident energy may be one of a very few instances where the consequences of *fast mechanical thermalization* has actually been demonstrated experimentally. These authors studied distributions of emerging CO using a pulse shape analysis that allowed them to separate signals arising from processes with widely differing time scales. For wide angle incidence they found angular distributions comprising a main lobe and a diffuse component *with no detectable time delay*. This suggests an interpretation along the lines given above, with effective thermalization occurring in just a few round trips as a consequence of mechanical chaos. These data were taken at a surface temperature of $T_s = 585$ K to prevent accumulation of the CO on the surface. If this scattering system is actually displaying mechanical chaos, a fast diffuse component would also have to be present at low T_s. Experimental verification might be difficult because of the high sticking coefficient, but it would seem to be worth some little effort to find out.

The influence of the rotational corrugation of the CO–surface interaction, and the multiple round trips to which this gives rise, on the sticking behaviour is illustrated in Figure 11b. The symbols refer to measured values[20] and the hatched area to calculations for the two-dimensional cluster referred to above. The theoretical results are quoted as an area rather than a single line because some of the particles scudded over the cluster before a clear cut decision as to sticking or re-emergence was made. The upper edge assumes all these stick, the lower one that none of them do. The remarkable thing about the data is the high value of the sticking coefficient at energies ≥ 2.0 eV and the slow fall off with energy. This cannot be a consequence of energy transfer to the lattice during a single round trip because even the 'well-depth corrected' Baule critical energy, equation (11), is only 1.1 eV for this system and this should serve as an upper limit for the trapping threshold in mechanical collisions. In fact, the trapping threshold for C-down orientation, normal incidence on-top collisions, the most favourable case in the absence of rotations, is about 0.8 eV. The inclusion of the rotational degree of freedom, on the other hand, explains the observed sticking adequately in terms of *trans→rot* conversion on initial collision and enhanced energy loss due to multiple round trips. At low energy, the theoretical sticking coefficient goes to unity while the experimental value is about 0.84. Calculations by Billing[23] at elevated temperature confirmed that this difference can be attributed to the surface temperature of 350 K used in the experiment. In addition, Billing used a PES that allowed for breaking of the CO bond, but found negligible dissociation up to incident energies as large as 8.0 eV.

Experimentally, no dissociation at all was detected up to the highest attainable incident beam energy of 3.5 eV. This again highlights the relative inefficiency of excitation of the molecular bond co-ordinate in mechanical collisions.

Experimentally, the system NO–Ag(111) has emerged as a prototype for study of the effects of rotational coupling and excitation. The probe molecule has readily accessible spectral bands that facilitate laser detection and its interaction with Ag is relatively weak, with a well depth of ~ 220 meV.[24] The first state-resolved measurements for this system, by Kleyn et al.,[25] revealed explicitly the very high level of rotational excitation with detection of a significant fraction where almost all the incident energy emerged as rotational energy. This and other work up to 1988 has been summarized by Rettner.[26] More recently, detailed studies of the velocity distributions of molecules emerging in specific rotational states have been reported[27] and data have become available that were taken with beams having the molecular axis *oriented* with respect to the surface normal, *i.e.* preferentially O-down or N-down orientations.[28] Most recently, these measurements have been extended to include also the distribution of rotational energy for different emergent angles.[29] Some of the theoretical work on this problem is described in Chapter 3. The initial interpretation of the Kleyn et al. data[30] was in terms of a 'stiff-lattice' potential whose orientation dependence gave rise to a double 'rainbow' (two points of inflection) and corresponded to the ground state for adsorbed NO having the N—O axis almost parallel to the surface. This would seem contrary to the elementary chemical considerations outlined above which suggest that the angular dependence should follow the overlap of the $2\pi^*$ orbital with metal orbitals (presumably monotonic with the largest binding energy for the N-down orientation). However, in view of the low binding energy, which suggests an almost physisorption-like interaction with the total electronic energy determined by the balance of individually large, partially cancelling, contributions with different angular dependences, a more complex orientational dependence cannot be ruled out. Nevertheless, agreement achieved with data on the basis of *any* stiff-lattice potential which ignores completely the very large energy transfers in this system[27] (in rough accordance with the Baule factor of 0.7) cannot be regarded, in any sense, as conclusive. The most extensive calculations reported to date, those of Muhlhausen et al., referred to earlier,[13] used a more conventional potential construction with a monotonic fall-off of the well depth with increasing bond inclination, and treated all salient aspects of the system within classical mechanics. Although the rotational distributions of Kleyn et al. were reproduced quantitatively, the model PES has a well depth more than three times larger than the experimental value. Since the rotational distributions were found to be very sensitive to the well depth[13] (and assuming the experimentally assigned well depth is correct), one would have to conclude that the achieved agreement is spurious. In a more recent paper on the anticorrelation which exists between energy transfer and rotational excitation, Kimman et al.[18] stated that a minor modification of the

PES of Muhlhausen et al. gave the correct well depth and also qualitative agreement with the salient features of all existing data. No reason for this apparent contradiction of the claim of Muhlhausen et al.—that a shallower interaction could *not* reproduce the data—was given and, as far as the present author is aware, details of the new PES and the angular couplings to which it corresponds, together with its relation to the earlier one, have not been published. At present, therefore, in spite of a wealth of data and a remarkably extensive theoretical literature on the NO–Ag(111) system, the nature of the fundamental interactions and the angular couplings remains a matter of speculation.

5 Scattering from a Quantum Lattice

Quantum effects in particle–surface scattering are the subject of Chapter 3 of this volume and we will here just complement the above discussion by focusing on the quantization of the *lattice* motion and its implications for the energy transfer. *En route* we will establish a criterion for the validity of the classical limit. Not particularly surprisingly, this will turn out to be that the energy transfer, $\delta\epsilon$, is large compared with the energies, ϵ_k, of the quantized lattice vibrations (phonons).[31] If $\delta\epsilon$ lies within the band of phonon energies, the scattering is in the quantum limit and classical mechanics breaks down completely. The most important consequence is the possibility of *elastic scattering* where exactly zero energy is exchanged.

The quantum mechanics of the *harmonic lattice* is particularly simple because the stationary states or eigenmodes are just the normal modes of the classical equations of motion. Thus, the displacement patterns for the normal modes of the linear chain, equation (2), are also *eigenfunctions* of the quantum mechanical Hamiltonian and correspond to stationary state energies $\epsilon_k = \hbar\omega(k)$, where $\omega(k)$ are the natural frequencies of the chain, also given in equation (2). The range of ϵ_k-values spans the *phonon band*, which therefore has width $\epsilon_0 = \hbar\omega_0$. Typically, ϵ_0 for solids is of order 25–40 meV. To illustrate how the quantum lattice responds to a collision, let us adopt the *forced oscillator model* within which the particle that scatters from the lattice is assumed to follow a classical trajectory.[32–35] The trajectory carries the particle towards the lattice, from which it rebounds and escapes. If $z_p(t)$ is the location of the particle at time t along the trajectory, we can define a force:

$$F(t) \equiv -\frac{\partial}{\partial z_0} V[z_0 - z_P(t)] \tag{20}$$

where z_0 is the co-ordinate of the outermost chain atom and V the pair interaction between this and the scattering particle (z_0 will in general depend on t unless we run the trajectory with a 'stiff-lattice' potential which is adequate for many purposes). $F(t)$ is the force acting on the outermost chain atom at time t during the collision and switches on and off so that $F(-\infty) = F(\infty) = 0$. While we consider the simplest possible case, the generalization to include forces on all the atoms of the chain (or of a three-dimensional lattice) is trivial. The influence of the particle on the chain is now reduced to the response of the chain to the time-dependent force, $F(t)$, which acts on the outer atom and gives rise to an additional potential $V(t) = z_0 F(t)$ in the chain's Hamiltonian. Suppose the chain starts off in its ground state (or no-phonon state), $|\phi_0\rangle$, and write for its state at time t under the action of perturbation $\hat{V}(t)$:

$$|\psi\rangle_t = \sum_k a_k(t) |\phi_k\rangle \tag{21}$$

where $|\phi_k\rangle$ is a state with one phonon having wavevector k excited. Then within first-order perturbation theory (which we crib from any elementary quantum mechanics book[36]) we have:

$$a_k(\infty) = -i \langle \phi_k | z_0 | \phi_0 \rangle F(\epsilon_k) \tag{22}$$

where

$$F(\epsilon) \equiv \int_{-\infty}^{\infty} dt\, F(t) e^{i\epsilon t} \tag{23}$$

is the Fourier transform of the time-dependent force. (For brevity, we adopt atomic units where $\hbar = 1$, lengths are in units of the first Bohr radius and masses in units where the electron mass $= 1$.) The matrix element is just the dipole matrix element for excitation of a harmonic oscillator from its ground state to its first excited state and we can crib this also from any quantum mechanics text:[36]

$$\langle \phi_k | z_0 | \phi_0 \rangle = \frac{1}{\sqrt{(2M\epsilon_k)}} u_0(k) \tag{24}$$

where M is the mass of the oscillator (in our case of the atoms in the chain) and $u_0(k)$ is the normalized displacement eigenvector of the oscillator [in our case, projected onto the outermost atom; viz., from equation (2), $u_0(k) \sim \cos(ka/2)$]. Thus, to lowest order in $\hat{V}$, the probability that the lattice left by the collision in one-phonon state $|\phi_k\rangle$ is:

$$|a_k(\infty)|^2 = \frac{|F(\epsilon_k)|}{2M\epsilon_k} |u_0(k)|^2 \tag{25}$$

Since the state $|\phi_k\rangle$ corresponds to lattice energy ϵ_k we can now write down the *energy gain* distribution of the lattice (which by correspondence is also the *energy loss* distribution for the particle:

$$P_1(\epsilon) = \sum_k |a_k(\infty)|^2 \delta(\epsilon - \epsilon_k) \qquad (26)$$

This is denoted $P_1(\epsilon)$ because it refers to energy transfer arising from *single phonon* processes. $P_1(\epsilon)$ can be factored conveniently to separate out the coupling to the particle via the force Fourier transforms and the availability of phonon states of requisite energy, determined by the *local density of phonon states*, $\rho(\epsilon)$, defined by:

$$\rho(\epsilon) \equiv \sum_k |u_0(k)|^2 \delta(\epsilon - \epsilon_k) \qquad (27)$$

We have,

$$P_1(\epsilon) = \frac{|F(\epsilon)|^2}{2M\epsilon} \rho(\epsilon) \qquad (28)$$

This result can be generalized trivially to a three-dimensional lattice, for which the force becomes a tensor with indices running over three spatial directions and those lattice sites involved in the direct collision. The relevant 'local density of phonon states' is also a tensor and the product in equation (28) is of the form $\hat{F} \cdot \hat{\rho} \cdot \hat{F}^*$ where the dots imply summation over three spatial directions and as many surface sites as the force tensor has non-zero components. Within our chain model, there is only one such site and one spatial dimension. We will need shortly to use two properties of the local density of states, which follow from the orthonormality and completeness of the phonon eigenvectors $|\phi_k\rangle$:

$$\int d\epsilon \rho(\epsilon) = 1 \qquad (29)$$

$$\langle \phi_0 | z_0^2 | \phi_0 \rangle = \sum_k |\langle \phi_k | z_0 | \phi_0 \rangle|^2 = \frac{1}{2M} \int d\epsilon \frac{\rho(\epsilon)}{\epsilon} \qquad (30)$$

Here, z_0 is the co-ordinate of the outermost atom of the chain and equation (30) links the mean square fluctuation in this co-ordinate with the chain in its ground state to the first inverse energy moment of the local density of states. These properties also can be appropriately generalized to three dimensions.

A very attractive feature of the harmonic lattice, and one that is unique to this particular many-body system, is that the *exact* distribution of energy extracted from a time-dependent force (*i.e.* valid to all orders in perturbation theory), $P(\epsilon)$, can be written in terms of the single-phonon distribution. The derivation is a bit wearisome and we just quote the result:[37]

$$P(\epsilon) = \int_{-\infty}^{\infty} \frac{dt}{2\pi} \exp\{i\epsilon t + \int d\epsilon' [e^{-i\epsilon' t} - 1] P_1(\epsilon')\} \qquad (31)$$

Unlike $P_1(\epsilon)$, which gives the energy distribution only when its overall weight is small and the dominant 'event' is zero-energy absorption (elastic scattering), $P(\epsilon)$ is properly normalized to unity and gives the distribution of energy drawn from the time-dependent force exactly and under all circumstances. A number of useful properties can readily be proved directly from equation (31):

$$\delta\epsilon \equiv \int d\epsilon\, \epsilon P(\epsilon) = \int d\epsilon\, \epsilon P_1(\epsilon) \tag{32}$$

$$\sigma^2 \equiv \int d\epsilon (\epsilon - \delta\epsilon)^2 P(\epsilon) = \int d\epsilon\, \epsilon^2 P_1(\epsilon) \tag{33}$$

$$P(\epsilon) \to e^{-2W} \delta(\epsilon) \quad \text{as} \quad \epsilon \to 0 \tag{34}$$

where

$$2W \equiv \int d\epsilon\, P_1(\epsilon) \tag{35}$$

is the *Debye–Waller factor* for the collision. These equations show that: (*a*) the *average energy transfer* absorbed by the lattice is the mean of the *single phonon distribution*, (*b*) the width of the energy distribution about its mean is the second energy moment of the single phonon distribution and, (*c*) that there is a non-zero probability, $P_{el} = e^{-2W}$, that the lattice absorbs *no energy at all*, i.e. that the collision is purely elastic.

Let us examine this aspect first. In classical mechanics, all collisions will excite the lattice to some degree so *elastic scattering* is an intrinsically quantum effect. To gain some insight into the likelihood of observing the effect for a given quantum system, we make use of equation (28) to write:

$$2W = \int d\epsilon\, \frac{|F(\epsilon)|^2}{2M\epsilon} \rho(\epsilon) \tag{36}$$

If the collision is fast, then the Fourier coefficients of the force will vary on an energy scale that is large compared with the phonon band width where the phonon density of states differs from zero. In this case, we may write, using equation (30):

$$2W \approx |F(0)|^2 |\langle \phi_0 | z_0^2 | \phi_0 \rangle|^2 = P_0^2 \langle z_0^2 \rangle \tag{37}$$

where $P_0 \equiv \int dt\, F(t)$ is the momentum transferred to the surface atom by the fast collision. The Debye–Waller factor is just the product of the square of the momentum transfer times the mean-square displacement of the atom that is struck. In the present case, this is a ground state expectation value and is due solely to the *zero-point* fluctuations of the lattice. At finite surface temperature, $\langle z_0^2 \rangle$ acquires an additional contribution due to thermal fluctuations, the Debye–Waller factor increases and the elastic scattering probability, P_{el}, falls off *exponentially* with T_s. The first prerequisite for observation of a substantial elastic fraction, therefore, is low T_s. Secondly, if we assume for simplicity that the collision essentially reverses the direction of the probe's velocity, the momentum transfer is just $P_0^2 = 8m_p \epsilon_i$, where ϵ_i is the incident energy and m_p the mass of the probe. Thus, for given incident energy, P_{el} falls off exponentially with the probe mass. The second prerequisite for observing a large elastic fraction, therefore, is to use *light mass* probes, *e.g.* H_2 and He. A third factor that can influence $2W$ is the *speed* of the collision. If the probe particle is heavy and traverses the interaction slowly equation (37) gives an overestimate of $2W$ because the force Fourier transform in equation (36) falls off rapidly across the phonon band and cuts off the contribution of higher energy phonons. Since heavy particles travel slower than light ones, this effect can counteract the larger momentum transfer and, in extreme cases, even reverse the trend. For the same reason, slow collisions are much more strongly affected by the surface temperature than fast ones because the energy transfer is dominated by low energy phonons, whose populations in a thermal ensemble are most strongly affected by T_s.

Turning now to the energy distribution, $P(\epsilon)$, we establish the classical limit by working out the integral for large energy. If ϵ is large the phase factor $i\epsilon t$ in the exponent of the exponential in equation (31) will vary rapidly compared with the integral term, whose t-dependence is on a time scale ϵ_0^{-1} [The single-phonon distribution, $P_1(\epsilon)$ is different from zero only within the phonon band $0 < \epsilon < \epsilon_0$]. Accordingly, we can expand the integral term in the exponent as a power series in t. Retaining terms linear and quadratic in t and using equations (32) and (33), we find:

$$P(\epsilon) \to \int \frac{dt}{2\pi} e^{i(\epsilon - \delta\epsilon)t} \exp[-\sigma^2 t^2/2] = \frac{1}{\sqrt{(2\pi\sigma)}} \exp[(\epsilon - \delta\epsilon)^2/2\sigma^2] \quad (38)$$

The energy distribution is therefore a Gaussian, normalized to unity, and centred symmetrically about the average energy transfer, $\delta\epsilon$, with width σ (Figure 12b). This result gives only the limiting behaviour of $P(\epsilon)$ for $\epsilon \gg \epsilon_0$. However, if $(\delta\epsilon - \sigma) \gg \epsilon_0$, the entire distribution falls in this high-energy region and so is the correct distribution at all energies. Details of the collision, via the force Fourier transforms, influence the parameters $\delta\epsilon$ and σ but not the Gaussian shape of the distribution. For a fast collision, where $F(\epsilon)$ varies little over the phonon band, we can replace $F(\epsilon)$ in all formulae by $F(0) = P_0$, the momentum transfer in the collision. From equations (28), (29), and (32), we then find:

$$\delta\epsilon \sim \frac{P_0^2}{2M} \int d\epsilon \rho(\epsilon) = \frac{P_0^2}{2M} = \Delta_{cl} \quad (39)$$

where Δ_{cl} is the *classical* energy transfer in an impulsive collision. Similarly:

$$\sigma \sim \sqrt{(\epsilon_0 \Delta_{cl})} \tag{40}$$

so the width of the distribution is just the geometric mean of the classical energy transfer and the phonon band width. The width increases with the energy transfer but as a square root, so that σ decreases *relative* to the energy transfer, i.e. $\sigma/\Delta_{cl} \to \sqrt{(\epsilon_0/\Delta_{cl})}$. This is the way in which the response of a quantum lattice approaches that of a classical lattice in the limit of high energy. Under conditions that prevail in molecular beam experiments σ is rarely negligible. For instance, for classical energy transfer 0.2 eV and the phonon band edge at ~ 40 meV, $\sigma \sim 0.09$ eV and the energy distribution is almost as broad as its average value (which means in this case that the Gaussian formula is not the correct energy distribution over the full range of energies). When classical mechanics is used to treat the response of the lattice, therefore, one should in principle assign to any given trajectory a Gaussian distribution of energy transfers centred on the classical energy transfer. This is rarely done in practice, though the smearing out of the discrete classical losses may be just as important as averaging over impact parameter or rotational angle.

The extreme quantum limit obtains when the energy transfer lies in the phonon band width, $\delta\epsilon \leq \epsilon_0$. In this limit we can expand the exponential in the integrand of equation (31) as a power series in P_1 and obtain:

$$P(\epsilon) \approx e^{-2W}\{\delta(\epsilon) + P_1(\epsilon) + \frac{1}{2}\int d\epsilon' P_1(\epsilon - \epsilon')P_1(\epsilon') + \ldots\} \tag{41}$$

whose terms represent successive processes involving more and more phonons. The first term corresponds to zero phonons and its weight gives the elastic scattering probability, $P_0 = e^{-2W}$. The second is the one-phonon distribution we calculated above scaled down by P_0 to take account of the unitarity. The third term describes the excitation of two phonons, and so on. The series convergence is governed by the Debye–Waller factor with the weight of the N phonon term given by $\sim (2W)^N/N!$. A phonon expansion converges rapidly, therefore, only when the dominant event (at low T_s) is elastic scattering. Sketches of the energy distributions that obtain in the classical and quantum limits are shown in Figure 12. In the classical limit, the distribution is Gaussian irrespective of the details of the collision, and its width is much smaller than the average energy transfer. In the quantum limit the distribution has a δ-function contribution at zero energy and an inelastic tail that is *broader* than the average energy transfer and depends in an essential way on the collision, unless this is fast, in which case the energy dependence just reflects that of the phonon density of states. Note that for a fast collision, the *average* energy transfer is given correctly by classical mechanics even in the extreme quantum limit. This refers, however, to a distribution that is completely different from the narrow Gaussian that characterizes the classical regime. The quantum form of the distribution has a crucial effect on all observables, in particular on the behaviour of the

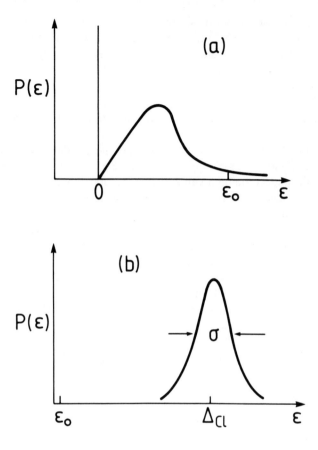

Figure 12 *Schematic energy distributions for a harmonic lattice whose surface atoms are subject to a time-dependent force that switches on and off. In the quantum limit (a) the distribution consists of a δ function contribution whose weight gives the probability for zero energy transfer (elastic scattering) and an inelastic distribution due to excitation of one or more phonons. The quantum limit is characterized by a distribution that is broader than its average value. In the classical limit (b) the distribution is a Gaussian centred on the classical energy transfer and with a width that is the geometric mean of this and the maximum phonon energy. In order for the classical limit to obtain, the distribution must be much narrower than its average value, $\Delta_{cl} \gg \epsilon_0$*

sticking coefficient, S, which, in any classical calculation, will always approach unity for incident energies that are small compared with the depth of the well. Quantum mechanically, however, the presence of elastic scattering places a limit on S of less than unity. Within the forced oscillator model the limiting value is $\sim (1 - P_0)$, where P_0 refers to the saturation value of the scattering probability at low energy. (This limit holds in the meV range of energies but is not the same as the true 'zero-energy limit' which depends on the quantum nature of the probe whose wavefunction tends to develop a node in the surface region. If the inelastic coupling can be treated in perturbation theory it can be shown that at exactly zero energy S is identically zero. To date, no exact non-perturbative treatment of this problem has been given and it remains quite intriguing.[32,38] However, the energy scale on which this particular quantum behaviour of the particle becomes crucial is many orders of magnitude smaller than any energy of relevance in beam scattering.)

Figure 13 shows the behaviour of the sticking coefficient in the extreme quantum limit. The measurements are for H_2 and D_2 incident normally at a Cu(100) surface and are shown as a function of incident beam energy.[39] The saturation of S at values of about 0.15 and 0.2 eV for H_2 and D_2 respectively, is clear and reflects the high value of the specular reflectivity for this system (*i.e.* the dominant scattering event is reflection of the H_2 as though from a mirror). The full lines in Figure 13 refer to calculations performed using the forced-oscillator model (FOM) considered above, and the distorted wave Born approximation (DWBA), where the entire problem of scattering particle plus lattice is treated quantum mechanically to first order in the particle–lattice coupling.[39,40] This means that only *single phonon* processes are included. Within the FOM, the sticking coefficient is interpreted in terms of the probability that the lattice absorbs more energy than the incident particle originally possessed:

$$S(\epsilon_i) = \int_{\epsilon_i}^{\infty} d\epsilon\, P(\epsilon) \qquad (42)$$

with the energy distribution, $P(\epsilon)$, given by equation (31) (appropriately generalized to three spatial dimensions). As can be seen, the theory gives a quite reasonable description of the fall-off of S over the entire energy range and agrees with the DWBA acceptably at low energy. This shows that the quantum nature of the particle is much less important than the quantization of the lattice vibrations, a point that has been stressed by Brenig.[41] For energies > 15 meV, which is only half of the bulk phonon band edge, the DWBA falls off far too rapidly as compared with the experiment, while the FOM remains reasonably accurate. The reason for this is the large contribution to the phonon density of states due to *surface phonons* (lattice waves whose displacement fields are localized at the surface). This contribution extends up to the surface phonon band edge, which for Cu is at about ~ 15 meV, and is sufficiently important that for energies that lie in the upper half of the phonon band, processes involving two surface phonons are much more important than single bulk phonon processes.

The behaviour of the sticking coefficient in Figure 13 is symptomatic of *direct* or 'normal' phonon processes whereby the incident energy of the particle is taken up by phonons during the initial collision with the surface. Quantum sticking can also occur in a two- or multi-step process where the particle is *trapped* at the surface on initial collision but at positive energy. Such particles may ultimately stick, or re-emerge depending on the balance of matrix elements governing these two competing channels. This is just the

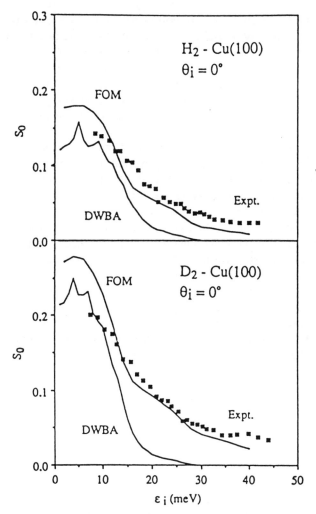

Figure 13 *Sticking coefficient in the quantum limit:* H_2 *and* D_2 *on* Cu(100) *for normal incidence and as a function of incident energy. The experimental data (filled symbols) are compared with the results of a calculation using the forced oscillator model (FOM, see text) and the distorted-wave Born approximation, which includes only single-phonon processes. The tops of the bulk and surface phonon bands for Cu lie at about 3 and 15 meV, respectively. Both theories give the low-energy behaviour reasonably, but at higher energies the DWBA fails because it does not take account of multi-surface phonon excitation which dominates over bulk phonon excitation because of the enhanced density of states*

quantum analogue of *trans→rot* conversion or the conversion of normal kinetic into parallel kinetic energy, as discussed in Section 3. However, the trapping step can also occur in quantum mechanics via an elastic, *resonant* process that has no classical analogue. The resonance arises because of a *degeneracy* between two quantum states, one describing a scattering particle and the other a particle that is bound to the surface but with enhanced parallel energy (or rotational energy). These states are degenerate for a flat surface but can be coupled by a small matrix element and can give rise to a large trapping probability. Resonant sticking has been observed for H_2 on Cu[39] and the matrix elements devolve from the rotational and/or spatial corrugation of the interaction and are characterized by the rotational energy and the vectors of the surface reciprocal lattice, respectively. The precise resonance condition is discussed more fully in Chapter 3. When the resonance condition is fulfilled, the sticking coefficient displays a sharp peak and the elastic scattering probability a corresponding trough.

Returning now to the question of the validity of the classical limit, one might suppose from the above discussion that this is guaranteed if P_0 is small (large Debye–Waller). This condition is indeed *necessary*, but it is not *sufficient*. For instance, the strong fall-off of P_0 with surface temperature does not mean that a lattice that at $T_s = 0$ is behaving quantum mechanically will start behaving classically at room temperature. An even more dramatic example is illustrated by the behaviour of our stalwart harmonic linear chain, of which for some time there has been little mention. The reason is that the quantum linear chain, in spite of its apparent simplicity, is a very strange beast indeed, as those readers who have taken the trouble to work through the above material explicitly for this case have found out. The phonon density of states defined by equation (27) can be worked out trivially from its definition, viz:

$$\rho(\epsilon) \sim \sqrt{\left[1 - \left(\frac{\epsilon}{\epsilon_0}\right)^2\right]} \tag{43}$$

and goes to a constant as $\epsilon \to 0$. This means that the Debye–Waller factor, basically the first inverse moment of ρ, *diverges* for any collision that corresponds to a non-zero change of momentum (*i.e.* for which $F(\epsilon = 0) \neq 0$). In view of equation (37) this means that the mean square fluctuation in the position of the atoms of the chain is infinite, a notion that requires some little digesting! This curious behaviour is an example of an *infra-red catastrophe* that arises in all one dimensional systems that have a linear excitation spectrum. Since the quantum linear chain has an infinite Debye–Waller factor, a phonon expansion of the form equation (41) does not exist and, in particular, the chain *cannot scatter elastically* however weak the coupling. Instead, in the weak coupling limit, $P(\epsilon)$ displays an *edge singularity* of the type encountered in some electron spectroscopies:

$$P(\epsilon) \sim \epsilon^{\nu - 1}$$

$$\nu \sim \frac{4\delta\epsilon_{cl}}{\pi\epsilon_0} \tag{44}$$

This is the only consequence of the 'catastrophe', however, and $P(\epsilon)$ behaves otherwise rather regularly and, in the limit of weak coupling, has the general form shown in Figure 12a, but with the δ-function replaced by half a sharp spike. For a three-dimensional quantum harmonic lattice the density of states vanishes quadratically with the energy, $\rho(\epsilon) \sim \epsilon^2$ as $\epsilon \to 0$ so the Debye–Waller factor is well defined and the mean square fluctuations of the atomic locations in the ground state are finite (and are typically of the order of a few percent of a lattice spacing). This will be of considerable comfort to experimentalists as an explicit confirmation of their implicit assumption that erratic data cannot be attributed to quantum fluctuations causing the target to slop about all over the vacuum chamber. And what a happy note on which to end!

Acknowledgements. I thank collaborators and colleagues too numerous to be listed individually for their unwitting contributions. Special thanks are due to Charlie Rettner for demonstrating the extraordinarily beneficial effect on the creative process of editorial threats of physical violence.

References

1. J. A. Barker and D. J. Auerbach, *Surf. Sci. Rep.*, 1984, **4**, 1.
2. J. W. Gadzuk, *Annu. Rev. Phys. Chem.*, 1988, **39**, 395; *Comments. At. Mol. Phys.*, 1985, **10**, 219.
3. R. W. Zwanzig, *J. Chem. Phys.*, 1960, **32**, 1173.
4. S. A. Adelman and J. D. Doll, *J. Chem. Phys.*, 1976, **64**, 2375.
5. See, for example, J. C. Tully, *J. Chem. Phys.*, 1980, **73**, 6383; *Surf. Sci.*, 1981, **111**, 461.
6. J. Harris, *Phys. Scr.*, 1987, **36**, 156.
7. E. K. Grimmelmann, J. C. Tully, and E. J. Helfand, *J. Chem. Phys.*, 1985, **83**, 2594.
8. J. E. Hurst, C. A. Becker, J. P. Cowin, K. C. Janda, L. Wharton, and D. J. Auerbach, *Phys. Rev. Lett.*, 1979, **43**, 1175.
9. K. C. Janda, J. E. Hurst, J. P. Cowin, L. Wharton, and D. J. Auerbach, *Surf. Sci.*, 1983, **130**, 395.
10. H. F. Winters, H. Coufal, C. T. Rettner, and D. S. Bethune, *Phys. Rev. B*, 1990, **41**, 6240.
11. D. M. Goodstein, R. L. McEachern, and B. H. Cooper, *Phys. Rev. B*, 1989, **39**, 13129.
12. H. Zacharias, *Int. J. Mod. Phys.*, 1990, **B4**, 45.
13. C. W. Muhlhausen, L. R. Williams, and J. C. Tully, *J. Chem. Phys.*, 1985, **83**, 2594.
14. C. T. Rettner, F. Fabre, J. Kimman, and D. J. Auerbach, *Phys. Rev. Lett.*, 1985, **55**, 1904.
15. J. W. Gadzuk and J. K. Nørskov, *J. Chem. Phys.*, 1985, **81**, 2828.
16. S. Holloway and J. W. Gadzuk, *J. Chem. Phys.*, 1985, **82**, 5203.
17. J. Harris and P. J. Feibelman, *Surf. Sci.*, 1982, **115**, 133.
18. J. Kimman, C. T. Rettner, D. J. Auerbach, J. A. Barker, and J. C. Tully, *Phys. Rev. Lett.*, 1986, **57**, 2053.
19. J. W. Gadzuk, 'Chaos in Surface Dynamics', in 'Chemistry and Physics of Solid Surfaces VIII', ed. R. Vanselow and R. Howe, Springer-Verlag, Berlin, 1990, pp. 159–181.

20. J. Harris and A. C. Luntz, *J. Chem. Phys.*, 1989, **91**, 6421.
21. B. N. J. Persson and J. E. Mueller, *Surf. Sci.*, 1986, **171**, 219.
22. L. K. Verheij, J. Lux, A. B. Anton, B. Poelsema, and G. Comsa, *Surf. Sci.*, 1987, **182**, 390.
23. G. D. Billing, *Chem. Phys.*, 1984, **86**, 349.
24. R. J. Behm and C. R. Brundle, *J. Vac. Sci. Technol.*, 1984, **A2**, 1040.
25. A. W. Kleyn, A. C. Luntz, and D. J. Auerbach, *Phys. Rev. Lett.*, 1981, **47**, 1169.
26. C. T. Rettner, *Vacuum*, 1988, **38**, 295.
27. C. T. Rettner, J. Kimman, and D. J. Auerbach, (to be published).
28. E. W. Kuipers, M. G. Tenner, A. W. Kleyn, and S. Stolte, *Surf. Sci.*, 1989, **211/212**, 819; *Nature*, 1988, **334**, 420.
29. F. Greuzebroek (private communication).
30. H. Voges and R. Schinke, *Chem. Phys. Lett.*, 1983, **100**, 245.
31. R. Sedlmeir and W. Brenig, *Z. Phys.*, 1980, **B36**, 245.
32. W. Brenig, *Z. Phys*, 1979, **B36**, 81.
33. J. Böheim and W. Brenig, *Z. Phys.*, 1981, **B41**, 243.
34. H. D. Meyer, *Surf. Sci.*, 1981, **104**, 117.
35. R. Brako and D. M. Newns, *Surf. Sci.*, 1982, **117**, 42.
36. See, for example, L. I. Schiff, 'Quantum Mechanics', McGraw-Hill, New York, 1949.
37. See, for example, M. Sunjic, and A. Lucas, *Phys. Rev. B*, 1971, **3**, 719.
38. Th. Martin, R. Bruinsma, and P. Platzman, *Phys. Rev. B*, 1989, **39**, 12411.
39. S. Andersson, L. Wilzén, M. Persson, and J. Harris, *Phys. Rev. B*, 1989, **40**, 8146.
40. M. Persson and J. Harris, *Surf. Sci.*, 1987, **187**, 67.
41. W. Brenig, *Phys. Scr.*, 1987, **35**, 329.

CHAPTER 2

Dynamics of Dissociative Chemisorption

A. E. DEPRISTO

1 Introduction

Molecule–surface reactions are important in a myriad of applications.[1-3] Studies can be found as far back as the turn of the century in relationship to heterogeneous catalysis. Thirty years later, the few dynamical treatments focused on simplistic one-dimensional models and qualitative descriptions[4] of the simplest chemical reaction at a surface, namely the dissociation of a diatomic molecule at a solid surface leading to the adsorption of the two atoms on the surface:

$$AB(g) \xrightarrow{\text{surface}} A(a) + B(a)$$

In the last two decades the microscopic details of this process have yielded to new experimental,[5-11] theoretical, and computational tools.[6,12-18]

I will present the recent conceptual, theoretical, and computational developments in dissociative chemisorption dynamics in this chapter, emphasizing the microscopic understanding provided by atomic and electronic level theories and computer simulations. The style will be closer to that of a textbook than a monograph, and the level will be consistently below that of a research level article. The chapter should be understandable to a college senior or a first year graduate student in chemistry with some exposure to quantum and classical mechanics. More rigorous discussions can be found in the literature.[14,16,17] In particular, the author has recently published two detailed reviews[16,17] on this topic, and the present chapter draws heavily on

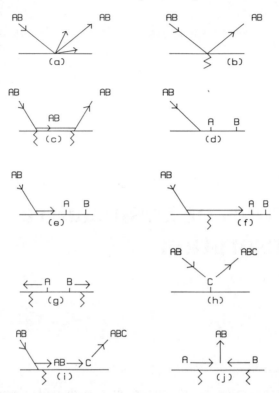

Figure 1 *Schematic of the different types of dynamical processes involved in molecule–surface systems. The jagged line indicates energy exchange with the surface. See text for further details*
(Reproduced with permission from *Adv. Chem. Phys.*, 1990, **77**, 163)

these. The article in *Advances in Chemical Physics* provides much more depth than the present chapter; the interested reader should also consult that review for more complete references.

To place the topic in perspective, Figure 1 shows some possible outcomes of the collision between a diatomic molecule (AB) and a surface (S). The z-axis is normal to the surface pointing into the vacuum (up in Figure 1). The translational (T) and internal rotational-vibration (R, V) energy of the molecule are given by $\hbar k$ and ϵ, respectively. The total energy, E, is just the sum of these two. Unprimed and primed quantities specify the initial (pre-scattering) and final (post-scattering) quantities, respectively.

Figures 1a–c involve physical processes. For *direct elastic scattering* in Figure 1a, the molecule scatters without exchanging energy with the surface. Two further divisions occur: for *specular scattering* the angles of k' and k with the surface normal are equal and the process occurs in a plane, while for *non-specular scattering* the angles differ or the process does not occur in a plane. The latter arises from *diffraction* due to surface corrugation and/or energy exchange between translational and internal ro-vibrational modes. The appropriate physical equation is:

$$AB(k, \epsilon, E) + S \rightarrow AB(k', \epsilon', E) + S.$$

For *direct inelastic scattering* in Figure 1b, the molecule exchanges some energy with the surface, symbolized by the equation:

$$AB(k, \epsilon, E) + S \rightarrow AB(k', \epsilon', E') + S.$$

For *indirect scattering* or *trapping-desorption* the molecule exchanges so much energy with the surface that it is trapped on the surface for a long time, or *adsorbed* AB(a), before returning to the gas phase, or *desorbing*. The equation is:

$$AB(k, \epsilon, E) + S \rightarrow AB(a) + S \rightarrow AB(k', \epsilon', E') + S.$$

Dissociative chemisorption is shown in Figures 1d–f in which the incoming molecule breaks apart and the two atoms bond to the surface. For *direct dissociation*, this occurs in the initial collision as in Figure 1d, with the chemical equation:

$$AB(k, \epsilon, E) + S \rightarrow A(a) + B(a) + S.$$

For *indirect dissociation* in Figure 1e, T→V energy transfer leads to a vibrationally excited complex which exists for a number of vibrational periods before the bond breaks, with the equation:

$$AB(k, \epsilon, E) + S \rightarrow AB(a; k', \epsilon', E') + S \rightarrow A(a) + B(a) + S.$$

For *dissociation via a precursor* in Figure 1f, the complex exists for enough vibrational periods to equilibrate with the surface before dissociation:

$$AB(k, \epsilon, E) + S \rightarrow AB(a) + S \rightarrow A(a) + B(a) + S.$$

It is difficult to quantify the difference between direct, indirect, and precursor mediated dissociative chemisorption since these depend upon the degree of equilibration between the molecule and the surface, a qualitative distinction.

The only processes that I will describe in detail in this chapter are direct and indirect dissociation. Complex processes such as oxidation, sputtering, and implantation in which the incident molecule, or some part of it, can react with or physically modify the surface are beyond the scope of this article.[13,15] However, for completeness, I do want to mention a few further events. First, as illustrated in Figure 1g, the adsorbates can move on the surface, termed diffusion. Secondly, as illustrated in Figure 1h, the incident molecule can collide with one of the previously formed adsorbates and form a new chemical species; this is the Eley–Rideal reaction mechanism for forming gaseous products. Thirdly, as illustrated in Figure 1i, the incident molecule can adsorb either molecularly or dissociatively and then react with one of the previously formed adsorbates, form a new chemical species, and desorb; this is the Langmuir–Hinshelwood mechanism for forming gaseous products. The final process, as illustrated in Figure 1j, is a particularly simple Langmuir–Hinshelwood reaction in which two adsorbates recombine and desorb.

I want to emphasize that the transfer of energy between the molecular and solid degrees of freedom may be significant for all the more complex processes 'd–j'. These low energy excitations involve motion of the nuclei (*phonons*), and for metals, excitation and de-excitation of electrons with energies close to the Fermi level (*electron-hole pairs* or e, h). I will consider only the former in this chapter since they are dominant. The reader should consult references 13, 17, 18, and 19 for more details.

I have intentionally left off a wide variety of topics in order to focus on dissociative chemisorption reactions. My goals are to introduce the concepts in this area and describe some of the current theoretical methodology. This chapter can provide an introduction to this topic, but is not intended to be at the research level. For that, the reader is referred to recent reviews and books.[1–3,9,13,15,18] I have not provided an exhaustive assessment of the current literature, but have referred to informative articles where necessary, in the same textbook spirit.

A little more detailed terminology about dissociative chemisorption dynamics is useful. When the dissociation occurs irrespective of the incident kinetic energy and angle and internal energy of the molecule, the process is labelled *non-activated*. By contrast, when the dissociation increases incident kinetic energy it is generally labelled *activated*. The dissociation probability on the clean surface, or 'the initial sticking coefficient', for non-precursor dissociation, universally denoted as S_0, depends on all three components of translational momentum. Two limiting cases are: S_0 depends only on $\hat{z} \cdot \boldsymbol{k}$ (*normal energy scaling*) and on $|\boldsymbol{k}|$ (*total energy scaling*).

2 Potential Energy Surfaces (PES)

General Characteristics

First, I must introduce the notation. For simplicity, I will limit considerations to a diatomic molecule interacting with a solid with only a single type of atom. Denote the positions of the gas molecule's atomic nuclei by $(\boldsymbol{X}_1, \boldsymbol{X}_2)$ and the solid's atomic nuclei by $(\boldsymbol{Y}_i, i = 1, \ldots, N_S)$ or $\{\boldsymbol{Y}_i\}$, with the equilibrium positions as $\{\boldsymbol{Y}_i^{(eq)}\}$. m_1, m_2, and m are the masses of the gas atoms and solid's atoms, respectively. The former are also located by the centre-of-mass, $\boldsymbol{X} = (m_1 \boldsymbol{X}_1 + m_2 \boldsymbol{X}_2)/(m_1 + m_2)$, and relative, $\boldsymbol{r} = (\boldsymbol{X}_2 - \boldsymbol{X}_1)$, coordinates, with the equilibrium value of the bond length, r, denoted as $r^{(eq)}$.

Associated with these co-ordinates are the total and reduced molecular masses, $M = (m_1 + m_2)$ and $\mu = m_1 m_2/(m_1 + m_2)$.

The basic physics of the interaction is described in a concise and useful model due to Nørskov and co-workers.[20,21] At distances far from the surface, the molecule–surface interaction is a van der Waals attraction of the form:

$$V_{LR}(X, r, \{Y_i\}) = -C_3/Z^3 \quad (1)$$

where C_3 depends upon the polarizability of the molecule and the electronic response of the solid. The distance $Z(\equiv \hat{z} \cdot X)$ is measured from the surface plane. As Z decreases, a repulsive term arises from the Pauli exclusion principle operating between the solid's and molecule's electrons, leading to the form:

$$V(X, r, \{Y_i\}) = V_{SR}(X, r, \{Y_i\}) - C_3/Z^3 \quad (2)$$

For a non-reactive system, V_{SR} is generally strongly dependent upon the orientation of the molecule, but only weakly dependent upon the bond length r and solid atom's positions. (The full AB–S PES, $V_{AB.S}$, is then just $V + V_{AB}(r)$ where the latter is the interaction in the isolated molecule). The effect of solid atom displacement in V_{SR} is typically repulsive in the separation between each gas atom and each nearby solid atom. Because of the weak dependence of V on r, variation of the bond length from $r^{(eq)}$ on the full AB–S PES increases the energy irrespective of the height of the molecule above the surface, thereby leading to very small coupling of r and X.

For reactive systems another effect occurs. As Z decreases, the affinity level (LUMO) of the molecule feels an image-charge type attraction, leading to an energy shift as:

$$E_a(Z) = E_a(\infty) - 1/4Z \quad (3)$$

At small Z, the affinity level gets pulled below the Fermi level of the solid, and the molecular orbital fills. If this orbital is antibonding, then the molecular bond may be weakened sufficiently to rupture, thereby leading to dissociative chemisorption. Filling of an antibonding level leads to an increase in the bond length and thus the PES will display a strong and non-monotonic dependence upon 'r' at small Z. A barrier to dissociation results when the Pauli repulsion increases more quickly than E_a decreases.

More evidence for the charge transfer mechanism comes from self-consistent calculation[20,21] of H_2 on Mg as illustrated in Figure 2. Various *possible* characteristics of a reactive PES are apparent:
1. a weak physisorption well at far distances from the surface ($Z \approx 7$ bohr here) with $r \approx r^{(eq)}$;
2. a barrier to molecular chemisorption closer to the surface ($Z \approx 5$ bohr here) with $r \approx r^{(eq)}$, denoted by point 'A';
3. a strongly bound molecular chemisorption well at closer distances ($Z \approx 3.5$ bohr here) with $r \geq r^{(eq)}$, denoted by point 'M';

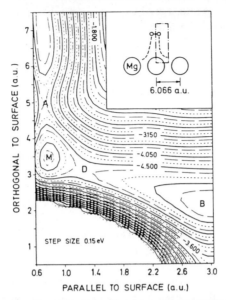

Figure 2 *Contour plot of a two-dimensional cut in the* $H_2/Mg(100)$ *PES for the configuration shown on the top right of the figure. The energies are in eV relative to those of the free atoms.* 'A' *is a barrier to molecular chemisorption.* 'M' *is a molecular chemisorption well.* 'D' *is a barrier to dissociation and* 'B' *is an atomic chemisorption well. The distance from the surface is measured from the first atomic layer and the distance parallel to the surface is measured from the top site towards the bridge site*
(Reproduced with permission from *Phys. Rev. Lett.*, 1981, **46**, 257)

4. a barrier to dissociative chemisorption slightly closer to the surface ($Z \approx 3$ bohr here) with $r > r^{(eq)}$, denoted by point 'D';
5. a strongly bound atomic chemisorption well at still closer distances ($Z \approx 2\frac{1}{4}$ bohr here) with $r \gg r^{(eq)}$, denoted by point 'B'.

Not all of these features will appear for every system. For example, a non-reactive system may have a very large value at either 'A' or 'D'. In addition, even for a reactive system all of these features may not appear in every geometry of a particular molecule–surface system. For example, if the H_2 bond was perpendicular to the surface, feature 'M' would either disappear or be much smaller; the barrier 'D' would disappear since point 'B' would lie considerably above (≈ 2.3–2.5 eV) the reactant H_2 molecule. In this configuration, the PES would appear similar to that of a non-reactive system.

I should point out that the barrier at 'A' is described as being in the entrance channel since the molecular bond length is 'nearly unchanged' from its gas phase value: $(r - r^{(eq)}) \ll r^{(eq)}$. Similarly, an exit channel barrier occurs when the molecular bond length is 'significantly stretched' from its gas phase value: $(r - r^{(eq)}) \approx r^{(eq)}$. These subjective criteria are useful as limiting cases since realistic PESs will generally fall in between each limit.

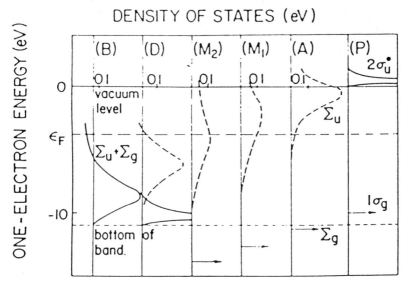

Figure 3 *One-electron density of states for* H_2 *approaching* Mg(100) *as in Figure 2* (Reproduced with permission from *J. Less-Common Met.*, 1987, **130**, 475)

The charge transfer corresponding to the various marked points in Figure 2 can be appreciated by inspection of the one-electron density of states diagrams shown in Figure 3. The antibonding $2\sigma_u^*$ level is not filled in the physisorbed ('P') and 'A' points on the PES; is slightly filled early in the 'M' well (denoted 'M_1') and nearly half filled late in the 'M' well (denoted 'M_2'); and is nearly fully filled by the dissociation barrier ('D').

The contour plot in Figure 2 also serves to illustrate a few general points about molecule–surface interactions. The points in the contour plot can be mapped onto a one-dimensional reaction path form in Figure 4. Such a one-dimensional plot illustrates a few important features in the contour plot concisely, and thus is often used to discuss real systems. This is not a good practice since such conciseness is associated with a lack of detail. For example, in the contour plot of Figure 2, it is apparent that point 'D' is associated with a stretched molecular bond, while in Figure 4, this is obscured and could just as easily arise from a barrier at short distances above the surface with an equilibrium bond length. Such different topologies of the PES will lead to greatly different dynamics.

Even the two-dimensional contour plot obscures some important facts about the true multi-dimensional PES, as illustrated by the following (erroneous) argument. Some of the features in the PES can be probed by experiments: the weakly physisorbed, molecularly and atomically chemisorbed species by surface spectroscopies; and the activation barriers by molecular beam scattering experiments. Thus, one can simply join these individual regions together and 'find' the PES!

The error is the neglect of the other degrees of freedom in the system. A

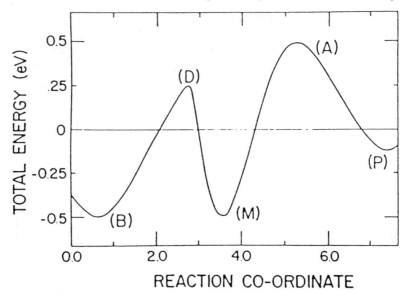

Figure 4 *One-dimensional representation of the PES in Figure 2*
(Reproduced with permission from *J. Less-Common Met.*, 1987, **130**, 475)

PES contour plot describes dissociative chemisorption for a particular location and orientation of the molecule above a particular geometry of the surface atoms. In real systems, all of the PES features will vary with orientation of the molecule $\hat{r}$, position of the molecule in the surface unit cell (X, Y), and displacement of the surface atoms $\{Y_i - Y_i^{(eq)}\}$. These other variables complicate the problem tremendously, and may even control the particular dynamical process. For example, the atomically and molecularly chemisorbed species may be stable at different positions in the surface unit cell. Spectroscopic probes of these species would then determine two completely different regions of the PES without any means for connecting them.

Clearly computations of full PES are needed. At the present time, however, the best available *ab initio* and density functional methods are limited to the calculation of just a few points on a full PES.[20,22–24] Moreover, the accuracy of the detailed calculations is still not sufficient for chemical accuracy, ≤ 1 kcal mol^{-1} for instance, and the computational expense still prohibits calculation of enough points to map out the full multi-dimensional PES, especially including surface atom displacement.

Construction of a Model Potential Energy Surface

In a dynamical system, the molecule can rotate and translate parallel to the surface, and the surface atoms can move, thereby distorting the PES. When the molecule rotates, the various barriers and even the gross topology of the PES may change. Similarly, the PES at different locations in the surface unit cell will be different. The dynamics in these other degrees of freedom may play a central role, and thus one must include them. It is necessary to develop computationally efficient global representations of PES which can be used to construct the full PES using all available information, experimental as well as theoretical. Such a synthesis plays a major role in this field and is based upon providing PES topologies similar to those in Figure 2 for appropriate sites and orientations.

A common construction of these global PESs is based upon a modified four-body LEPS form that was initially developed by McCreery and Wolken.[25] This was later modified and quantified to be more accurate for metal surfaces.[26,27] The basic idea is to utilize valence bond theory for the atom–surface interactions, V_{AS} and V_{BS}, along with V_{AB} to construct $V_{AB.S}$. For each atom of the diatomic, we associate a single electron. Since association of one electron with each body in a three-body system allows only one bond and since the solid can bind both atoms simultaneously, two valence electrons are associated with the solid. Physically, this reflects the ability of the infinite solid to donate and receive many electrons. The use of two electrons for the solid body and two for the diatomic leads to a four-body LEPS potential[28] that is convenient mathematically, but contains non-physical bonds between the two electrons in the solid. These are eliminated, based upon the rule that each electron can only interact with an electron on a different body, yielding the modified four-body LEPS form. One may also view this as an empirical parametrized form with a few parameters which have well controlled effects on the global PES.

The explicit form is:

$$V_{AB.S} = Q_{AS} + Q_{BS} + Q_{AB} - [J_{AB}(J_{AB} - J_{AS} - J_{BS}) + (J_{AS} + J_{BS})^2]^{1/2} \quad (4)$$

where Q and J are 'coulomb' and 'exchange' integrals, respectively, for each constituent. One important feature of this form is the non-additivity of the interaction potentials. For example, the AB interaction alone is given by:

$$V_{AB} = Q_{AB} - |J_{AB}| \quad (5)$$

which evidently does not determine Q_{AB} and J_{AB} individually. Another important feature is the incorporation of all four asymptotic limits: AB(g) + S; A(a) + B(g) + S; A(g) + B(a) + S; A(a) + B(a) + S. For example, note that if the terms with A–S and B–S vanish, then $V_{AB.S} = Q_{AB} - |J_{AB}| = V_{AB}$, and if the terms with A–B vanish, then $V_{AB.S} = (Q_{AS} - |J_{AS}|) + (Q_{BS} - |J_{BS}|)$. The precise division into Q and J for each interaction in equation (4) controls the topology and energies of the full PES for the reaction.

The A–B interaction is:

$$Q_{AB} + J_{AB} = V_{AB}$$
$$= D_{AB}\{\exp[-2a_{AB}(r - r_{AB})] - 2\exp[-a_{AB}(r - r_{AB})]\} \quad (6a)$$

$$Q_{AB} - J_{AB} = \tfrac{1}{2}[(1 - \Delta_{AB})/(1 + \Delta_{AB})] \times$$
$$D_{AB}\{\exp[-2a_{AB}(r - r_{AB})] + 2\exp[-a_{AB}(r - r_{AB})]\} \quad (6b)$$

The A–B interaction potential is represented by the Morse potential in equation (6a) with the parameters D_{AB}, a_{AB}, and r_{AB} characterizing the bond energy, range parameter, and equilibrium bond length ($= r^{(eq)}$ for the A—B bond), respectively. The A–B 'antibonding' potential in equation (6b) is represented by the anti-Morse form. Δ_{AB}, controls the division of the Morse–anti-Morse forms into Q_{AB} and J_{AB} and is called a Sato parameter.

Now consider the A–S interaction (with the equations for the B–S interaction derived by simply changing A to B everywhere). This interaction is:

$$Q_{AS} + J_{AS} = V_{AS}$$
$$= D_{AH}\{\exp[-2a_{AH}(r_{AS} - R_{AH})] - 2\exp[-a_{AH}(r_{AS} - R_{AH})]\}$$
$$+ \Sigma D_{AS}\{\exp[-2a_{AS}(R_{A\beta} - R_{AS})]$$
$$- 2\exp[-a_{AS}(R_{A\beta} - R_{AS})]\} \quad (7a)$$

$$Q_{AS} - J_{AS} = \tfrac{1}{2}[(1 - \Delta_{AS})/(1 + \Delta_{AS})] \times$$
$$[D_{AH}\{\exp[-2a_{AH}(r_{AS} - R_{AH})] + 2\exp[-a_{AH}(r_{AS} - R_{AH})]\}]$$
$$+ \Sigma D_{AS}\{\exp[-2a_{AS}(R_{A\beta} - R_{AS})]$$
$$+ 2\exp[-a_{AS}(R_{A\beta} - R_{AS})]\}] \quad (7b)$$

where

$$R_{A\beta} = |X_A - Y_\beta| \quad (8)$$

is the distance between gas atom 'A' and solid atom 'β'. The summation extends over all of the atoms in the solid. The two general terms in equation (7) describe the bonding between a molecule and a metal.

First consider the terms in D_{AH}. *These represent the interaction between A and the valence electrons of the metal.* This is modelled by the interaction of A with jellium (a uniform electron gas with compensating positive background). The density of this jellium is provided by the metal's valence electrons at the position of A. The parameters are thus:

(i) D_{AH} = strength of the interaction between atom A and jellium;
(ii) a_{AH} = range of the above interaction;
(iii) $R_{AH} = (3/4\pi n_0)^{1/3}$ where n_0 is the density at the minimum of the atom–jellium binding curve.

These are determined from the self-consistent-field local density (SCF-LD) functional calculations[29] on the embedding energy of an atom in jellium. (Illustrative curves are shown in Figure 5 for H and N atoms.) The values as a function of jellium densities are represented by a Morse-like form in the density. The variable, r_{AS} is defined as:

$$r_{AS} = (3/4\pi n(X_A))^{1/3} \tag{9}$$

where $n(X_A)$ is the density of the solid at the position of A. The additional assumption that the solid's density is well represented by the sum over the individual atomic densities is made:

$$n(X_A) = \Sigma n_\beta(R_{A\beta}) \tag{10}$$

where $n_\beta(R_{A\beta})$ is the atomic density of the solid atom β at the position of the gas atom A and again the summation extends over all of the solid's atoms.

Next consider the terms in D_{AS}. *These represent the interaction between atom A and the localized electrons and nuclear charges of the metal atoms.* These are described by a two-body interaction between A and each solid atom, β. The particular form used is the Morse potential with parameters:
(i) D_{AS} = strength of the localized two-body interaction between atom A and the solid's atoms
(ii) a_{AS} = range of the above interaction
(iii) R_{AS} = position of the minimum of the two-body interaction

These parameters can be determined from either experimental or theoretical information on the atom–surface interaction potential.

The parameters for the two types of interactions can be assumed fixed by the binding curves for the asymptotic fragments and the known SCF-LD embedding energy for an atom in jellium. However, the remaining variables

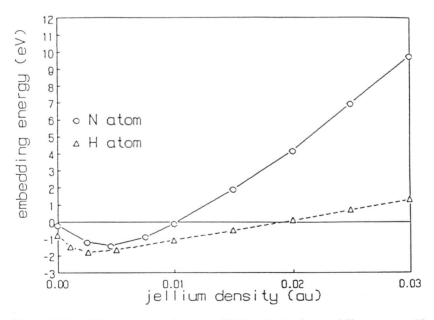

Figure 5 *Embedding energy as a function of jellium density for two different atoms. The data are from reference 29*
(Reproduced with permission from *Adv. Chem. Phys.*, 1990, **77**, 163)

in the PES, the so called Sato parameters, Δ_{AB}, Δ_{AS}, and Δ_{BS}, are undetermined and available for flexible representation of the full molecule–solid reactive PES. In particular, these can adjust the location and size of activation barriers.

At this point, I want to re-emphasize the basic physics and chemistry: (i) interactions with localized and delocalized metal electrons; (ii) non-additive chemical bonding. I should also note that the representation of an atom's interaction with metals in terms of an embedding function (in jellium) plus two-body terms is identical in spirit to the embedded atom method (EAM).[30,31] The distinction here is that the EAM is not used for the A–B interaction and non-additive energies are explicitly incorporated via the LEPS prescription, both of which are important for the accurate representation of the reactive PES.

3 Dynamical Theories

As indicated in Section 2, a one-dimensional representation of the molecule–surface PES is often used to provide a qualitative description of experimental data; a schematic is shown in Figure 6 following the more complex form in Figure 4. In Figure 6, the molecular well for physisorption, W_p, and energy for desorption, E_d, correspond to point 'P' in Figure 4. The barrier to chemisorption, E_c, corresponds to point 'D' in Figure 4 while the chemisorption well, W_c, corresponds to point 'B'. This one-dimensional representation is simplified in order to eliminate the possibility of a molecularly chemisorbed species.

For indirect dissociative chemisorption, W_p plays two important roles: (i) acceleration of the incoming molecule to allow for significant energy exchange with the lattice; and, (ii) stabilization of the trapped molecular species for a long enough time to allow significant energy exchange among the molecular degrees of freedom. For a precursor mechanism, W_p plays the same roles as in the indirect mechanism but the stabilization is long enough to allow for equilibration between the physisorbed molecule and the lattice.

As an aid to understanding this case, I consider a chemical kinetic description. k_d and k_c are the rate constants for desorption and chemisorption, respectively, while a is the physisorption probability for an incident gas phase molecule. Under a steady-state assumption for physisorbed species, this yields the initial sticking coefficient for dissociation as:

$$S_0 = ak_c/(k_c + k_d) \tag{11}$$

Since increasing the incident kinetic energy, E_i, of the molecule decreases a, S_0 will decrease with increasing kinetic energy in a precursor mechanism. This is one signature of a precursor mechanism often looked for experimentally. To derive another, consider the simple Arrhenius rate forms:

$$k_d = v_d \exp(-E_d/k_B T) \tag{12a}$$

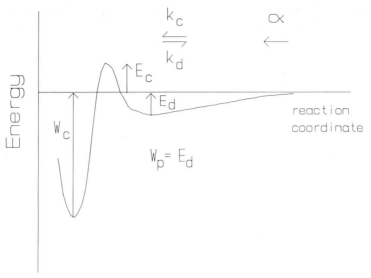

Figure 6 *A one-dimensional representation of a molecule–surface interaction. The depth of the physisorption and chemisorption wells are W_p and W_c, respectively. The barrier to chemisorption is E_c and the barrier to desorption is E_d. The probability of physisorption is a, while the rate constants for desorption and chemisorption out of the physisorbed state are k_d and k_c, respectively* (Reproduced with permission from *Adv. Chem. Phys.*, 1990, **77**, 163)

$$k_c = v_c \exp[-(E_d + E_c)/k_B T] \qquad (12b)$$

where T is the surface temperature. Combining equations (11) and (12) yields:

$$S_0 = a[1 + (v_d/v_c)\exp(E_c/k_B T)]^{-1} \qquad (13)$$

For activated chemisorption (*i.e.* $E_c > 0$), S_0 increases with increasing T; this behaviour is not unique since it can also occur for activated reactions in both direct and indirect mechanisms. By contrast, for non-activated chemisorption (*i.e.* $E_c < 0$), S_0 decreases with increasing T, which is a unique signature.

I should emphasize that increasing the kinetic energy enough will lead to direct dissociation for either activated or unactivated dissociative chemisorption. Thus, at high kinetic energy, one should expect S_0 to increase with E_i. The value at which S_0 stops decreasing and starts increasing with E_i will depend upon the value of E_c and the detailed dynamics of the dissociation.

For direct and indirect dissociative chemisorption the problem is to surmount E_c. The theoretical difficulty is to determine the dissociative sticking probability for a gas molecule to directly surmount the chemisorption barrier, E_c. Due to the rather complicated nature of the many dimensional reactive PES, it is not easy to predict such a probability.

I will present some methods to make such a prediction, assuming the PES is available. I begin with the general quantum mechanical approach to solution of the dynamics assuming a rigid substrate. The time-dependent Schrödinger equation is:

$$i\hbar \frac{\partial \Phi}{\partial t} = \mathcal{H}\Phi \qquad (14a)$$

$$= [\mathcal{T}(X, r) + V_{AB.S}(X, r, \{Y_i^{(eq)}\})]\Phi \qquad (14b)$$

$\mathcal{T}$ is the nuclear kinetic energy operator of the gas atoms. Solution of this equation is now feasible, under certain restrictions.[32] To illustrate the method, I shall use a simpler notation, letting 'q' stand for the set of co-ordinates (X, r). Then equation (14) has the formal solution:

$$\langle t + \Delta t; q' | \Phi \rangle = \int \langle q' | \exp(-i\mathcal{H}\Delta t/\hbar) | q \rangle \langle t; q | \Phi \rangle dq \qquad (15)$$

This is a low dimensional integral which upon evaluation at time 't' yields the full spatial wavefunction at time '$t + \Delta t$'. Repetition for many time steps, Δt, provides a full propagation of the wavefunction.

The problem is to evaluate the exponential operator in equation (15). This is complicated for two reasons. First, the kinetic and potential operators do not commute. However, if Δt is small enough then one can utilize the approximation:[33]

$$\exp(-i\mathcal{H}\Delta t/\hbar) \approx \exp(-i\mathcal{V}\Delta t/2\hbar) \exp(-i\mathcal{T}\Delta t/\hbar) \exp(-i\mathcal{V}\Delta t/2\hbar) \qquad (16)$$

which is accurate to order $(\Delta t)^3$. Secondly, $\mathcal{T}$ is a local operator in momentum space and $\mathcal{V}$ is a local operator in co-ordinate space. Thus, insert a complete set of momentum states, $|p\rangle \langle p|$, between the $\mathcal{V}$ and $\mathcal{T}$ parts and substitute equation (16) into (15) to get the final working equation:

$$\langle t + \Delta t; q' | \Phi \rangle = \int \langle q' | \exp(-i\mathcal{V}\Delta t/2\hbar) | q' \rangle$$

$$\langle q' | p \rangle \langle p | \exp(-i\mathcal{T}\Delta t/\hbar) | p \rangle$$

$$\langle p | q \rangle \langle q | \exp(-i\mathcal{V}\Delta t/2\hbar) | q \rangle \langle t; q | \Phi \rangle dp\, dq \qquad (17)$$

The transformation between co-ordinate and momentum space is accomplished via the fast fourier transform (FFT) algorithm. This transforms the co-ordinate space evaluation of $\exp(-i\mathcal{V}\Delta t/2\hbar)\langle t|\Phi\rangle$ into momentum space; the effect of $\exp(-i\mathcal{T}\Delta t/\hbar)$ is determined in this space to yield the momentum space function; the momentum space function is transformed back to a co-ordinate space function via another FFT; and, finally the new co-ordinate space function is further evolved by the local operator $\exp(-i\mathcal{V}\Delta t/2\hbar)$ to yield the co-ordinate space wavefunction at time $t + \Delta t$. This is efficient because the FFT algorithm is highly developed and the values of $\langle q'|\exp(-i\mathcal{V}\Delta t/2\hbar)|q'\rangle$ and $\langle p|\exp(-i\mathcal{T}\Delta t/\hbar)|p\rangle$ are evaluated at the grid points (in the FFT) once at the beginning of the calculation and stored.

The propagation procedure proceeds in the following manner. Start with a wavepacket, $\Phi(t=0; X, r)$, which is the product of an internal state eigenfunction and a localized translational packet at large Z with an average velocity directed towards the surface. The wavepacket at time $t + \Delta t$ is then generated via equation (17) with $q = (X, r)$. Next, use $\Phi(\Delta t; X, r)$ to generate the packet at $2\Delta t$ via equation (17). This procedure is repeated many times until the final wavefunction, $\Phi(t \to \infty; X, r)$, is obtained. From this function, all scattering information is extracted via projection onto plane waves and appropriate eigenstates.

There are a number of advantages to such a time-dependent wavepacket approach. First, one specific initial internal state of the reactant molecule can be treated. By contrast, time-independent methods generally calculate the values for all internal states at once, even though this information is often not desired. Secondly, this scattering information can be obtained at a number of translational energies in a single calculation since the initial translational packet is composed of large numbers of initial plane waves. Thirdly, the method is applicable to a non-rigid surface by letting $\{Y_i^{(eq)}\}$ in equation (14b) vary classically. Fourthly, the method is applicable to non-periodic surfaces.

The major limitation is computational: for each time step, the computational time is proportional to $N_g \ln N_g$ where N_g is the number of grid points used in the evaluation of the FFT. Assuming for simplicity an equivalent and low number of grid points of $2^4 = 16$ in each degree of freedom, then in M degrees of freedom we have $N_g = 16^M$ which limits treatments to $M \leq 3$ on current supercomputers unless only a single calculation must be performed in which case $M = 4$ may be feasible. However, even a rigid rotor–rigid surface collision entails $M = 5$ (unless further simplifying assumptions are made). The FFT approach is only rigorously applicable at present to atom–surface scattering unless the molecular internal co-ordinates, r, are not important in the dynamics.[34,35]

Under the approximations that some degrees of freedom can be ignored, the FFT solution will allow for detailed investigation of quantal effects in four degrees of freedom. For example, recent work by Nørskov, Holloway, and co-workers[36] assumes that the scattering occurs in a plane perpendicular to the surface (*e.g.* the X–Z plane). This reduces the number of variables to four (r, θ, Z, X). I expect that this will capture much of the effect of the full six degrees of freedom systems, especially if the calculations are performed for many different planes.

The most direct treatment of reactive scattering for molecule–surface systems ignores completely the quantum mechanical nature of the molecular internal and translational motions, and includes the motion of the solid's atoms. The Newtonian equations of motion are:

$$\mu d^2 r/dt^2 = - \nabla_r V_{AB.S}(X, r, \{Y_i\}) \tag{18a}$$

$$M d^2 X/dt^2 = - \nabla_X V_{AB.S}(X, r, \{Y_i\}) \tag{18b}$$

$$m d^2 Y_j/dt^2 = - \nabla_{Y_j} V_{AB.S}(X, r, \{Y_i\}) - \nabla_{Y_j} V_{SS}(\{Y_i\}) \tag{18c}$$

where μ is the reduced mass and M is the total mass of the gas molecule. These are rather standard molecular dynamics simulation equations and can be used directly providing enough computer time is available to treat the large number of the solid's atoms needed to model a surface.

However, the treatment of the motion of the solid is not generally included simply by integration of Newton's equations. This is because the collision process, and $V_{AB.S}$, are quite strongly dependent upon the exact location (and hence motion) of only a small subset of the entire atoms of the solid, called the primary zone. These primary zone atoms are strongly disturbed by the collision, while the remainder of the solid remains near equilibrium at temperature T. The role of the remainder of the solid is to allow for dissipation of the collision energy into the bulk and to provide energy from the bulk into the primary zone atoms via thermal fluctuations. The basic idea is then to treat the motion of these few strongly perturbed atoms explicitly, and to incorporate the remainder of the solid's atoms by methods based upon statistical mechanics.

The most accurate method for accomplishing the above is the generalized Langevin equation (GLE) formalism for gas–surface processes. Introductions can be found in a number of articles.[16,37–40] Solving Newton's equations for the motion of the harmonic secondary atoms and substitution of this result into the equations of motion of the primary zone atoms yields a GLE for the latter:

$$\begin{aligned} m d^2 \tilde{Y}/dt^2 = & - \nabla_{\tilde{Y}} V_{AB.S}(X, r, \tilde{Y}, \tilde{Y}_Q^{(eq)}) \\ & - \nabla_{\tilde{Y}} V_{SS}(\tilde{Y}) + m M(0) \tilde{Y} - m M(t) \tilde{Y}(0) \\ & - \int m M(t - t') d\tilde{Y}(t')/dt' \, dt' + F(t) \end{aligned} \tag{19}$$

The notation $\tilde{Y}$ signifies all the co-ordinates of $\{Y_i\}$ which are included in the primary zone, while the co-ordinates $\tilde{Y}_Q$ are those remaining from $\{Y_i\}$ that are not included in $\tilde{Y}$. The memory function and random force are defined by:

$$M(t) = \Omega_{PQ}^2 \cos(\Omega_{QQ} t) \Omega_{QQ}^{-2} \Omega_{QP}^2 \tag{20a}$$

$$F(t) = - m\Omega_{PQ}^2 \cos(\Omega_{QQ} t) \tilde{Y}_Q(0) - m\Omega_{PQ}^2 \sin(\Omega_{QQ} t) \Omega_{QQ}^{-1} d\tilde{Y}_Q(0)/dt \tag{20b}$$

where 'P' signifies a projection onto the primary zone and 'Q' signifies a projection on the secondary zone. (Such projections simply keep track of the atoms' locations.) The frequency matrix of the original solid is $\boldsymbol{\Omega}$. Note that the interaction $V_{AB.S}$ depends upon the full set $\{Y_i\}$ but with the secondary zone at equilibrium, *i.e.* $\{\tilde{Y}, \tilde{Y}_Q^{(eq)}\}$.

The kernel in equation (20a) involves the response of the full many-body system and thus retains 'memory' of previous velocities. This differs from a standard Langevin equation which replaces the memory function by a Dirac delta function, leading to a temporally local friction. The random force, $\tilde{F}(t)$, and the memory kernel are related via the fluctuation–dissipation theorem:

$$\langle \tilde{F}(t)\tilde{F}(0)^T \rangle = mkTM(t) \qquad (21)$$

where the $\langle \rangle$ indicate an average over initial conditions of the secondary atoms. The derivation of equation (21) uses the equilibrium properties of the positions and velocities of the secondary zone atoms.

Equations (19)–(21) are equivalent to the original set of molecular dynamics equations in equation (18c). Indeed, the GLE equations are no easier to solve in their present form. The advantage of the GLE approach is that it is possible to approximate the memory kernel and random force[16,39–41] to provide a reasonably accurate description of both the short-time and long-time (actually long wavelength) response of the primary zone atoms to an external perturbation (*e.g.* a collision).

One expects a decaying and oscillating function on physical grounds, which can be represented by:

$$M(t) = M_0^{1/2} \exp(-\gamma t)[\cos(\omega_1 t) + \tfrac{1}{2}\gamma\omega_1^{-1} \sin(\omega_1 t)]M_0^{1/2} \qquad (22)$$

This is the multi-dimensional generalization of the position autocorrelation function of a Brownian oscillator and provides such a function with a number of unknown parameters. Using this approximate memory function, allows replacement of the GLE in equation (19) by:

$$m d^2 \tilde{Y}/dt^2 = -\nabla_Y V_{AB.S}(X, r, \{Y_{ij}\}) - \nabla_Y V_{SS}(\tilde{Y}) + mM_0^{1/2}\omega_0\tilde{s} \qquad (23a)$$

$$m d^2 \tilde{s}/dt^2 = m\omega_0 M_0^{1/2} \tilde{Y} - m\omega_0^2 \tilde{s} - m\gamma d\tilde{s}/dt + \tilde{f}(t) \qquad (23b)$$

where $\tilde{f}(t)$ is a gaussian white noise random force obeying:

$$\langle \tilde{f}(t)\tilde{f}(0)^T \rangle = 2mkT\gamma\delta(t) \qquad (23c)$$

$\delta(t)$ is a Dirac delta function. The fictitious particles obeying the equations of motion of $\tilde{s}$ are commonly referred to as 'ghost' atoms, and thus this formulation is referred to as the GLE-'ghost' atom method.

One can determine these parameter matrices by two fundamentally different methods. The first,[39] solves equation (19) at $T = 0$ without the AB-S interaction and with harmonic primary zone interactions and then uses this solution to find the phonon density of states. The parameters are adjusted to achieve agreement with the experimental results. The second,[40] assumes a microscopic interaction model for the forces among the solid's atoms, finds Ω, and substitutes this into equation (20a) to define the exact $M(t)$; then equation (22) is forced to agree with the exact $M(t)$ at short times. In both methods, the low frequency density of states is determined by the relationship $\gamma = 1\pi\omega_D/6$.

The two methods will provide very similar parameters if an accurate frequency matrix, Ω, is available. Both will capture the essential features of the GLE, namely frictional energy loss from the primary atoms to the secondary atoms and thermal energy transfer from the secondary atoms to the primary atoms. And both will provide a reasonable description of the bulk and surface phonon density of states of the solid. Neither will provide the exact time-dependent response of the solid due to the limited number of parameters used to describe the memory function.

It is worthwhile to make a few points about the GLE formalism. First, it is not possible to implement an (atomic based) GLE in any *practical* and systematic way if the primary–secondary or secondary–secondary interactions are anharmonic. Secondly, a complication arises within the GLE formalism due to the localized nature of the primary zone atoms. In a molecule–surface collision, the initial localized interaction specifies a set of primary zone atoms as illustrated for a diatomic interacting with a b.c.c.(110) surface in Figure 7. (There is nothing fundamental about this surface or the number of primary zone atoms; the same argument holds irrespective of the lattice and primary zone size.) When the molecule moves outside of these thirteen atoms, a new set of primary zone atoms must be defined, a process termed switching of the primary zone.[39] There is no *a priori* method to consistently define new primary zone atoms since there is no information on the flow of energy from the original primary zone into the specific secondary zone atoms (*i.e.* the exact $M(t)$ is no longer available). The current practice is to assume that the motion of the molecule across the surface is slow compared to thermalization of the surface atoms. Then the new primary zone atoms are re-initialized from a thermal distribution. Such a method will break down if the motion of the molecule is very fast and/or the distortion of the lattice is sufficiently great to inhibit thermalization on a fast enough time scale.

Thirdly, the GLE is not a particularly simple method to implement. One must choose the parameter matrices for the memory function, which will differ for each material and surface face, a procedure which is intensive of human time. Changing the number of primary zone atoms requires re-determination of these parameters, and major modifications of a computer code to implement the switching process. If a treatment of a particular solid and a surface face is to be the focus of investigation for an extended period of time, the investment in human time is definitely worthwhile. (If computational resources are limited to non-supercomputers, it may be the only viable approach.) However, if a number of faces and materials are to be treated, still simpler methods combining LE and molecular dynamics (MD)

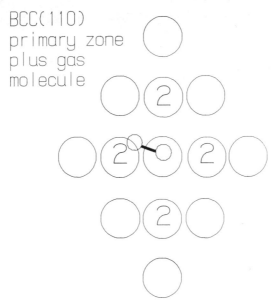

Figure 7 *Diatomic molecule and primary zone atoms used for GLE simulations of dissociative chemisorption on a* b.c.c.(110) *surface. The atoms labelled by '2' are in the second layer*
(Reproduced with permission from *Adv. Chem. Phys.*, 1990, **77**, 163)

are desirable, although more expensive computationally.[42-46]

In these theories, the solid's atoms are divided into three classes: inner active, edge active, and fixed. The *inner active* atoms follow classical dynamics with:

$$m\mathrm{d}^2 Y_j/\mathrm{d}t = - \nabla_{Y_j} V_{\mathrm{AB.S}}(X, r, \{Y_i\}) - \nabla_{Y_j} V_{\mathrm{SS}}(\{Y_i\}) \qquad (24)$$

The *edge active* atoms follow Langevin dynamics in which local (isotropic) frictional and gaussian white noise random forces are added to the potential forces:

$$m\mathrm{d}^2 Y_j/\mathrm{d}t = - \nabla_{Y_j} V_{\mathrm{AB.S}}(X, r, \{Y_i\}) - \nabla_{Y_j} V_{\mathrm{SS}}(\{Y_i\}) - m\gamma \mathrm{d} Y_j/\mathrm{d}t + f_{Y_j}(t) \qquad (25\mathrm{a})$$

$$\gamma = \pi \omega_\mathrm{D}/6 \qquad (25\mathrm{b})$$

$$f_{Y_j}(t) = (2\gamma k_\mathrm{B} T m/h)^{1/2} \xi_{Y_j} \qquad (25\mathrm{c})$$

Here, ω_D is the Debye frequency; equation (25b) ensures that long time energy transfer into the bulk system is given correctly; the time step in the integration is 'h'; and ξ_{Y_j} is a vector of gaussian random numbers. Equation (25c) ensures that the frictional and random forces obey the second fluctuation–dissipation theorem:

$$\langle f_Y(t)f_Y(0)^T\rangle = \delta(t)2\gamma mk_B T\,\mathbf{1} \tag{26}$$

at least over the time step of the numerical integration. The fixed atoms surround the active atoms and are not allowed to move. They provide a structural template for the surface and smooth out the potential energy of the active atoms.

Local Langevin theories are much easier to apply than the GLE. They are also more expensive computationally since they require a larger number of moving atoms than the GLE. This occurs because the proper memory kernel for the real friction forces have been replaced by a simple local friction. The edge atoms must be far enough away from the strongly perturbed region so as to appear to exhibit damping forces with memory. The ease of applicability and generality of these local LE methods are so compelling that they have essentially replaced the GLE methods.

I do want to point out that the damping coefficient in the local LE methods is still tied to the elastic properties of the solid. This is important since the rate of energy transport in the solid is controlled by these properties. Hence, the local LE dissipates energy on the proper time scale. Not all methods do this. The common approach of simply rescaling all the velocities to thermostat the temperature is designed to provide an equilibrium situation but will not provide the proper dynamical response unless the velocity rescaling time is somehow tied to the elastic properties of the solid. There is no real advantage to this velocity rescaling since gaussian random numbers can be generated very efficiently; the major time in any simulation involves the computation of the forces; and, in any event, it is more important to describe the system accurately than do the computations quickly.

At this point, I have presented the global dynamical methods which, in principle, require knowledge of the entire AB–S and/or S–S interactions. There are a number of methods which involve only local dynamics, in the sense of using only localized properties of the full PES. As such, they will be much less dependent upon the large (and unknown) regions of configuration space which must be predicted by any PES form. If only small regions are needed, then these might even be provided by high quality *ab initio* or density functional calculations. This can be a mixed blessing since one must make *a priori* assumptions about which regions are most important without a full multi-dimensional dynamical simulation as a guide. In the worst case, this will yield dynamical results which are completely incorrect and inappropriate for the system under study.

The most common assumption is one of a reaction path in hyperspace in which a saddle point on the PES is found and the steepest descent path (in mass-weighted co-ordinates) from this saddle point to reactants and products is defined as the reaction path.[47] The information needed, except for the path and the energies along it, is the local quadratic PES for motion perpendicular to the path. The reaction path Hamiltonian is only a weakly local method since it can be viewed as an approximation to the full PES and since it is possible to use any of the previously defined global dynamical methods with this potential. However, it is local because the approximate PES restricts motion to lie around the reaction path. The utility of a reaction path formalism involves the convenient approximations to the dynamics which can be made with the formalism as a starting point.

The most important such method is classical canonical transition state theory (CTST). This approximation is based upon three assumptions: (i) all motion is classical; (ii) the reactants and activated complex are at equilibrium; (iii) all activated molecules with momentum along the reaction path in the direction of products will proceed to products (*i.e.* no recrossing). This leads to:

$$k_{TST}(T) = (k_B T/h)(Q^{\ddagger}/Q_r)\exp(-\beta V^{\ddagger}) \qquad (27)$$

for the rate of reaction. Here, $Q^{\ddagger}$ and $V^{\ddagger}$ are the partition function (including all degrees of freedom except 'u') and barrier height, respectively, at the saddle point, *i.e.* the transition state. In the more accurate and modern versions of this theory, the location of the saddle point is varied to provide a minimum value for k_{TST}, termed canonical variational TST (CVTST) and, in addition, tunnelling corrections are explicitly incorporated in order to incorporate some quantum mechanical features.[48–50] These corrections lead to the form:

$$k(T) = \Gamma(T)k_{TST}(T) \qquad (28)$$

where $\Gamma(T)$ is a ubiquitous factor including all such corrections.
Applications of this formula can utilize some or all corrections, ranging from simple CTST to very sophisticated CVTST with least action paths for tunnelling.

In all cases, TST provides a relatively simple method for the prediction of the rate constant. However, one must always keep in mind the assumption of TST, especially the short-time and positive momentum criteria, which prohibit recrossing of the saddle point. The fundamental idea that the dynamical process follows a reaction path must also be critically examined.

4 Selected Illustrations

I now turn to an illustration of the theoretical results. Since these are for clean, perfect surfaces, it is important to indicate when this assumption is valid. Serri *et al.*[51] developed a kinetic model for NO (molecular) chemisorption on Pt which incorporated adsorption, desorption, and diffusion of NO on terraces, diffusion from the terraces to the steps, and escape from the steps to the terraces. Adsorption and desorption directly from the steps were not included. The major conclusion was that, under the physically realistic condition of fast diffusion at low coverage and moderate to high temperature, the presence of steps plays a dominant role in thermal desorption. In dissociative chemisorption via either an intrinsic or extrinsic precursor the

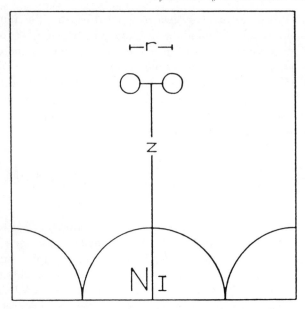

Figure 8 *Geometry and co-ordinate system for* $H_2/Ni(100)$
(Reproduced with permission from *J. Chem. Phys.*, 1987, **87**, 5497)

presence of steps may also play a dominant role, especially when the dissociative chemisorption probabilities are small.[8,52,53]

Consider H_2 colliding with Ni and Cu surfaces. The light mass of H_2 precludes significant energy transfer to the lattice. This indicates that for S_0 of substantial size, a direct mechanism should apply and thus S_0 and the dissociation probability, P_d, are equal. Since H_2 is so light it is important to identify possible quantum mechanical effects, such as tunnelling, zero-point energy, and diffraction.

Start with a simple model in which the H_2 orientation is fixed parallel to a rigid linear chain of Ni atoms [representing a row of Ni(100)] and is located above a Ni atom as shown in Figure 8. Due to the symmetry, only Z and r can vary, yielding a two-dimensional dynamical problem for which the exact solution of the time-dependent Schrödinger equation is feasible using the FFT algorithm described in Section 3. This has been accomplished on the PES shown in Figure 9.[54] This PES is a model for the dissociation over an atop atom towards the bridge sites on Ni(100), but this should not be taken too seriously since, as will be discussed later, the likely dissociation process does not even involve a symmetric site or orientation of H_2. This work is important because it provides a detailed investigation of quantum mechanical effects.

The results are shown in Figure 10 for four isotopes of hydrogen with masses of 1, 2, 3, and 7. The last is fictitious but useful in delimiting the importance of quantum effects. The classical and quantum values are in general agreement with the former slightly exceeding the latter, except at the

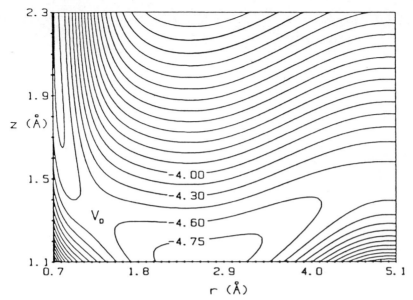

Figure 9 *A contour plot of the PES for H_2/Ni(100) in the configuration of Figure 8* (Reproduced with permission from *J. Chem. Phys.*, 1987, **87**, 5497)

lowest kinetic energies. Even for H_2, the classical results are clearly not quantitative but are not terrible. It is interesting that the classical results are too large at low kinetic energies. This is due to two effects. The first is the occurrence of quantum mechanical reflection (at the second atop position) of the part of the wavepacket which surmounts V_0. This effect will be negligible on a real three-dimensional surface. The second, and more important, is the inability of the classical simulations to enforce zero-point energy restrictions throughout the trajectory, thereby allowing the initial H_2 zero-point vibrational energy to be converted into motion along the reaction co-ordinate, which increases the classical S_0. Both effects will become less important for larger masses, as can be seen for the $m = 7$ results in Figure 10.

This incorrect treatment of V→T energy transfer is not only important in dissociation. For example, the stretching of the H_2 as it approaches the surface can give rise to a trapped molecular physisorbed species.[55] Due to the lowering of the vibrational frequency as H_2 approaches the surface, some vibrational zero-point energy is released into translational energy which can then be transferred to a non-rigid surface. Loss of even a small fraction of this local translational energy will trap a molecule with low initial kinetic energy since the zero-point energy is typically much larger than the initial low kinetic energy. This is the vibrational analogue of the Beeby mechanism for translations:[56] acceleration by the physisorption well provides a large local speed when the molecule hits the surface, and a small fractional loss of this local translational energy leads to trapping. The vibrational zero-point effect is independent of the well depth of the molecule–surface PES, except,

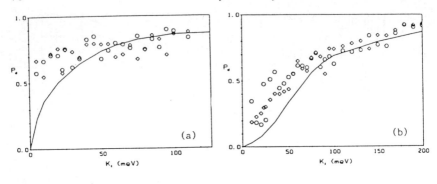

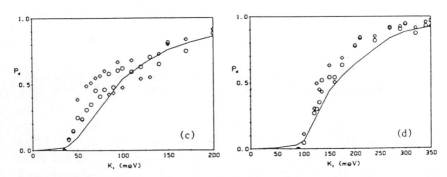

Figure 10 (a) *Dissociation probability as function of the initial kinetic energy for* $H_2/Ni(100)$. *The solid curves are from the quantum calculation, and the circles and diamonds correspond to the quantum-weighted classical and quasiclassical approaches, respectively;* (b) *same as* (a) *except the gas atom mass is 2;* (c) *same as* (a) *except that the gas atom mass is 3;* (d) *same as* (a) *except that the gas atom mass is 7*
(Reproduced with permission from *J. Chem. Phys.*, 1987, **87**, 5497)

of course, that a well must exist to support a physisorbed species. Classical mechanical treatments are intrinsically unable to describe this vibrational enhanced trapping.

It is also interesting that the shape of the P_d vs. kinetic energy curves is not a step function, even though only a single barrier occurs. This is easy to understand classically. Each trajectory is initiated with a different division between kinetic and potential vibrational energy (*i.e.* particular vibrational phase) as well as with the same translational kinetic energy. Thus each trajectory oscillates in r as it decreases in Z. A trajectory which overshoots the reaction path and bounces off the hard wall (*i.e.* small Z and $r \approx r^{(eq)}$) will not dissociate because its motion will not lie along the reaction co-ordinate. At low kinetic energy it will be most important to have precisely the correct phase to ensure motion over the minimum barrier. At higher kinetic energies, enough energy is available to surmount more than the minimum barrier making the vibrational phase less important, and thus yielding larger P_d.

The above results demonstrate that even a single barrier located in the exit channel will not give rise to a step function change in S_0. They also demonstrate that even when the kinetic energy is not enough to surmount the minimum barrier, the ability of the PES to transform vibrational zero-point energy into motion along the reaction path can also lead to $S_0 > 0$. Thus, one should not ascribe these two characteristics to quantal tunnelling without knowing the shape of the PES.

Now consider models which are of higher dimensionality. The first is still fully quantum mechanical but in which a special form for the PES is used that is most appropriate for the H_2/Ni and H_2/Cu systems.[34,35,57] The basic assumptions are that a barrier to chemisorption exists in the entrance channel, much like point 'A' in Figure 2; this barrier is nearly independent of the molecular orientation and vibrational bond length; and, all molecules which pass over this barrier dissociate (*i.e.* there is a 'sink' in the PES). The justification for the orientation independence is the free rotational motion in the physisorbed species. The justification for the independence of bond length is the absence of a significantly stretched molecular bond at the position of the activation barrier 'A'. The irreversible dissociative behaviour after surmounting a point like 'A' in Figure 2 assumes that either no barrier, or at most an extremely small barrier, exists in the exit channel, unlike point 'D' in that figure.

The distinguishing feature of this work, however, is that the activation barrier is allowed to vary with different positions of the H_2 in the surface unit cell. This approach captures much of the influence of the corrugated surface. Moreover, under these approximations the dynamical equations become only three dimensional and, in fact, are identical to those of an atom scattering from a rigid, corrugated surface but with a potential with a 'sink'. The 3-D Schrödinger equation is solved by the FFT procedure. Hence, it is possible to treat diffraction, tunnelling, and dissociative chemisorption by accurate quantum techniques, at least on this simple type of PES.

Calculations of the diffraction intensities were performed for three different PESs. The first had the minimum activation barrier when the H_2 was located over a Ni atom (*i.e.* atop site) while the second had the minimum over a four-fold (*i.e.* centre) site. The third did not allow for dissociation at all since it did not have a barrier (*i.e.* the PES continually increased as Z decreased). The first is shown in Figure 11a as a function of Z and position of H_2 centre-of-mass along a line connecting atop and centre sites. A different representation is shown in Figures 11b–d, providing a graphic illustration of the energy dependence of the size of the hole. The PES with the minimum barrier over the centre site was very similar.

Diffraction intensities are shown in Figure 12 with the notation [00], [10], and [20] indicating the specular, first, and second order diffracted beams, respectively. It is immediately apparent from the contrast between the three figures that the availability of the PES 'sink' has a marked effect on all the

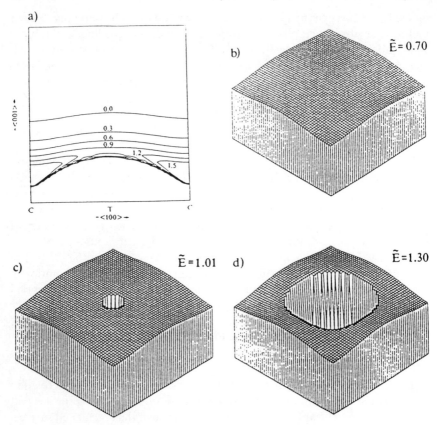

Figure 11 (a) *Contour map of a model potential for an adsorbing* H_2 *with points 'C' and 'T' referring to centre and atop sites, respectively;* (b)–(d) *surfaces of constant* H_2 *surface energy plotted across the unit cell. See text for further description of the model*
(Reproduced with permission from *Surf. Sci.*, 1987, **179**, L41)

beams. The specular beam does not go through a minimum at ≈ 350 meV because as the energy is raised, dissociation becomes more probable, and thus more of the flux is channelled into this dissociation channel instead of into the reflected specular channel. Similarly, the peak values of the [10] and [20] beams are significantly reduced. In a non-reactive system, it is well known that the corrugation of the potential increases with energy, thereby leading to larger diffracted intensities at high energy. In the reactive system, however, the competition for the flux at high energy by the dissociative channel significantly lessens the peak of the diffracted beams. Similar results were also found for D_2 scattering. The conclusion is that the energy dependence of the diffracted beams may contain significant information on the reactive PES, at least when the barrier lies in the entrance channel.

I now consider the results of classical dynamics simulations of dissociative chemisorption. These will utilize the full dimensionality of the problem.

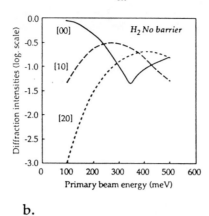
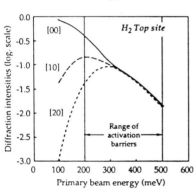
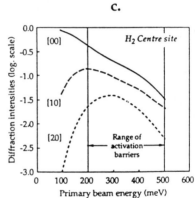

Figure 12 (a) H_2 intensities for the first three diffraction states as a function of initial energy for a primary beam at normal incidence to a potential having no reactive channel. The sharp minimum occurring in the specular beam at 370 meV arises as a consequence of destructive interference between molecules scattered at top and centre sites; (b) same as (a) but here the minimum value of the activation barrier is above an atop site; (c) same as (a) but here the minimum barrier value is located above a centre site
(Reproduced with permission from *J. Chem. Phys.*, 1988, **88**, 7197)

Again, consider the H_2/Ni(100) system for which a LEPS PES has been constructed along the lines presented in Section 2.[58] Both *ab initio* calculations[22] of the H_2/Ni(100) system and experimental data[59] on S_0 were used to determine the parameters. The contour plots in Figure 13 orient the H_2 parallel to the surface with its centre-of-mass positioned either over the bridge or atop sites. These demonstrate that there are small activation barriers of ≈ 0.035 eV and ≈ 0.045 eV in the entrance channel of the bridge-to-centre and atop-to-centre PES, respectively. The dependence of the entrance channel barrier on the orientation of H_2 has not been determined quantitatively, but the dependence on r is quite weak. The variation with position in the unit cell is also weak. In addition, for the atop-to-centre

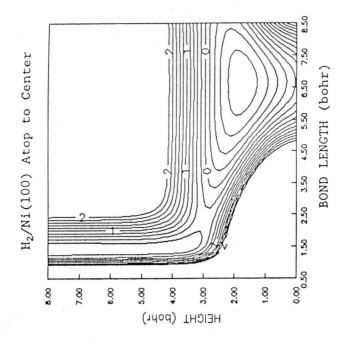

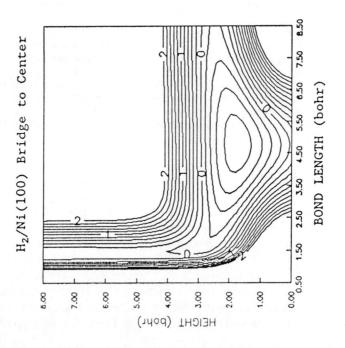

ZOOMING THE ENTRANCE CHANNEL

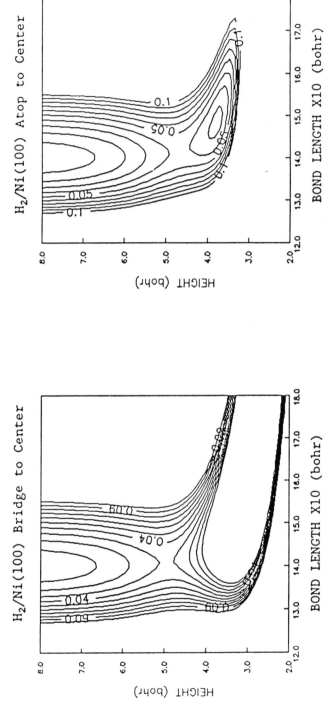

Figure 13 *Two-dimensional contour plots for a LEPS potential of* $H_2/Ni(100)$. (Reproduced with permission from *J. Chem. Phys.*, 1989, **92**, 5653)

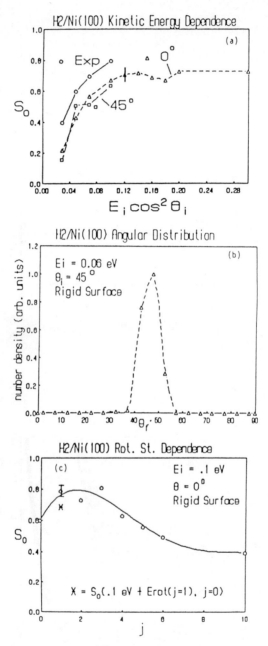

Figure 14 (a) *Variation of the initial sticking coefficient with initial normal kinetic energy (circles represent experimental results from reference 59 while triangles and squares represent classical GLE calculations at incident angles of 0° and 45°, respectively)*; (b) *calculated number of density of scattered* H_2 *vs. final polar angle;* (c) *calculated initial rotational state dependence of the dissociation probability*
(Reproduced with permission from *J. Chem. Phys.*, 1989, **92**, 5653)

$H_2/Ni(100)$

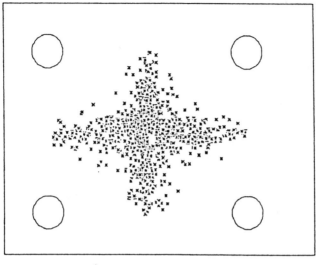

a) $E_i = .14$ eV

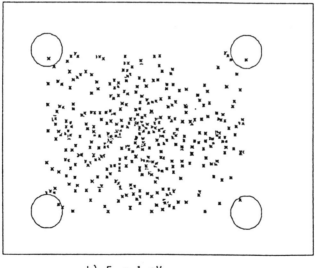

b) $E_i = 1$ eV

Figure 15 *Location within the unit cell of the centre-of-mass of the molecule at the occurrence of molecular dissociation in the $H_2/Ni(100)$ system, for normal incidence and initial kinetic energies of 0.14 eV in (a) and 1 eV in (b)* (Reproduced with permission from *J. Chem. Phys.*, 1989, **92**, 5653)

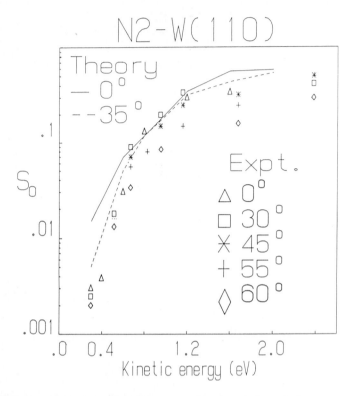

Figure 16 *The initial sticking coefficient as a function of the initial kinetic energy for the $N_2/W(110)$ system; experimental data are from reference 60 and the theoretical curves are from reference 27*
(Reproduced with permission from *Surf. Sci.*, 1988, **193**, 437)

dissociation there is also a barrier in the exit channel. The exit channel barrier obviously has a strong dependence upon bond length. Thus, the PES is complex, but many features are similar to the model described above.

GLE-'ghost' atom simulations showed that surface motion had negligible effect on the dynamics.[58] The results shown in Figure 14 are for a rigid surface with all six degrees of freedom of the H_2 allowed to vary. First, note in Figure 14a that S_0 is clearly a function of normal kinetic energy, *i.e.* $S_0(E_i, \theta_i) = S_0(E_i \cos^2\theta_i, 0)$. I want to emphasize that such scaling behaviour does not imply a barrier in the entrance channel. To see why, note first that the shape of the PES shown in Figures 2 and 13 arises from two general factors: (i) the molecular physisorption well is located further from the surface than the atomic chemisorption well; and, (ii) the bond must stretch in transforming the reactant molecule into the separated atomic adsorbates. Smooth contours drawn between these limits will yield a PES with a topology similar to those shown previously. Thus most PESs should be able to transform translational motion normal to the surface into motion along the reaction path with con-

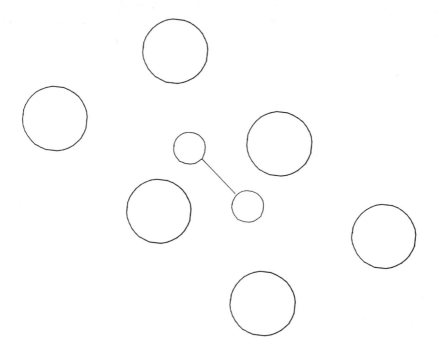

Figure 17 *Configuration of the N_2/W(110) system with only a few W atoms shown for clarity*
(Reproduced with permission from *Surf. Sci.*, 1988, **193**, 437)

siderable efficiency. This is the physical explanation of why many systems with exit channel barriers obey near normal energy scaling.

Secondly, note that there must be some dynamical features of orientation and bond length since S_0 does not become unity at $E_i > 0.1$ eV, which exceeds all the entrance channel barriers in Figure 13. However, S_0 is quite large (0.6–0.7) at such energies and so most of the molecules do dissociate after surmounting the entrance channel barrier. Thirdly, the angular distribution of scattered molecules is quite narrow, indicating small translational to rotational energy transfer. Fourthly, the rotational state dependence is not insignificant at high kinetic energies, as in Figure 14c, but will be much less important at low kinetic energies. Indeed, I would expect that the neglect of all quantum effects inherent in the classical simulation would be a much more severe approximation at low kinetic energies. Hence, the reduced dimensionality quantal model[34,35,57] may have real utility at low kinetic energy. This will depend upon the detailed pattern of reactivity, *i.e.* whether an energy dependent 'hole' in Figure 11b–d is a good description.

Determination of the pattern of reactivity cannot be done unless the bond

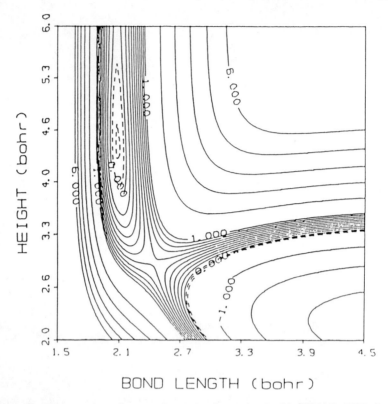

Figure 18 *Contour plot of a two-dimensional cut in the N_2/W(110) PES for the configuration in Figure 17 which yielded total energy scaling*
(Reproduced with permission from *Surf. Sci.*, 1988, **193**, 437)

stretching to dissociation is fast compared to the motion of the molecular centre-of-mass across the surface. For the H_2/Ni(100) system this was the case[58] since the molecular bond length oscillated around its gas-phase value until increasing quickly and monotonically during the irreversible dissociation. The rapid increase in the bond length occurred much more quickly than any translational motion of the molecule's centre-of-mass. (The rotation of the molecule occurred on the same time scale as the increase in bond length, which precluded a precise definition of the molecular orientation at dissociation.) Figure 15 shows the distribution of dissociation impacts. The reaction exhibits site selectivity at $E_i = 0.14$ eV even though $S_0 \approx 0.8$, only losing the selectivity at $E_i = 1.0$ eV where $S_0 \approx 0.96$. While the reaction zone is very distorted from the near-circular one shown in Figures 11b–d, it could clearly be modelled along the same lines as done in references 34, 35, 37.

A different type of system is N_2/W(110) since it exhibits a scaling with the total kinetic energy, $S_0(E_i, \theta_i) = S_0(E_i, 0)$ according to experiment.[60] It is also possible to duplicate this behaviour using full GLE-'ghost' atom simulations

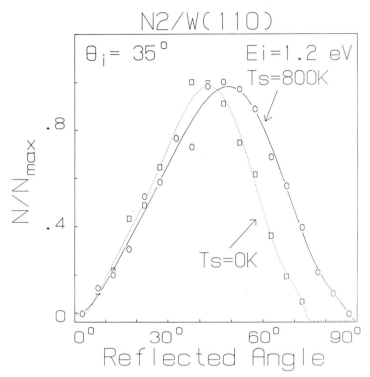

Figure 19 *Number distribution of final angles for N_2 molecules scattered from the W(110) surface*
(Reproduced with permission from *Surf. Sci.*, 1988, **193**, 437)

as shown in Figure 16. As before, the simulations used a four-body LEPS PES constructed according the methods in Section 2.[27,61,62] For the configuration of N_2/W(110) in Figure 17, a contour plot of the PES is shown in Figure 18. The very constricted activation barrier region enforces a much more stringent steric dependence on the molecule before reaction can occur, which does not allow for the indiscriminate transfer of translational to vibrational energy to be efficient at surmounting the barrier.

The angular distribution of scattered molecules is shown in Figure 19. The surprising features are the narrowness and the peak at a slightly supraspecular angle. Both agree with the experimental data of reference 60. The reason for these features is that the molecules which scatter back into the gas phase nearly always undergo only a single impact with the surface, because of an improper orientation during the initial collision. These scattered molecules do not sample the dissociative part of the PES but instead act just like the scattered molecules for inelastic scattering. Further evidence for this is seen in Figure 20 for the distribution of rotational, vibrational, kinetic, and total energies for all the scattered molecules, each summed over all final angles. The surprising result is the negligible change in the vibrational energy even though E_i is large enough to excite $N_2(n = 4)$. This demonstrates that the part

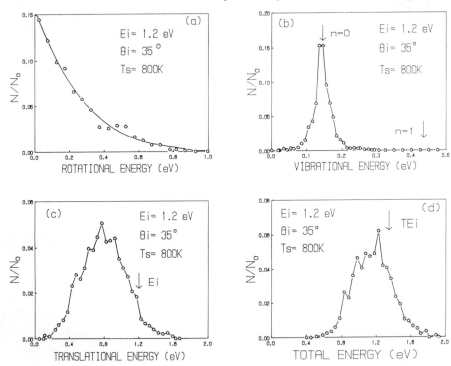

Figure 20 (a) *Distribution of final rotational energy for N_2 molecules scattered from the W(110) surface;* (b) *same as* (a) *except for vibrational energy;* (c) *same as* (a) *except for kinetic energy;* (d) *same as* (a) *except for total molecular energy*
(Reproduced with permission from *Adv. Chem. Phys.*, 1990, **77**, 163)

of the PES which stretches the N_2 bond is not sampled by the scattered molecules, corroborating the above idea. The scattering is not elastic however. The energy loss to W(110) is indicated by the average of the final total energy being less than the initial value by about 0.3 eV. By contrast, the change in E_i is about 0.5 eV, which demonstrates that about 0.2 eV is transferred from translational into rotational energy.

Another interesting feature of the dissociative chemisorption is the dependence upon initial vibrational and rotational excitation of the N_2. The results in Figure 21 demonstrate that translational and vibrational energy are equally efficient at increasing S_0. This may be expected for a system in which total energy scaling is found. By contrast, a complicated dependence upon rotational state is exhibited in Figure 22. At each E_i there is a fast increase of S_0 at $j = 2$ followed by a nearly constant value. The increase is much larger than the variation of S_0 with E_i since the rotational energy of $j = 2$ is only ≈ 0.0015 eV.

In the N_{23}/W(110) system, S_0 remains less than unity even at kinetic energies well above those of the minimum barrier. By contrast, for the H_2/Ni(100) system considered previously S_0 approaches unity at high kinetic

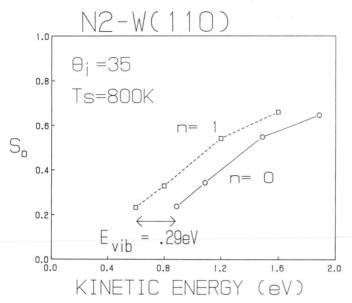

Figure 21 *A comparison of the efficiency of vibrational and translational energies in the dissociative chemisorption of N_2 on W(110)*
(Reproduced with permission from *Adv. Chem. Phys.*, 1990, **77**, 163)

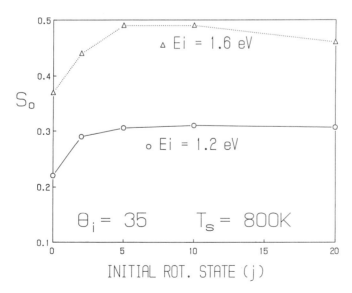

Figure 22 *Initial rotational state dependence of the dissociative chemisorption probability of N_2 on W(110)*
(Reproduced with permission from *J. Chem. Phys.*, 1988, **88**, 5240)

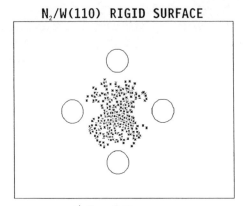
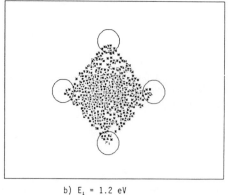
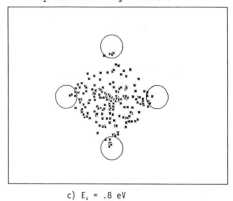

Figure 23 *Location within the unit cell of the centre-of-mass of the molecule at the occurrence of molecular dissociation in the $N_2/W(110)$ system, for normal incidence. (a)–(b) Results are for a rigid surface at different kinetic energies; (c) results are for a non-rigid surface*
(Reproduced with permission from *J. Chem. Phys.*, 1989, **92**, 5653)

energies. To understand this behaviour more fully, consider the location of the N_2 molecule in the surface unit cell when dissociation occurred. This was possible for the same reasons as given for the $H_2/Ni(100)$ system. Figure 23 shows the distribution of dissociation impacts for $N_2/W(110)$ at low and high kinetic energy. Two surprising features are:
(i) the absence of site selectivity in $N_2/W(110)$, for the non-rigid surface at $E_i = 0.8$ eV in Figure 23c, even when $S_0 \approx 0.15$;
(ii) the change from site selective to non-site selective behaviour for the $N_2/W(110)$ system at $E_i = 0.8$ eV, an effect which requires $E_i = 1.2$ eV in the rigid surface.

Rationalization of these results depends upon the influence of molecular orientation on the PES. In the $N_2/W(110)$ case, molecular orientation is much more important than position in the surface unit cell. By contrast, for the $H_2/Ni(100)$ case, the opposite relationship holds. This can be understood in terms of the variation of the underlying atom–surface PES and the size of the molecules. Rotation of the N_2 relative to the surface changes the N–W distance substantially on the scale which the N–W(110) interaction changes. By contrast, rotation of the H_2 relative to the surface changes the H–Ni distance only a small amount on the scale which the H–Ni(100) interaction changes. Thus rotation of the molecule is more important in the former case. In addition, location in the unit cell is more important in $H_2/Ni(100)$ because the atop site does not bind H strongly enough to dissociate H_2 (*e.g.* atop, bridge and four-fold sites bind H by $\approx$ 2.2, 2.5, and 2.8 eV, respectively). By contrast, the relevant quantities for N–W are 5.8, 6.8, and 7.2 eV. In other words, the atom–surface PES is much more corrugated for the H–Ni(100) than N–W(110) systems.

More work will need to be performed to ascertain the generality of these conclusions, but I expect that it will be found necessary to discuss dissociative chemisorption reactions in a much higher dimensionality than the simple one- or two-dimensional models. This lack of dominance by high symmetry dissociation sites implies that *ab initio* and density functional calculations of PES must sample a wide variety of sites in order to provide detailed information about the dissociation process.

Acknowledgement. This work has been supported by the National Science Foundation, division of Chemical Physics. Partial support by the Petroleum Research Foundation administered by the American Chemical Society has also been received.

References

1. G. A. Somorjai, 'Chemistry in Two Dimensions', Cornell University Press, Ithaca, New York, 1981.
2. A. Zangwill, 'Physics at Surfaces', Cambridge University Press, New Rochelle, New York, 1988.
3. R. P. H. Gasser, 'An Introduction to Chemisorption and Catalysis by Metals', Clarendon Press, Oxford, 1985.

4. J. E. Lennard-Jones, *Trans. Faraday Soc.*, 1932, **28**, 333.
5. S. L. Bernasek, *Adv. Chem. Phys.*, 1980, **41**, 477.
6. J. A. Barker and D. J. Auerbach, *Surf. Sci. Rep.*, 1984, **4**, 1.
7. D. W. Goodman, *Acc. Chem. Res.*, 1984, **17**, 194.
8. G. Comsa and R. David, *Surf. Sci. Rep.*, 1985, **5**, 145.
9. M. J. Cardillo, *Langmuir*, 1985, **1**, 4.
10. S. T. Ceyer, *Annu. Rev. Phys. Chem.*, 1988, **39**, 479.
11. C. T. Rettner, *Vacuum*, 1988, **38**, 295.
12. J. C. Tully, *Acc. Chem. Res.*, 1981, **14**, 188.
13. R. B. Gerber, *Chem. Rev.*, 1987, **87**, 29.
14. J. W. Gadzuk, *Annu. Rev. Phys. Chem.*, 1988, **39**, 395.
15. D. Brenner and B. J. Garrison, *Adv. Chem. Phys.*, 1989, **66**, 281.
16. A. E. DePristo, in 'Interactions of Atoms and Molecules with Solid Surfaces', ed. A. Bortolani, N. March, and M. Tosi, Plenum Press, New York, 1989.
17. A. E. DePristo and A. Kara, *Adv. Chem. Phys.*, 1990, **77**, 163.
18. A. Bortolani, N. March, and M. Tosi, in 'Interactions of Atoms and Molecules with Solid Surfaces', ed., A. Bortolani, N. March, and M. Tosi, Plenum Press, New York, 1989.
19. G. D. Billing, *Chem. Phys.*, 1987, **116**, 269.
20. J. K. Nørskov, A. Houmoller, P. Johansson, and B. I. Lundqvist, *Phys. Rev. Lett.*, 1981, **46**, 257.
21. J. K. Nørskov and F. Besenbacher, *J. Less-Common Met.*, 1987, **130**, 475.
22. P. Siegbahn, R. A. Blomberg, and C. W. Bauschlicher, *J. Chem. Phys.*, 1984, **81**, 2103.
23. T. H. Upton, P. Stevens, and R. J. Madix, *J. Chem. Phys.*, 1988, **88**, 3988.
24. H. Yang and J. L. Whitten, *J. Chem. Phys.*, 1988, **89**, 5329.
25. J. H. McCreery and G. Wolken, *J. Chem. Phys.*, 1977, **67**, 2551.
26. C. Y. Lee and A. E. DePristo, *J. Chem. Phys.*, 1986, **85**, 4161.
27. A. Kara and A. E. DePristo, *Surf. Sci.*, 1988, **193**, 437.
28. H. Eyring, J. Walter, and G. E. Kimball, 'Quantum Chemistry', Wiley, New York, 1944.
29. M. J. Puska, R. M. Nieminen, and I. Manninen, *Phys. Rev. B*, 1981, **24**, 3037.
30. M. S. Daw, *Phys. Rev. B*, 1989, **39**, 7441.
31. M. S. Daw and M. I. Baskes, *Phys. Rev. B*, 1984, **29**, 6443.
32. R. B. Gerber, R. Kosloff, and M. Berman, *Comput. Phys. Rep.*, 1986, **5**, 59.
33. J. A. Fleck, Jr., J. R. Morris, and M. D. Feit, *Appl. Phys.*, 1976, **10**, 124.
34. M. Karikorpi, S. Holloway, N. Henriksen, and J. K. Nørskov, *Surf. Sci.*, 1987, **179**, L41.
35. D. Halstead and S. Holloway, *J. Chem. Phys.*, 1988, **88**, 7197.
36. U. Nielsen, D. Halstead, S. Holloway, and J. K. Nørskov, *J.Chem.Phys*, 1990, **93**, 2879.
37. S. A. Adelman and J. D. Doll, *Acc. Chem. Res.*, 1977, **10**, 378.
38. S. A. Adelman, *Adv. Chem. Phys.*, 1980, **44**, 143.
39. J. C. Tully, *J. Chem. Phys.*, 1980, **73**, 1975.
40. A. E. DePristo, *Surf. Sci.*, 1984, **141**, 40.
41. D. J. Diestler and M. E. Riley, *J. Chem. Phys.*, 1987, **86**, 4885.
42. M. Berkowitz and J. A. McCammon, *Chem. Phys. Lett.*, 1982, **90**, 215.
43. C. L. Brooks, III and M. Karplus, *J. Chem. Phys.*, 1983, **79**, 6312.
44. (a) R. Lucchese and J. C. Tully, *Surf. Sci.*, 1983, **137**, 1570; (b) R. Lucchese and J. C. Tully, *J. Chem. Phys.*, 1984, **80**, 3451.

45. M. E. Riley, M. E. Coltrin, and D. J. Diestler, *J. Chem. Phys.*, 1988, **88**, 5934.
46. A. E. DePristo and H. Metiu, *J. Chem. Phys.*, 1989, **90**, 1229.
47. W. H. Miller, N. C. Handy, and J. E. Adams, *J. Chem. Phys.*, 1980, **72**, 99.
48. J. G. Lauderdale and D. G. Truhlar, *Surf. Sci.*, 1985, **164**, 558.
49. J. G. Lauderdale and D. G. Truhlar, *J. Chem. Phys.*, 1986, **84**, 1843.
50. D. G. Truhlar, A. D. Isaacson, and B. C. Garrett, in 'The Theory of Chemical Reaction Dynamics', Vol. 4, ed. M. Baer, CRC, Boca Raton, Florida, 1986, pp. 1.
51. J. A. Serri, J. C. Tully, and M. J. Cardillo, *J. Chem. Phys.*, 1983, **79**, 1530.
52. H. P. Steinruck, M. P. D'Evelyn, and R. J. Madix, *Surf. Sci.*, 1986, **172**, L561.
53. K. D. Rendulic, *Appl. Phys. A*, 1988, **47**, 55.
54. C.-M. Chiang and B. Jackson, *J. Chem. Phys.*, 1987, **87**, 5497.
55. J. E. Muller, *Phys. Rev. Lett.*, 1987, **59**, 2943.
56. F. O. Goodman and H. Y. Wachman, 'Dynamics of Gas Surface Scattering', Academic Press, New York, 1976.
57. S. Holloway, in 'Interactions of Atoms and Molecules with Solid Surfaces', ed. A. Bortolani, N. March, and M. Tosi, Plenum Press, New York, 1989.
58. A. Kara and A. E. DePristo, *J. Chem. Phys.*, 1989, **92**, 5653.
59. A. V. Hamza and R. J. Madix, *J. Phys. Chem.*, 1985, **89**, 5381.
60. H. E. Pfnur, C. T. Rettner, J. Lee, R. J. Madix, and D. J. Auerbach, *J. Chem. Phys.*, 1986, **85**, 7452.
61. A. Kara and A. E. DePristo, *J. Chem. Phys.*, 1988, **88**, 2033.
62. A. Kara and A. E. DePristo, *J. Chem. Phys.*, 1988, **88**, 5240.

CHAPTER 3

Quantum Effects in Gas–Solid Interactions

S. HOLLOWAY

1 Introduction

Surface science is a subject area that is fast reaching maturity. Within the past three decades it has been shown that it is possible to adsorb foreign species on well characterized single crystal surfaces and perform spectroscopic investigations to determine atomic positions, vibrational properties, and electronic structure.[1] Within the area of basic research this has given rise to a series of natural subject areas forming within surface science which may be characterized as being either static, electronic, or dynamic properties. While this division is in no way meant to be unique (there will of course be endeavours which overlap such boundaries) it does focus the attention onto the primary sub-groupings existing at this time. The articles in this volume are concerned with dynamical properties and in particular those which arise from the interaction of gas atoms and molecules with the bare surface.

The ancestry of dynamics goes back to the time just following the discovery of quantum mechanics. At this time there was a lively atmosphere which was fuelled by the notion that the forces experienced by reacting species may be calculated using the new mechanics and then the motion of the reactants could be followed in time. This spirit appears particularly clear in the article of Eyring:[2]

> 'Now a system moving on this (potential energy) surface will have kinetic energy that may be quantized for the different degrees of freedom in a variety of ways, consistent with the particular energy and the particular position on the surface. Low places in the potential energy surfaces correspond to compounds. If a particular low lying region is separated from all other low places by regions higher

than twenty-three kilocalories, the compound will be stable at and below room temperature. The higher the lowest pass the higher is the temperature at which the compound is still stable. A reaction corresponds to a system passing from one low region to another'.

This is the way in which the majority of dynamical processes are envisaged even to the present day! A potential energy surface describing a particular reaction of interest is found (in principle) by determining a total energy from an electronic structure calculation. Using this, an initial-value boundary condition problem is solved for the evolution of the nuclear dynamics. It is only comparatively recently that alternative schemes have been developed based on the density functional theory which allow both electronic and nuclear motion to be followed simultaneously.[3] Even in this case it is, however, still not possible to treat electronic non-adiabatic effects accurately since the density functional scheme generally only applies to ground state properties.

In addition to electronic non-adiabaticity, the surface presents a vibrational continuum to the incident molecule which, depending on the surface temperature, can supply or remove energy in a scattering encounter. Classically this may be incorporated in a very direct fashion by employing molecular dynamic simulation methods which have a long history in the area of dynamical phenomena. For a detailed review, Chapter 2 of this volume should suffice. If the (classical) energy exchange with the surface falls within the phonon band then a classical description of the collision is inadequate and quantum mechanics needs to be considered.[4] In this current article for reasons that will become clear, we will not deliberate at great length on phonon induced non-adiabaticity in the quantum limit; this was well covered in Chapter 1. Instead the major thrust of this article is to address the problem of the internal degrees of freedom in the gas-phase molecule and how they become coupled via the interaction with the solid surface.

The organization of this chapter will commence by introducing the background to the construction of, and nomenclature associated with, gas–surface potential energy surfaces (PES) in Section 2. Both adiabatic and diabatic surfaces will be presented and their various merits discussed. For each case, the problem of electronic non-adiabaticity will be introduced. Following this we shall present the pros and cons of classical vs. quantum scattering in surface dynamics. The problem of how to solve the time dependent Schrödinger equation for a given multi-dimensional potential will be addressed in Section 3. This will, by necessity, be a rather technical exercise and both the split operator and the Chebyshev methods will be examined. Section 4 will be concerned with the inelastic scattering of molecules from surfaces. In this case, no dissociation occurs and molecules can either trap onto the surface resonantly before undergoing inelastic encounters with the phonon sub-system (H_2/Cu) or scatter promptly, transferring rotational energy directly with the phonons (NO/Ag). For the former case, some recent experimental data[5] have been analysed which have necessitated a re-examination of the physisorption interaction while in the latter case progress is being made on the extremely complicated interaction between an open shell molecule and a metal surface.[6] In addition to rotational–translational coupling (R–T) this section will conclude with a short discussion of vibrational–translational (V–T) coupling in the system NO/Ag. This is a particularly intriguing problem because of the unexpected strong surface temperature dependence observed for the probability of vibrational excitation in the scattered beam.[7] As in the parallel field of gas-phase inelastic collisions, it is this area of surface dynamics which appears to be coming closest to unravelling potential energy surfaces by modelling dynamical interactions.

The following section, which will be the largest, contains a discussion of the reactive scattering of small molecules from metal surfaces. To reduce the coupling with the phonons, and to amplify the quantum effects, it is informative to study the interaction of hydrogen with surfaces. Paradoxically while this is advantageous on theoretical grounds, it is also of particular interest from an experimental viewpoint where the reaction of hydrogen on metal surfaces has wide ranging applications in industry. This has given rise to a model system which has been (and continues to be) extensively studied both theoretically and experimentally: H_2/Cu (see the following article by Hayden). This system exhibits a barrier to the dissociation reaction and as a consequence shows a variety of interesting dynamical phenomena. In particular, attention will be focused upon (i) the question of early and late barriers in the dissociation reaction and the relative merits of translational *vs.* vibrational energy in overcoming them, (ii) dimensionality in the scattering problem and in particular the important question of how to reduce the number of degrees of freedom in the problem but retain the essential dynamics, and (iii) the information content in the rotational state distributions for the scattered fraction in a chemisorption experiment. The scattering of heavier diatomics will also be discussed in this section along with some background to the dissociation of the polyatomic CH_4 molecule which shows some interesting isotope effects perhaps meriting a quantum interpretation. Much of the work presented here will employ fairly rudimentary potential energy surfaces, highly modelistic in nature. It is now broadly accepted that it is a useful exercise to develop a conceptual understanding of reactivity based upon PESs that display particular topological features.

The final section will contain a discussion of future directions in the topic area. It will be necessary to obtain a better understanding of the electronic part of the problem in order to generate PESs and understand more about electronic non-adiabatic effects. If the dissociation of diatomics is to be pursued, then what hope is there of including more (quantum) degrees of freedom in a scattering calculation? If surface reactions are really to be studied then what would be suitable systems (*i.e.* Langmuir–Hinshelwood or Eley–Rideal) and how might this be tackled from a theoretical point of view without becoming overwhelmed with uncertainties?

2 Potential Energy Surfaces

Formal Derivation

Potential energy surfaces are the *sine qua non* for theoretical studies of dynamic processes. Once obtained, it is possible to follow the time evolution of a system by integrating the equations of motion either classically or quantum mechanically. These aspects will be discussed in detail in the next section but first we shall present a brief account of what potential surfaces are, how they may be calculated, and the role of non-adiabatic terms in coupling together excited electronic states of the system.

In general, the aim is to solve for the combined motion of the electrons and nuclei when a gas-phase species approaches a solid surface; a formidable problem. For the combined system, suppose the co-ordinates of the electrons are given by $r \equiv (r_1, r_2, r_3, \ldots, r_n)$, and the nuclei by $R \equiv (R_1, R_2, R_3, \ldots, R_N)$. The total Hamiltonian may be expressed as:

$$\mathcal{H} = \mathcal{T}_R + \mathcal{H}_0 \tag{1}$$

where

$$\mathcal{T}_R = \sum_{M=1}^{N-1} -\left(\frac{1}{2\mu_M}\right)\nabla_M^2 \tag{2}$$

is the nuclear kinetic energy operator and

$$\mathcal{H}_0 = \sum_{i=1}^{n} -\tfrac{1}{2}\nabla_i^2 + \sum_{i=1}^{n-1}\sum_{j>i}^{n} \frac{1}{|r_i - r_j|} - \sum_{i=1}^{n}\sum_{M=1}^{N} \frac{z_M}{|r_i - R_M|}$$

$$+ \sum_{M=1}^{N-1}\sum_{M'>M}^{N} \frac{z_M z_{M'}}{|R_M - R_{M'}|} \tag{3}$$

is the electronic Hamiltonian for *fixed* nuclear positions. By selecting a suitable basis set for the electronic wave functions, $\phi_k(r; R)$, the total system wavefunction $\Psi(r, R)$ may be expressed as the linear combination:

$$\Psi(r, R) = \sum_{k} \phi_k(r; R)\chi_k(R) \tag{4}$$

where $\chi_k(R)$ describes the motion of the nuclei on the PES associated with the electronic state k. Substituting equations (1)–(4) into the Schrödinger equation:

$$[\mathcal{H} - \mathcal{E}]\Psi(r, R) = 0 \tag{5}$$

one obtains the infinite set of coupled equations:

$$[\mathcal{T}_R + U_{kk} - \mathcal{E}]\chi_k = -\sum_{k' \neq k}[\mathcal{T}'_{kk'} + U_{kk'}]\chi_{k'} \qquad (6)$$

where

$$U_{kk'} = \langle \phi_k | \mathcal{H}_0 | \phi_{k'} \rangle \qquad (7)$$

and

$$\mathcal{T}'_{kk'} = \sum_{M=1}^{N-1} -\left(\frac{1}{2\mu_M}\right)\langle \phi_k | \nabla_M | \phi_{k'} \rangle \cdot \nabla_M \qquad (8)$$

It is the diagonal elements of $U_{kk'}$ which are the potential energy surfaces governing the nuclear motion in a particular electronic state, k. The off diagonal terms $\mathcal{T}'_{kk'}$ and $U_{kk'}$ give rise to transitions between the various electronic states and are usually referred to as non-adiabatic couplings. They depend critically upon the velocity of the nuclei as can be seen from equation (8).

Adiabatic Representation

This is the most frequently used choice of representation in gas–surface problems and uses as basis states the eigenfunctions of $\mathcal{H}_0$

$$[\mathcal{H}_0 - V_k]\phi = 0 \qquad (9)$$

From equation (7) it is then seen that the eigenvalues V_k ($= U_{kk'}\delta_{kk'}$) themselves are the required PES. Since the off diagonal elements $U_{kk'}$ are identically zero, the only terms to couple motion between different electronic states are $\mathcal{T}'_{kk'}$. The states $\phi_k(r; R)$ satisfy the variational principle and may be calculated using, for example, density functional methods. There exist in the literature many examples of adiabatic ground state total energy calculations for surface problems. Perhaps the most popular methods involve either small clusters[8,9] or slabs [10,11] to represent the surface although more recently embedding methods[12] have been used to treat a single ad-species interacting with a semi-infinite substrate.

Diabatic Representation

Electronic non-adiabaticity occurs mainly at regions in 'R-space' where potential energy surfaces corresponding to different electronic states, k, approach one another. From the point of view of bonding, these regions accord to a change in electronic configuration of the interacting system. To illustrate this, consider Figure 1 which schematically illustrates the dissociation of a molecule on a surface. The bonding may be described by two simple diabatic states,[13] molecular and atomic in nature.[14] Far from the

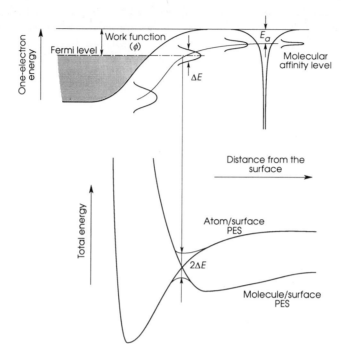

Figure 1 *The one-electron picture and the corresponding total-energy curves for a diatomic molecule approaching a metal surface. The molecule is taken to be of fixed orientation and the abscissa measures the distance from a notional surface plane. The molecular diabatic curve is for the electronic state where the affinity level is always empty. For large gas–surface separations this is a weakly attractive van der Waals interaction while closer in, a strongly repulsive force is felt. The atomic curve represents the interaction of the dissociated molecular fragments and is the well known chemisorption potential.[97] The adiabatic potential curves have been indicated and constitute a 'rounding off' of the diabatic curves near the crossing point the magnitude of which is related to the width of the antibonding resonance as indicated*

surface the molecular affinity level is above the Fermi level and only a weak dispersion force is experienced. Closer to the surface, the affinity level is lowered in energy and broadened (arising from image interactions) and it becomes 'partially' occupied. In the region when the affinity and Fermi levels are degenerate, the two diabatic states cross and it is here that the non-adiabatic effects are greatest.[15] The adiabatic PESs, which are suitable linear combinations of these diabatic states, in general will not cross[16] but will exhibit an 'avoided crossing' as the molecular state gradually evolves into an atomic one. The strength of the splitting is related directly to the width of the molecular affinity resonance which for the case of H_2 outside a simple metal surface is ~ 1 eV.[17] For molecular adsorbates there will be a continuous series of avoided crossings as, for example, the molecular bond is extended or the molecule rotates.[18] The locus of these points gives rise to 'crossing seams' in the PESs where activation barriers are most often found. It should be kept in mind that the low energy excitations in the substrate may also be mapped as a densely packed network of PESs above the adiabatic ground state which will totally fill up all of the possible configuration space. The question as to whether or not this is a particularly useful description, however, is a moot one because of the paucity of information regarding the coupling strengths $\mathcal{T}_{kk'}$.

Problematically, there exists no unique definition of the diabatic representation. One possibility of obtaining diabatic potentials is to simply extrapolate potentials in the neighbourhood of avoided crossings![15] Sometimes adsorbates are known to be ionic and density functional methods may be employed. Another possibility is to construct the basis functions in such a way as to minimize the $\mathcal{T}'_{kk'}$ terms which has the consequence that the off diagonal terms $U_{kk'}$ promote transitions between the various states.[19]

The diabatic representation is of most use when fast moving adsorbates have curve crossing points far from the surface,[14] since in this case the avoided crossing region remains well localized and the relative velocity is large [see equation (8)]. For the two state system in Figure 1, an initially prepared molecule will have a probability of $\exp(-\gamma_M)$ of remaining so,[20] where:

$$\gamma_M \propto \frac{E_a^2}{v_n |\nabla(V_1 - V_2)|} \qquad (10)$$

an approximation originally due to Massey. v_n is the nuclear velocity of the molecule normal to a crossing seam, E_a is the width of the antibonding state and V_1 and V_2 are the molecular and atomic adiabatic PESs defined by equation (9).

Quantum vs. Classical Mechanics

There is a quite general question regarding the validity of classical mechanics that is frequently pigeon-holed with the question of how important is tunnelling in chemical reactions. This is particularly misleading since even the Eyring paper,[2] referred to in the introduction, identified no less than three areas where quantum characteristics would manifest themselves in reactive encounters: (i) zero-point energy effects and quantized states in general, (ii) below barrier tunnelling and above barrier reflection, and (iii) coherence effects.

Assume that a molecule prepared in some particular ro-vibrational state is approaching a surface with an activation barrier; it experiences the potential and reacts accordingly. In the classical sense, the barrier functions like an on–off switch.[21] If the initial kinetic energy is greater than the barrier height then the trajectory will in general cross into the product state having no memory of its encounter.[22] Depending upon the position of the barrier along the reaction co-ordinate[23] there may be some slight R–T or V–T coupling in the region of approach to the surface which will result in a lower 'effective' barrier. Quantum mechanically this is quite different, with substantial reflection occurring for even quite modest barrier heights.[21] In addition, because of the lack of quantization in either the product or the reactant portions of the PES it is quite possible for a large number of initial trajectories to scatter either forward or backwards with vibrational energies below the zero-point threshold value. This is clearly nonsensical and, for the case of H_2, can give rise to large differences in results obtained for either quantum or classical simulations on the *same* PES. Proponents of classical molecular dynamical methods claim that if quantum calculations could be extended to include many more degrees of freedom, then results from the two methods would be similar due to the increased degree of averaging occurring. Thus far there appear to be no concrete grounds to support this point of view.

Coupling quantum and classical degrees of freedom is a problem with a rich history. Thus far there does not appear to be any consensus on how to treat this problem although the literature contains several attempts to do this. In particular, coupling quantum wavepackets with a classical solid has been investigated and produced interesting results.[24,25] In parallel to this, the driven oscillator model, which combines a classical projectile with a bath of quantum oscillators, has also produced results of considerable importance (see Chapter 1 of this volume). Philosophically it seems most sensible to combine an 'exact' treatment of the projectile with a semi-classical treatment of the surface[26] since, for most problems, detailed information on the substrate degrees of freedom is totally unnecessary. At this moment in time, however, there exist no theoretical methods to accomplish this.

3 Solving the Quantum Dynamics

Propagation Methods

Given a PES, the conventional approach to a full solution of the dynamics is to solve the *time-independent* Schrödinger equation by expansion of the wavefunction in a complete set of orthogonal basis functions. The accuracy of the method depends on the quality of the expansion scheme used, while numerical efficiency is usually limited by diagonalization of the Hamiltonian matrix, which scales with the number of basis states cubed. For surface collision problems the number of final states may become prohibitively large; for example, in problems involving dissociation, there may be difficulties in matching the wavefunction between the entrance and exit channel.

An alternative approach is to solve the *time-dependent* Schrödinger equation, by representing the incoming wavefunction as a wavepacket, thereby restricting the solution to a localized region of space. There are several review articles devoted to time-dependent quantum mechanical methods, to which the interested reader is referred.[27–30] The problem is then of the initial-value type—rather than the boundary-value problem time-independent solution—starting from an initial state, with repeated applications of a short-time propagator to yield the final state a time T later. An advantage of this approach is that an intuitive insight into the dynamics is gained, since the wavepacket follows a path similar to that of a classical trajectory. The method also gives the on-shell elements of the S-matrix and is thus highly efficient.

All time-dependent methods involve a representation of the initial non-stationary wavepacket on a pre-defined spatial grid; it being assumed that interpolation of the wavefunction between the grid points will be valid. This initial wavepacket will, in general, be confined within the space defined by the grid. The problem is then to solve the time dependent Schrödinger equation:[31]

$$i \frac{\partial \psi(\mathbf{R}; t)}{\partial t} = \mathcal{H} \psi(\mathbf{R}; t) \tag{11}$$

where $\mathcal{H}$ is the Hamiltonian for the problem:

$$\mathcal{H} = -\frac{\nabla^2}{2M} + V(\mathbf{R}) \tag{12}$$

This involves evaluating both a first-order differential with respect to time- and a second-order differential with respect to $\mathbf{R}$ in the Laplacian, ∇^2. The method used to evaluate the time development involves expanding the time evolution operator, $\mathcal{U}$, given by the formal solution to equation (11):

$$\psi(\mathbf{R}; t + \Delta t) = \mathcal{U} \psi(\mathbf{R}; t) = \exp[-i\Delta t \mathcal{H}] \psi(\mathbf{R}; t) \tag{13}$$

An efficient method to evaluate the second order spatial derivative in the Laplacian employs the Fast Fourier Transform (FFT) algorithm.[32] A spatial derivative is mapped into reciprocal space as a multiplication by ik, where k is the wavevector. Similarly, the second derivative is mapped as a multiplication by $-k^2$. Therefore, by transforming the wavefunction from its real-space to its momentum-space representation with a FFT, the kinetic energy operator simply becomes multiplicative. Since the FFT algorithm is extremely efficient, great savings in computer time may be made with its implementation.

A further improvement in the time propagation was made in a method which also utilizes the FFT algorithm. From equation (13) it may be seen that if the evolution operator is split up in some way into factors containing only the kinetic or potential energy operators, it is possible to apply each operator in the representation in which it is diagonal. For example:

$$\mathcal{U} \psi(\mathbf{R}; t) \approx \exp[-i\Delta t \mathcal{T}] \exp[-i\Delta t \mathcal{V}] \psi(\mathbf{R}; t) \tag{14}$$

Errors will arise from the non-commutability of the kinetic and potential energy operators which are of second order in the time step:

$$\Delta \mathcal{U}_1 = \tfrac{1}{2}[\mathcal{T}, \mathcal{V}]\Delta t^2 \qquad (15)$$

since an expansion of the resultant evolution operator gives terms only in $\mathcal{TV}$. Alternatively, other forms for the splitting may be formulated that reduce the error further. If equation (13) is written as:[33]

$$\mathcal{U}\psi(\mathbf{R}; t) \approx \exp\left[-i\frac{\Delta t}{2}\mathcal{T}\right]\exp[-i\Delta t\,\mathcal{V}]\exp\left[-i\frac{\Delta t}{2}\mathcal{T}\right]\psi(\mathbf{R}; t) \qquad (16)$$

the second-order errors are eliminated, and the error over each step is of third order:

$$\Delta \mathcal{U}_2 = \frac{i}{24}\{(\mathcal{T}, [\mathcal{T}, \mathcal{V}]) + 2(\mathcal{V}, [\mathcal{T}, \mathcal{V}])\}\Delta t^3 \qquad (17)$$

Operationally, there is almost no difference between using equation (14) and using equation (16), since the evolution operator will generally be applied repeatedly N times until a time T. The half time steps of the factors containing the kinetic energy operator will then run back to back, and may be combined to a full time step. The only difference is then at the start and at the end of the calculation, where a half time step operation is applied. Equation (16) is commonly called the 'symmetrically split operator method', and has been widely applied to problems ranging from the blooming of lasers by the atmosphere to wave-guide calculations, as well as for quantum dynamics.

It is no trivial exercise to use the split operator methods for Hamiltonians containing terms that mix space and momentum co-ordinates, an example being that for a rigid rotor in spherical polar co-ordinates:

$$\mathcal{H} = -\left(\frac{1}{2M}\right)\frac{\partial^2}{\partial z^2} - \left(\frac{1}{2I}\right)\left\{\frac{\partial^2}{\partial \theta^2} + \cot\theta\frac{\partial}{\partial \theta} + \left(\frac{1}{\sin^2\theta}\right)\frac{\partial^2}{\partial \phi^2}\right\} \qquad (18)$$

where I is its moment of inertia and M its mass. It is, however, still possible to use a grid based FFT method to evaluate the Hamiltonian in such a case using an alternative propagation scheme[34] that involves a polynomial expansion of the evolution operator. The most commonly used method is to use a Chebyshev polynomial expansion of the exponential in equation (13):

$$\exp[-i\mathcal{H}t] \approx \sum_{n=0}^{M} a_n T_n(-i\mathcal{H}t) \qquad (19)$$

The operations of both the translational and rotational energy operators of the Hamiltonians on the wavefunction are again evaluated using the FFT algorithm.

For either the split operator or the Chebyshev method, the computation time required to evaluate the effect of the Hamiltonian on the wavepacket scales as $K \log K$, where K is the number of grid points used. The grid will generally be chosen so that the initial wavepacket may be placed in an asymptotic region of space, and the sides of the grid will either be bounded by high potential walls or asymptotic arrangement channels. Once the wavefunction has left the region of strong interaction it may be no longer necessary to follow part of it; in this case the size of the grid may be greatly reduced by using an absorbing boundary.[35] With a suitable choice of an imaginary potential covering a strip near the edge of the grid, the wavefunction may be removed from this region smoothly, without reflection back onto the grid. In this way the wavefunction never builds up appreciable weight at the edge of the grid, and the problems associated with the periodicity of the FFT are therefore avoided. There is a slight problem arising from the use of an absorbing boundary in the Chebyshev scheme since the eigenvalues of the Hamiltonian are no longer real and the domain of the evolution operator in equation (9) is out of the range of definition of the Chebyshev polynomials. This can lead to dramatic instabilities. However, since the complex part of the eigenvalues of the Hamiltonian is very small compared to the real part, the error introduced by the optical potential very much depends on the propagation time step. Instead of using a single time step, as used in most calculations employing this method, it is possible to use multiple time steps which for most purposes appears to be satisfactory.

Choice of the Initial Parameters and Final State Projection

For a typical scattering event the initial wavepacket will be a product of the eigenfunctions for internal and translational degrees of freedom, since in the initial (prepared) state the various degrees of freedom are essentially uncoupled. The choice of the grid parameters will be strongly influenced by the requirements of the initial wavepacket. Because a Fourier transform forces a periodicity of length L_Z—the length of the grid in the Z co-ordinate—onto any non-periodic function, the initial eigenfunction for the Z co-ordinate must be either confined completely within L_Z, or periodic in Z. Vibrational eigenfunctions are of the first type, confined to a limited region of space, while eigenfunctions for free space translation parallel to the surface are of the second type. Translation perpendicular to the surface, however, is not defined by a periodic function because of the presence of the surface. An appropriate choice of wavefunction in this case is then a distribution of plane waves having the form:

$$g(Z - Z_i, p_Z) = \frac{1}{\sqrt[4]{(2\pi\delta^2)}} \exp[-(Z - Z_i)^2/4\delta^2] \exp[-i p_Z Z] \quad (20)$$

which is centred on Z_i, with root mean square deviation δ. The average translational energy in the Z-direction is given by:

$$\langle E_{\text{trans}} \rangle = \frac{1}{2\mu}\left(p_Z^2 + \frac{\delta^2}{4}\right) \tag{21}$$

The choice of suitable parameters for the grid is governed by the dual requirements of the real- and momentum-space properties. The periodic nature of the FFT algorithm makes it necessary that the wavepacket remain on the grid throughout the calculation, in both real- and momentum-space. A suitable choice of the Z space grid resolution must be made in order that the maximum momentum component, which will be present in the wavefunction during the collision, may be represented on the k_Z-space grid. For example, if the length of the Z-space grid is L_Z, and the number of points is N_Z, the Z-space grid spacing dZ is L_Z/N_Z. The wave vector of a specific component is $2\pi/\lambda$, where λ is the wavelength, and it is clear that the maximum wavelength which may be represented on the grid is L_Z, with smaller possible values given by L_Z/n, where n is an integer. This gives for the k_Z grid spacing the value of $2\pi/L_Z$. The range of positive and negative k_Z values that can be represented on the k_Z-space grid of N_Z points is then:

$$-\frac{2\pi}{L_Z}\left(\frac{N_Z}{2}\right) \le k_Z \le \frac{2\pi}{L_Z}\left(\frac{N_Z}{2} - 1\right) \tag{22}$$

which gives a condition for dZ:

$$|k_Z| \le \frac{\pi}{dZ} \tag{23}$$

Additionally, dZ must be small enough that the potential be smoothly represented and interpolation of the wavefunction between the grid points is valid. The momentum grid spacing, $dk_Z = 2\pi/L_Z$, must also be fine enough to give the required degree of resolution in the final state. These considerations will be governed, to some extent, by the size of L_Z, which must be large enough to span both the potential surface and the initial wavepacket. This relates to the choice of the initial width of the wavepacket in equation (20): a very wide packet will have a well defined momentum and energy resolved results will be obtained, while for a narrow initial packet it may be possible to obtain several rows of the S-matrix from a single calculation.[36]

For internal degrees of freedom essentially the same conditions apply to the choice of a grid. In this case however the initial wavefunction would be a product state corresponding to a particular ro-vibrational state. The rotational eigenfunctions of a diatomic are simply the spherical harmonics while the vibrational degree of freedom is in most cases chosen to be a Morse oscillator. This is a particularly convenient choice since the eigenfunctions can be expressed in an analytic form.[37] The projection of the final wavefunction onto these stationary states is then a relatively straightforward exercise in numerical integration.

4 Inelastic Scattering

To begin investigating the form of the PES for a gas–solid interaction it is informative to address that class of systems where there is no bond-breaking but instead, an energetic rearrangement within the various internal degrees of freedom. This is generally referred to as inelastic scattering and within the past ten years there has been significant progress both in understanding PES topologies and the quantum dynamics.

The Trapping of $H_2/Cu(100)$

This is a particularly interesting problem that has received substantial experimental and theoretical interest over the past five years.[38] In general the form of the laterally averaged physisorption PES that the H_2 experiences on approaching a *static* surface is a combination of attractive and repulsive parts:[39]

$$V_{\text{phys}}(Z, \theta) = V_R(Z, \theta) + V_{vW}(Z, \theta) \quad (24)$$

where

$$V_R(Z, \theta) = V_R \exp(-a_R Z)\{1 + \beta_R P_2(\cos \theta)\} \quad (25)$$

and

$$V_{vW}(z, \theta) = \frac{C_{vW}}{|z - z_{vW}|^3} \{1 + \beta_{vW} P_2(\cos \theta)\} f(k_c[z - z_{vW}]) \quad (26)$$

V_R and a_R are the strength and inverse range of the repulsive potential and β_R its anisotropy in θ, which is the angle between the molecular bond and the surface normal. $P_2(x)$ is the Legendre polynomial. C_{vW} is the strength of the asymptotic van der Waals attraction and Z_{vW} the position of the dynamical image plane. β_{vW} is the anisotropy in the polarizability tensor of the free H_2 molecule ($=0.05$) and $f(x)$ accounts for the saturation of the van der Waals potential due to the finite size of the molecule.[39] For the H_2/Cu system, theoretical results for the parameters appearing in the repulsive part have been obtained from calculations employing a Hartree–Fock approach to calculate the ground state energy appearing in equation (7): the value of β_R found was 0.2.[40]

When a low energy H_2 molecule scatters from a surface, there are several possibilities for a final state: (i) specular elastic scattering, (ii) diffraction, (iii) selective adsorption into a bound state of equation (24) via the surface corrugation (CMSA) or the rotational corrugation (RMSA), (iv) scattering with phonon creation into either free or bound states with net positive final energy, and (v) scattering with phonon creation into bound states with negative final energy (trapping). Wilzén *et al.* have measured the position of the bound state energies ϵ_n, and the CMSA and RMSA for low energy beams of H_2 scattering from a Cu(100) surface at ~ 20 K.[5] At a surface the $(2J + 1)$ degeneracy of the Jth rotational state in the nth bound state is lifted by an amount:[41]

$$\Delta\epsilon_n^{(J,m_J)} = \frac{3\langle n|\beta_R|n\rangle}{(2J+3)}\left\{\frac{J^2 - M^2}{2J - 1} - \frac{J}{3}\right\} \qquad (27)$$

and by measuring the probability for rotational transitions (either by CMSA or RMSA) it is possible to obtain an estimate of the matrix element $\langle n|\beta_R|n\rangle$. Wilzén et al. measured the splittings of the CMSA resonances and found that $\langle n|\beta_R|n\rangle$ was negative for $n = 1, 2$ which is at variance with the theoretical estimates mentioned above and implies that, at equilibrium, the molecule prefers to be oriented with its axis perpendicular to the surface.[5]

In order to reconcile theory and experiment, it is necessary to probe a little deeper into the calculated value of β_R. The repulsive part of the potential arises (to first order) from the overlap between the molecular orbitals and those of the conduction electrons. In the original calculation,[40] only the bonding σ_g state of the molecule was included in the calculation while the antibonding σ_u state was neglected. Wilzén et al. have shown that this omission is a serious deficiency and, on the basis of a simple three-level model to describe the interaction, have shown that the contributions to β_R from the σ_g and σ_u states approximately cancel one another for surface electronic states of high symmetry.[5] This implies that the majority of the anisotropy in V_{phys} arises from the van der Waals interaction which, having a negative sign, rationalizes the experimentally determined value.

This is a particularly interesting study since it represents one of the very few occasions in surface dynamics where it really has been possible to test the predictions of a theoretical PES to its limit. As a consequence, it has been necessary to go 'back to the drawing board' and critically examine the approximations that went into the theoretical description. The physisorption model was developed for the He/surface interaction and the experiments have shown in an unequivocal manner that, even at the lowest energies, effects arising from the unfilled antibonding state of the molecule cannot be totally neglected. Ultimately this will have significant repercussions on the PES for the dissociation of H_2 discussed later in this chapter.

T–R: the Scattering of NO/Ag

Internal state resolved experimental studies of the scattering of NO from metal surfaces were among the first to be performed, taking advantage of the possibility to directly measure the rotational state distribution of the scattered molecules using laser-induced fluorescence.[42] Kleyn et al.[43] scattered rotationally cold NO from Ag(111) over a range of kinetic energies from 0.1–1.7 eV and at various angles of incidence. The observed rotational state population distributions showed differing behaviour at low and high rotational energies. For low J, the distribution was linear in a Boltzmann plot. However, the 'rotational temperature' obtained was linearly dependent on the translational energy normal to the surface and independent of surface temperature, suggesting a direct-inelastic scattering mechanism. For high J, the scattering showed marked deviation from Boltzmann-like behaviour particularly for high incident energies where the appearance of a broad peak was interpreted as a rotational rainbow, arising from a singularity in the classical cross section for rotational excitation.

The effect of energy exchange with the surface on rotational excitation was investigated[7] using time-of-flight detection to measure the final velocity of individual rotational states of NO scattered from Ag(111). An 'anticorrelation' was found between rotational and phonon excitation; high rotational excitation being accompanied by less energy transfer to the surface. In general the collisions were found to be highly inelastic particularly for high initial translational energies, with transfer of up to one half of the available kinetic energy to the surface! The anticorrelation was described as being due to an 'effective mass' effect[7,44] whereby collisions producing high rotational energy (the molecule being inclined with respect to the surface) transfer less energy to the surface because of a reduction in the effective mass normal to the surface. A similar anticorrelation effect was found for the direct–inelastic component of NO scattered from germanium.[45]

A theoretical study to rationalize the above results was performed by Barker et al.[46] Modelling the collision as a two-dimensional rigid rotor striking a rigid surface, both classical trajectories and a coupled-channels quantum formalism were used to obtain results for rotational scattering probabilities. The peak at high J was reproduced by using a repulsive potential which was anisotropic in θ, explicitly containing a term in the second-order Legendre polynomial, $P_2(\cos\theta)$, in the repulsive part of the potential. Thus the potential was symmetric about $\theta = \pi$, containing only terms in $\cos^2\theta$. While the quantum solution improved agreement with experiment by smoothing the classical singularity at the rainbow, the low J distribution was not well described by this potential. As an alternative, Voges and Schinke[47] devised an empirical potential containing both attractive and repulsive terms in $P_1(\cos\theta)$ as well as $P_2(\cos\theta)$. This extra asymmetry of the PES ($\sim \cos\theta$) models the difference in chemical interaction of the oxygen and nitrogen ends of the molecule with the metal surface. Again under the assumption of a rigid surface, in a quantum study they found two rainbow maxima, one at low and one at high J. This PES significantly improved the agreement with the low rotational energy part of the experimental distributions.

In a study performed using the rotationally sudden approximation, Schinke and Gerber[48] determined a Debye–Waller factor which was an increasing function of the rotational energy transfer. In a model of NO scattering from Ag(111) they found that variation of the surface temperature had only a minor effect on the rotational distribution, and that the mean rotational energy transfer showed a slight linear increase with temperature,

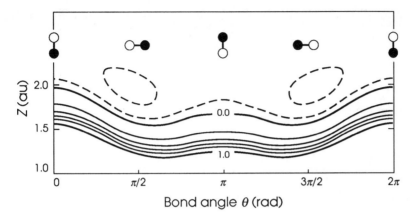

Figure 2 *A contour plot of the Voges–Schinke potential energy surface[47] [equation (28)] for the two-dimensional model of NO interacting with Ag(111) described in the text. The Z direction measures the distance of the molecular centre-of-mass above the surface and θ the angle between the NO bond and the surface normal. The energies shown are in eV*

in agreement with experiment. A coupled channels study of rotational excitation within a three-dimensional model was performed by Brenig et al.[49] In this calculation a two-dimensional PES with terms in $P_1(\cos\theta)$ and $P_2(\cos\theta)$ was coupled to a surface 'atom' represented by a harmonic oscillator. It was found that the main effect of the coupling was to shift the rainbow maxima towards lower rotational energy. A semi-classical treatment of the scattering of NO from GeO[50] reproduced the experimentally observed anticorrelation between phonon and rotational excitation.

This system provides a good testing ground for the time-dependent quantum methods discussed in Section 3.[51] The PES chosen is that empirically derived by Voges and Schinke[47] to give a 'best fit' to the scattering experiments of Kleyn et al.[43] The potential has the form:

$$V(Z, \theta) = \frac{A}{(Z - \alpha \cos\theta - \beta \cos^2\theta)^8} - \frac{B}{(Z - \delta \cos\theta)^3} \quad (28)$$

and is shown in Figure 2. The optimum value for the constants appearing in equation (28) were found to be:[47] $A = 10$ eV a.u.^{8}, $B = 1$ eV a.u.^{3}, $\alpha = \beta = 0.16$ a.u., and $\delta = 0.20$ a.u. The first term describes a short range repulsive interaction due to the overlap of the molecular orbitals with the metal conduction band states, while the second term describes the long range van der Waals attraction. The potential contains both a P_1 term, describing deviations from sphericity of the molecule, and a P_2 term which differentiates the two ends of the molecule:

$$V(Z, \theta) = \frac{A}{Z^8} - \frac{B}{Z^3} + CP_1(\cos\theta) + DP_2(\cos\theta) \quad (29a)$$

where

$$C = \left[\frac{8Aa}{Z^9} - \frac{3B\delta}{Z^4}\right] \quad (29b)$$

and

$$D = \frac{4}{3Z}\left[\frac{4A\beta}{Z^8} + \frac{1}{Z}\left(\frac{16Aa^2}{Z^8} - \frac{3B\delta^2}{Z^3}\right)\right] \quad (29c)$$

This latter asymmetry is the result of two factors: (i) arising from a difference in the chemical interaction of the oxygen and nitrogen ends of the molecule with the metal surface, and (ii) from the centre-of-mass transformation for a heteronuclear molecule. As mentioned in the introduction, the inclusion of the asymmetric terms was necessary to explain the low J part of the observed rotational distributions.

As in the calculation by Brenig et al.,[52] the inclusion of surface motion may be modelled by coupling the two-dimensional PES (Figure 2) to a surface atom represented as an Einstein oscillator with mass, M, equal to that of a Ag atom, 108 a.m.u., and frequency, $\omega \sim 20$ meV (an average phonon energy). The total potential is then:

$$\tilde{V}(Z, \theta, y) = V(Z - y, \theta) + \tfrac{1}{2}M\omega^2 y^2 \quad (30)$$

where y is the displacement of the surface atom.[53] This simple form is justified by the fact that, for the large energy transfers expected,[53] the shape of the final energy distribution becomes insensitive to details of the phonon spectrum.[52]

The initial wavepacket is the product state:

$$\psi(Z, \theta, y; t = 0) = g(Z - Z_i)\, e^{ik_Z^0 Z}\, \phi_n(y)\, e^{ij\theta} \quad (31)$$

where $g(Z - Z_i)$ is a Gaussian envelope centred at Z_i, k_Z^0 is the initial mean wavevector in the Z direction, and $\phi_n(y)$ is a harmonic oscillator eigenfunction with quantum number n. Because we employ a two-dimensional representation of the rigid rotor, the rotational wavefunctions are that of a particle on a ring.

In Figure 3a are shown the rotational scattering probabilities for transitions $P(j)$ obtained using a rigid-surface model where M has been set to infinity. The rotor is initially in the $j = 0$ state and the initial translational energy, E_{trans}^i, is 0.5 eV. The plot shows two distinct rainbow maxima at low and high j, arising from the points of inflection in the PES shown in Figure 2. Plotted in Figure 4a is the classical deflection function for the rigid surface which links the final angular momentum j to the initial rotor angle θ. At the two stationary points, the derivative $dj/d\theta$ vanishes, resulting in singularities in the classical scattering probability:

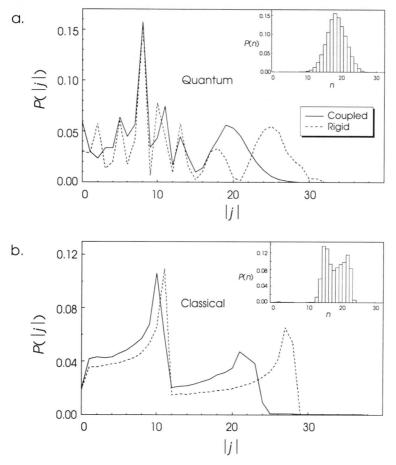

Figure 3 *Plots of the absolute rotational distributions for the scattering of oriented NO, as calculated for the rigid surface (dashed line) and recoiling surface (solid line).[51] Plotted in (a) and (b) are results for the quantum and classical calculations respectively. The insets show the post-collision surface oscillator energy distributions*

$$P(j) = P(\theta) \left| \frac{dj}{d\theta} \right|^{-1} \qquad (32)$$

seen quite clearly in the classical $P(j)$ shown in Figure 3b. Quantum mechanics smooths this classical catastrophe,[54] and a semi-classical analysis has shown[55,56] that the high energy rainbow becomes an Airy function which has its maximum at a slightly lower value of j and extends beyond the classical threshold. The shift of the maximum to a lower energy is due to quantum interference, which also results in the supernumary rainbows seen in the classically flat parts of the distribution. The tail to high j arises as a consequence of the penetration of the wavefunction into the repulsive part of

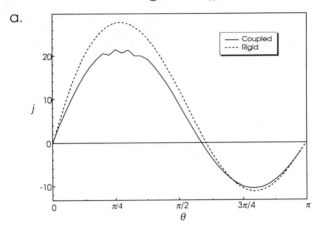

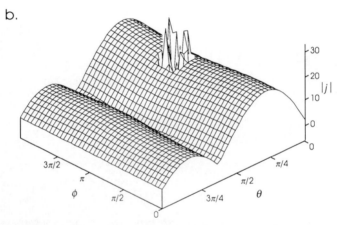

Figure 4 (a) *The dashed line shows the rigid-surface rotational excitation function, linking the final classical angular momentum (in units of the rotational quantum number) to the polar angle of the molecular axis before collision.*[51] *This shows clearly the way in which the Jacobian [equation (32)] vanishes at the points of inflection seen in Figure 2. Plotted in the full line is the excitation function calculated for the recoiling surface. This shows a reduction in the magnitude of the maxima and some new structure arising from the inclusion of surface motion.* (b) *The recoiling-surface excitation function shown in Figure 4(a), plotted as a function of the initial rotor and oscillator phase. The onset of chaotic behaviour over a small range of phases near the rainbow maxima is clearly seen*

the potential, such that it experiences a greater corrugation in θ, and thereby achieves a greater rotational energy.[49]

Also plotted in Figure 3 are the integrated probabilities for the rotational excitation when the surface atom mass is $M = 108$ a.m.u. In this calculation this atom is initially taken to be in its vibrational ground state. It is apparent that the main effect observed as a consequence of energy transfer to the

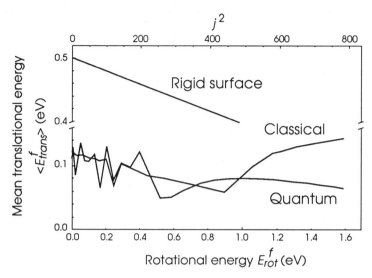

Figure 5 *A graph of the mean final translational energy of the scattered beam as a function of rotational energy.[51] The solid straight line is the result obtained with no energy transfer to the surface. Results obtained including surface recoil are shown lower in the figure (note the axis break). It is interesting to note that the structure in these curves is not due exclusively to quantum behaviour (see text for details)*

surface is a shift of the rainbow maxima to lower values of j.[49,50] In the classical calculation the lower energy rainbow is shifted less than the high energy rainbow, while for the quantum calculations it is not shifted at all. This compression of the rainbows also has the effect that the number of supernumary rainbows seen in the quantum distributions is reduced.

Plotted in the insets of Figure 3a and b are the post-collision oscillator vibrational populations for the quantum and classical calculations, $P(n) = \sum_j P_n(j)$. These are centred about the kinematic energy transfer for impulsive collisions, $\Delta E \approx 0.36$ eV (the 18th n state).[57] The double maxima observed in the classical calculations arises from singularities in the cross section for inelastic scattering from a classical oscillator as has been discussed elsewhere.[55,58] Any effects caused by an 'effective' mass dependence on the initial angle θ are generally washed out in the final summation over j.

In Figure 5 the mean final translational energy is plotted as a function of the final rotational energy for the parameters corresponding to Figure 3. The results for a rigid surface correspond to the condition $E^f_{trans} = E^i_{trans} - E^f_{rot}$. As has previously been observed both experimentally and theoretically,[7,50] results for the recoiling surface have a reduced slope, due to an anticorrelation between rotational and phonon excitation, such that greater rotational excitation is accompanied by less energy transfer to the surface. For an impulsive collision between a non-rotating rigid rotor and a stationary hard cube Hand *et al.*[51] have shown that:

$$\left.\frac{dE_{trans}^f}{dE_{rot}}\right|_{\lim E_{rot} \to 0} = \frac{\mu - 1}{\mu + 1} \qquad (33)$$

where μ is the rotor-surface atom mass ratio. In this case, equation (33) evaluates to -0.57, comparing favourably with the slope of ~ -0.58, obtained from the classical simulation. The state resolved probabilities calculated using the hard cube model are in good agreement with the classical trajectory calculation, and an examination of the two-parameter deflection function, $j(\theta, \phi)$ shown in Figure 4b, gives additional insight into the results obtained. The maximum in the oscillator excitation probability occurs for $\phi = \pi$, with the surface atom moving away from the incoming molecule, while the minimum occurs for $\phi = 0$, with the surface atom moving towards the incoming molecule. It is thus apparent that maximum

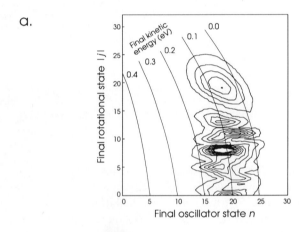

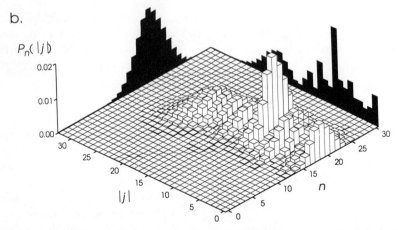

Figure 6 (a) *The quantum differential scattering probability, $P_n(j)$ as calculated for NO scattering from Ag(111) with translational energy 0.5 eV, in the form of a contour plot. The curved lines link areas of equal translational energy.* (b) *$P_n(j)$ in a three-dimensional representation*[51]

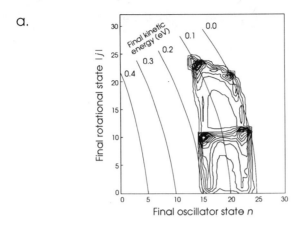

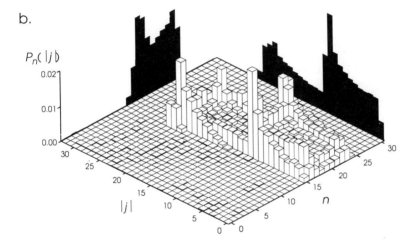

Figure 7 *The classical equivalent of Figure 6*[51]

energy transfer to the oscillator is accompanied by less rotational excitation, because the relative translational energy is accordingly reduced.

The oscillations in the low j part of the quantum results are attributable to the interference structure seen in the rotational distributions.[50] Although this oscillatory structure is, of course, absent in the classical result, the classical slope agrees well with quantum calculations. There are two pronounced dips in the classical results however, the origins of which become apparent upon closer examination of the state resolved scattering probabilities. Figures 6a and 7a show the quantum and classical probabilities, $P_n(j)$, in the form of contour plots. The curved lines indicate the dependence of E^f_{trans} on the final j and n which from simple energy conservation is:

$$j = \sqrt{[2I(E^i_{trans} - E^f_{trans} - n\omega)]} \qquad (34)$$

It can be clearly seen in Figure 7a that the dip in the classical final kinetic energy occurs because the low and high n maxima in the oscillator populations occur for differing values of j corresponding to the low energy rotational rainbow. The peak at high n lies close to the line where $E^f_{trans} = 0$, thus weighting the final kinetic energy to a lower value, and causing a dip in the mean final translational energy for the $j = 11$ rotational state.

In concluding this subsection it is important to reflect upon the usefulness of such calculations in increasing our general understanding of the gas–surface interaction. The available data base is most certainly not large enough to merit a unique inversion to obtain a definitive PES and so it is left to models of the form described above to offer insight into various scattering mechanisms. While this is perhaps not the ideal way to progress, it most certainly has had some degree of success in interpreting the T–R scattering phenomena. On a cautionary note, there have been quantum calculations performed on NO/Ag potential surfaces which attempt to include effects arising from the open-shell character of the NO.[6] The main concern is whether the existing experiments are sensitive to such details and to provoke this discussion Figure 8 compares results for NO and N_2 taken for the same initial experimental conditions.[59] In both cases a high J rainbow is observed, and at low J the data has a steep negative gradient. Are the origins of the similarities merely coincidental and do they require quite separate interpretations based upon potential energy surfaces of totally different character? The simple answer is that at this moment we do not know and what is required are more experimental results and, in addition, more theoretical iterations around the modelling loop.

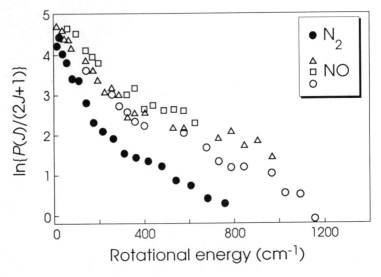

Figure 8 *Comparison of Boltzmann plots for N_2 and NO scattered from Ag(111) under similar conditions*[59]

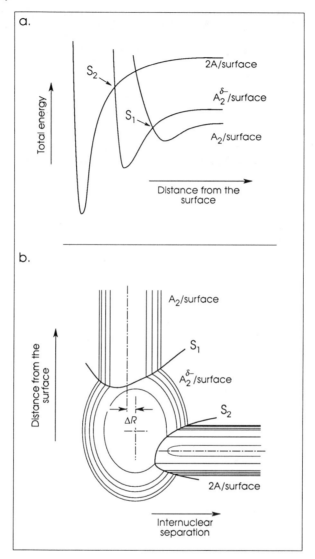

Figure 9 (a) *A more general form of Figure 1 where the presence of a third electronic state, corresponding to a negative molecular ion, is interposed between the molecular and chemisorption diabatic states. Depending upon the relative position of the states, an activation barrier can occur at one of two places, S_1 or S_2. (b) A contour plot for the lowest energy diabatic states where both the molecule–surface distance and the molecular bond length have been included. The geometry corresponds to a molecule approaching with its axis parallel to the surface ('elbow plot'). S_1 and S_2 now appear as crossing seams rather than unique points in space*[20]

T–V: the Scattering of NO/Ag

In a series of papers, Gadzuk and Holloway[60-65] investigated the conversion of translational into vibrational energy for a diatomic molecule scattering from a surface when some degree of charge transfer occurs. For the purposes of illustration, the form of the expected PES for such an event is illustrated in Figure 9 where a modification of the dissociation PES (Figure 1) has been made which incorporates a negatively charged molecular adsorption state located between the neutral molecule and the dissociated atomic states.[20] In the two-dimensional plot in Figure 9b the variation of the PES with internuclear separation is explicitly shown and this highlights the major topological features of the PES. For the case shown, the electron transfer from the surface is accommodated in the antibonding orbital of the A_2 molecule which with respect to the neutral results in (i) an increased equilibrium separation (ΔR), and (ii) a lower intramolecular vibrational frequency.

In this restricted dimensionality model the dynamical consequences of the PES are that A_2 molecules prepared in some vibrational state, on crossing the seam S_1, experience a net force in the vibrational co-ordinate. If surface recoil is ignored, then energy is taken from the translational degree of freedom and repartitioned into vibrational energy. After residing in the molecular state for some time the molecule either exits across S_1 back into the gas phase or across S_2 and is dissociated. Those which emerge from the surface have a high probability for exhibiting vibrational excitation as they retain some memory of the time spent in the resonant state.[61,63]

The experimental verification of this model came with the study of $n = 0$ NO molecules scattering from Ag(111).[66] In this study it was found that the degree of vibrational excitation depends both upon the *normal* translational energy and on the surface temperature T_s. Furthermore, time-of-flight measurements for the scattered molecules showed conclusively that these are not accommodated to the surface temperature. In Figure 10 are shown results for this system taken from the work of Rettner *et al.*[66]

On the basis of the model described above, the strong surface temperature dependence will arise from the fact that as T_s increases, the Fermi tail will broaden and the length of time spent in the molecular resonance will change. In turn this will alter the $n = 1$ scattering probability. Choosing PES parameters consistent with experimentally determined values, Gadzuk and Holloway[65] were able to describe the strong dependence on the normal translational energy. Newns pursued a similar idea from an adiabatic viewpoint[67] and within the framework of a model Hamiltonian study calculated the $n = 1$ excitation probability in the first Born approximation. This work has recently been extended to include also the de-excitation probability on the outward leg of the scattering trajectory[68] and it has been suggested that the original results considerably overestimate the excitation probability particularly at high kinetic energies.

It would seem that on the basis of the very limited amount of data available, there is currently no detailed microscopic model capable of explaining all of the observed findings. There is a general consensus that a

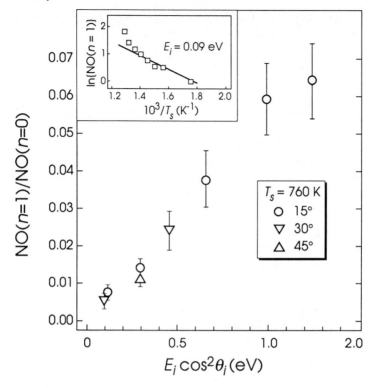

Figure 10 *The effect of incident energy and angle on the NO(n = 1)/NO(n = 0) scattering ratio. In the inset is shown the dependence upon surface temperature for a beam with initial energy 0.09 eV*[66]

partially charged molecular intermediate drives the vibrational degree of freedom but exactly how the electron-hole pair subsystem couples, remains a problem for future study.

5 Reactive Scattering

In this section we shall address the problem of dissociative adsorption of small molecules on surfaces. This is probably the most elementary surface reaction and its understanding will represent an important step in the dynamics of surface processes.

H_2 Scattering

The hydrogen–copper system is rapidly becoming a 'benchmark' dynamical problem in surface dynamics; Chapter 4 in this volume sets the scene from an experimental point of view. For the present purposes we will present details of some recent calculations on a series of model PESs and try to make contact with recent experimental findings when appropriate.

Classical calculations for the scattering of H_2 from metals have been performed for many years but it was with the advent of the calculation by Jackson and Metiu[21] that the 'quantum age' arrived. In this work a PES was constructed of the LEPS form with a restricted geometry corresponding to the two co-ordinates shown in Figure 9b (an 'elbow' PES); the molecule has its axis constrained to be parallel to the surface and is incident normally. For this choice, the ensuing PES resembles the co-linear dynamical problem beloved by the gas-phase scattering community. In their study they used the time-dependent wavepacket method to explore the dependence of the dissociation probability of H_2 on a variety of PESs having barriers representing (i) adsorption, (ii) dissociation, and (iii) surface diffusion. Calculations for the translational energy dependence of the dissociation showed that there was considerable V–T coupling near the surface although in this work only vibrationally ground-state molecules were considered.

Surface Recoil Effects. Jackson and Metiu did not consider the effects of non-adiabaticity from the electronic or vibrational degrees of freedom present in the surface. Hand and Harris[53] investigated the latter problem in a study where an elbow potential $V(Z, x)$ was coupled to a surface oscillator in a similar fashion to that discussed in Section 4. $V(Z, x)$ was a parametrized version of the PES obtained from a total energy calculation of a H_2/Cu_2 cluster.[9] Although a Cu_2 cluster is only a poor approximation to a semi-infinite surface, this model PES has been used extensively over the past five years both for quantitative calculations and qualitative discussion.[22,37,69,70] Its general form is given by combining the two diabatic PESs for the H_2/surface and the 2H/surface interactions via the Landau–Zener prescription:[15]

$$V_{H_2}(Z, x) = V_{Morse}(x) + W + V_0 \exp(-\lambda Z) \qquad (35a)$$

$$V_{2H}(Z, x) = V_1 \exp\{\sqrt{[A_x(x - x_0)^2 + A_z(Z - Z_0)^2]}\}$$
$$+ V_2 \exp\{-2\lambda_m x\} \qquad (35b)$$

$$V_{tot}(Z, x) = \tfrac{1}{2}(V_{H_2} + V_{2H} - \sqrt{[(V_{H_2} - V_{2H})^2 + \chi^2]}) \qquad (35c)$$

Table 1 *Potential parameters for the H_2/Cu system given in equation (35)*[53]

	V_{H_2}		V_{2H}	
Morse	$\begin{cases} W = 4.75 \\ a_M = 1.03 \end{cases}$		V_1	$= 0.15$
			A_x	$= 0.9$
	$V_0 = 18.0$		A_z	$= 0.6$
	$\lambda = 1.0$		x_0	$= 3.2$
			Z_0	$= 1.1$
			V_2	$= 15.0$
			λ_m	$= 1.03$

Coupling parameter: $\varkappa = 1.5$

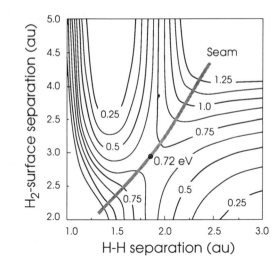

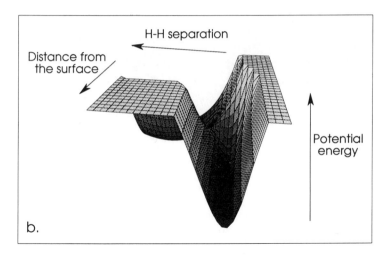

Figure 11 *The H_2/Cu interaction potential used in the calculation of Hand and Harris.[53] (a) A contour plot showing the main topological features: a structureless entrance channel representing the intact molecule, an activation barrier which is located at a molecular separation in excess of the neutral molecule, and an exit channel corresponding to atomic adsorption (c.f. Figure 1). This PES has been adapted from a cluster calculation which is described in the text. Energies are in eV and are measured with respect to a molecule infinitely far from the surface. (b) A perspective view which corresponds to looking down the entrance channel*

$V_{\text{Morse}}(x)$ is the gas-phase H_2 PES and W the well depth, the final term in equation (35a) is the repulsive physisorption wall alluded to in Section 4. Parameters for this PES are given in Table 1 and Figure 11 shows a contour map and perspective view of the surface. The topology is quite simple; two valleys representing entrance (reactant) and exit (product) channels separ-

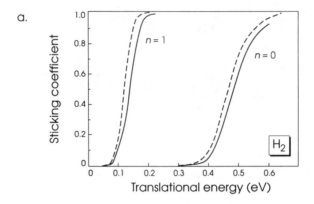

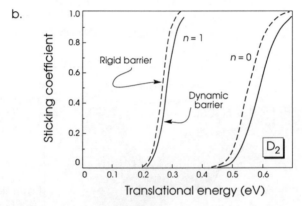

Figure 12 *Sticking coefficient as a function of initial translational energy for (a) H_2 and (b) D_2 scattering from a Cu surface.[53] The curves refer to incidence with the molecule in either its vibrational ground state or the first excited state. The solid lines are for the case when the surface can recoil while the dashed lines are for a rigid surface. The differences between the two arise because of the large mass mis-match between hydrogen and copper*

ated by a ridge, the lowest point of which is a saddle point with energy 0.72 eV with respect to the isolated systems. The surface atom had a mass equal to that of Cu (63 a.m.u.) and a frequency corresponding to an energy of 12 meV.

The initial state was again represented by a product state:

$$\psi(Z, x, y; t = 0) = g(Z - Z_i)\, e^{ik_Z^0 Z} \chi_n(x)\, \phi_0(y) \qquad (36)$$

where $\chi_n(x)$ is the nth eigenstate of the H_2 molecule and again it was assumed that the surface oscillator was in its ground state $\phi_0(y)$. The motion of the wavepacket was followed in time until the dissociated fraction and the back-scattered fraction reached their respective asymptotic values. Absorbing boundaries were employed to avoid having an excessively large grid and the final probabilities obtained by the method first suggested by Bringer and

Harris.[71] Figure 12 shows the results for the dissociation probability for H_2 and D_2 as a function of translational energy and vibrational state. For the case of H_2 it is clear that the inclusion of the surface degree of freedom makes essentially no difference to the probability for dissociation. The inclusion of energy in the vibrational co-ordinate, however, has a dramatic effect on the dissociation probability. This effect will be discussed in some detail in the following subsections. For the case of D_2 the effects of recoil are more noticeable as the mass mis-match is that much less. The direction of the effect is explained by the fact that as the hydrogen approaches the surface, the atom begins to recoil, which acts to raise the effective barrier height. This is in spirit similar to the anticorrelation effect seen in the rotational problem discussed previously. The mean value for the final surface atom oscillator probabilities, $P(n)$ is generally in good agreement with the kinematic energy transfer for impulsive collisions. The results from this calculation add credibility to the following calculations where the effect of surface motion has not been included.

Vibrational Excitation and Barrier Location. As was shown in the above calculation, by placing energy into the vibrational co-ordinate of the hydrogen molecule, the apparent translational energy requirement for the dissociation reaction is diminished by a certain fraction. In this subsection this feature will be pursued at some length and an attempt will be made to link the vibrational efficiency in promoting dissociation, to various topological features of the PES.[23] In spirit this is similar to the development of the gas-phase reaction propensity rules discussed by Polanyi.[72]

Again the principle vehicle for this study is a PES in two co-ordinates, Z and x; an elbow potential. Figure 13a shows a breakdown of such a potential and also serves to define the various regions which are characterized by their differing electronic structure.

(i) *The entrance channel* ($x \sim 0$, Z large) which is the gas phase molecular potential (A_2 in Figure 9) with a weak Z-dependence arising from van der Waals forces (here neglected). The potential in this region is again represented by a Morse potential:

$$V_1(x, Z) = V_{\text{Morse}}(x) \qquad (37)$$

where the origin in x is taken to be the H_2 equilibrium separation. For an H_2 reduced mass of 0.5 a.m.u., this potential yields a zero-point energy $\epsilon_0 = 0.269$ eV and a vibrational excitation energy $\epsilon_1 - \epsilon_0 = 0.514$ eV.

(ii) *The exit channel* (x large, $Z \sim 0$) which describes the electronic state corresponding to two chemisorbed hydrogen atoms (2A in Figure 9). For the assumed structureless surface under study, the PES in this region also takes the form of a Morse potential but this time in Z, with no dependence on x. As with the exclusion of the van der Waals force, this omission is by no means crucial, and any conclusions drawn are not affected by its absence. The potential in this region is then:

$$V_2(x, Z) = V_{\text{Morse}}(Z) \tag{38}$$

with the origin of Z being arbitrarily set at the potential minimum. For simplicity the Morse parameters are again taken to be those given in Table 1, which gives an H atom adsorption energy of -2.38 eV. Since the molecular mass, which governs the motion in the Z-direction, is a factor of four times that of the reduced mass, the vibrational states in the exit channel are one half of their values in the entrance channel,[73] $\epsilon_0 = 0.134$ eV and $\epsilon_1 - \epsilon_0 = 0.257$ eV.

(iii) *The reaction zone* ($x \sim 0$, $Z \sim 0$) links regions (i) and (ii) and corresponds to that region where molecular bonds are being severed and chemisorption bonds formed. *Ab initio* calculations for the interaction of H_2 with metals show that this curved portion of the PES is particularly sensitive to the substrate electronic structure.[9] In one-electron language, this region reflects the competition between an attractive interaction, arising from a charge transfer from the metal into the

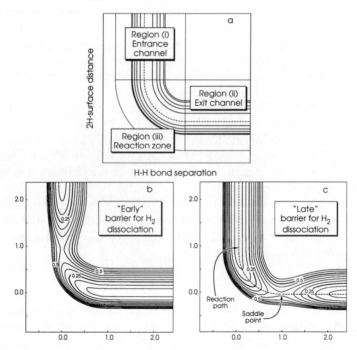

Figure 13 (a) *A contour diagram showing the essential features of this model potential energy surface. This template is then combined with a Gaussian barrier whose potential is varied in order to create surfaces with two different topologies.* (b) *Here the maximum of the dissociation barrier lies at the beginning of the reaction zone—an early barrier,* (c) *The activation barrier maximum is placed at the beginning of the exit channel—a late barrier. In this figure is shown the reaction path defined by equation (40). Energies are in eV and are measured with respect to a molecule infinitely far from the surface*[23]

molecular antibonding orbitals, and a repulsive interaction resulting from the overlap of the doubly occupied molecular $1\sigma_g$ state with the conduction electrons. For free-electron-like metals it is here that activation barriers are located, while for transition metals the reaction zone may simply show a smooth featureless region of decreasing energy. For this present study, the reaction zone PES is composed of two parts. To link the entrance and exit channels in the simplest way, we join the contours in these regions by a quarter-circle.[74,75] Thus the potential along any radius in this region is again a Morse potential (Figure 13a). There will be a gradual decrease in the zero-point energy as the corner is turned and the effective mass of the representative point changes from 1/2 to 2 a.m.u., consequently the PES is only thermo-neutral in the classical sense.

Superimposed upon this trough, is an activation barrier whose shape is fixed, and position along the minimum energy path serves to define the final potential surface. This barrier is Gaussian in the angular co-ordinate ϕ, which is taken to be zero at the junction with the entrance channel. The PES is then:

$$V_3(r, \phi) = V_{\text{Morse}}(r) + V_0 \, e^{-\beta(\phi - \phi_0)^2} \tag{39}$$

where again the Morse parameters are the same as for the entrance and exit channels. $\beta = 1.0$, defines the width of the activation barrier and ϕ_0 its location within the reaction zone, $r_{eq} = 1.0$ a.u. The height of the barrier V_0 has been chosen to be 0.538 eV (twice the zero-point energy of H_2), which is slightly greater than the $0 \to 1$ vibrational excitation in the free H_2 molecule. The choice of ϕ_0 serves to define a particular topology for the PES; $\phi_0 = 0$ corresponds to an *early barrier* (Figure 13b), and $\phi_0 = \pi/2$ a *late barrier* (Figure 13c).

Figure 13c also shows the reaction path which is defined as the path of steepest descent from the saddle point:

$$\frac{d\mathbf{r}'}{ds} = -\frac{\nabla V(\mathbf{r}')}{|\nabla V(\mathbf{r}')|} \tag{40}$$

where $\mathbf{r}' = (x, Z')$ and s is the arc length along the path. In order to calculate s, the variable transformation $Z' = 2Z$ is employed which makes the mass a scalar quantity. This is equivalent to making a skewed axis representation.[76] It is now convenient to define a co-ordinate ρ which is locally orthogonal to s. The PES $V(x, Z')$ can then, quite generally, be expressed as $V(\rho, s)$. Furthermore, it is useful to write:

$$V(\rho, s) = V_{\text{RP}}(s) + v(\rho; s) \tag{41}$$

where $V_{\text{RP}}(s)$ is the potential *along* the reaction path and $v(\rho; s)$ is the potential in the ρ direction for a *given* value of s. At each point along the

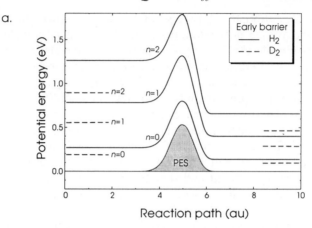

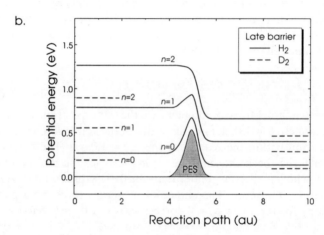

Figure 14 (a) *The early barrier one-dimensional effective potentials, $V_{eff}^n(s)$, for the first three vibrational states.*[23] *The shaded area shows the adiabatic barrier $V_{RP}(s)$, when the mass transformation, described in the text, is included. In order to generate $V_{eff}^n(s)$, the vibrational energy, calculated from the potential normal to the reaction path $v(\rho; s)$, is added to $V_{RP}(s)$. It should be noted that for each $V_{eff}^n(s)$, the barrier is of equal height. The dashed lines show the asymptotic energies of the D_2 states which behave in an identical manner to H_2 in the reaction zone.* (b) *The same plot, but for the late barrier PES. In this case the barrier height is markedly dependent on the vibrational quantum number, being negative for the $n = 2$ state!*

reaction path the bound state eigenvalues, $\epsilon_n(s)$, of $v(\rho; s)$ can be obtained, and it is then possible to define a set of one-dimensional vibrationally adiabatic potentials:

$$V_{eff}^n(s) = V_{RP}(s) + \epsilon_n(s) \tag{42}$$

In Figure 14, $V_{eff}^n(s)$ is plotted for the two surfaces described above. For the

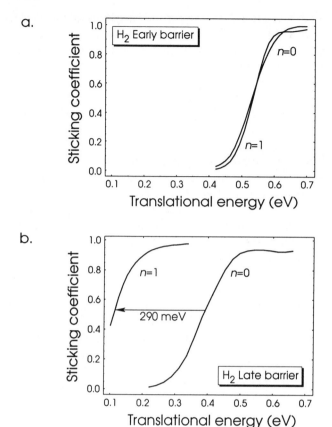

Figure 15 *A comparison of the dissociation probability for a H_2 molecule in its ground and first vibrationally excited state. (a) Shows the results for the early barrier PES and indicates that there is a minimal effect resulting from the additional vibrational energy. (b) In this late barrier case, where the barrier lies wholly in the vibrational co-ordinate (bond extension of 1.0 a.u.), the shift of 290 meV still only represents a usage of 56% of the available vibrational energy. For both potentials, it is found that the de-excitation probability for vibrationally excited molecules is < 10%, lending weight to the concept of vibrational adiabaticity in the scattering dynamics*[23]

early barrier case, there is no decrease in its height due to zero-point energy expenditure and thus $E_{act}^n = 0.54$ eV for all values of n. This is in stark contrast to the late barrier case where the barrier height depends markedly on n with $E_{act}^0 = 0.40$ eV, $E_{act}^1 = 0.15$ eV with E_{act}^2 being negative! In addition, the width of the barrier decreases as ϕ_0 increases. This occurs because of the co-ordinate transformation $Z \rightarrow Z'$, which shrinks the x co-ordinate as the reaction zone is traversed. Thus all late barriers are perceived as being approximately half of the width of their early barrier counterpart.

In Figure 15a are plotted the results for the $n = 1$ dissociation probability as a function of translational energy for the early barrier together with the $n = 0$ data. It is clear that the extra 514 meV of vibrational energy has resulted in a negligible change in the position of the curve, the main effect being a slight smearing in its profile. This illustrates the effects of a 'spectator co-ordinate' mentioned by Eyring,[2] in this case the molecular vibration, whose energy cannot be converted into the reaction co-ordinate despite the fact that the available energy is 1.5 times the height of the activation barrier! The ground state population of the reflected flux remained at less than 0.01% of the total flux for all energies, showing that almost no de-excitation occurs, confirming that the scattering is vibrationally adiabatic.

For the late barrier, the efficiency of vibrational energy in enhancing dissociation is greatly increased as seen in Figure 15b. Now there is a shift in the $n = 1$ curve of ~ 290 meV to lower energies when compared to the $n = 0$ result. Since the zero-point energy contribution has been included in the $n = 0$ sticking curve, the conversion efficiency arising from the extra 514 meV of vibrational energy is $\sim 56\%$. It is interesting to note that if all of the vibrational energy were available for barrier traversal, then there would be a reduction in height of the effective potential barrier of > 640 meV, (783/2 meV from the change in mass in the $n = 1$ state in the curvature region and an additional 257 meV if it then de-excited into the ground state of the exit channel) making the system non-activated! This is simply not the case and the observation of a finite, but much reduced barrier, is best understood by examining the $n = 1$ effective potential for the late barrier shown in Figure 14b. The height of this barrier is 146 meV relative to a zero kinetic energy $n = 1$ molecule, whereas if the diatomic were vibrationally de-excited before striking the barrier, its total energy would lie well above that of the ground state barrier as can be seen in Figure 14b. This again indicates the inefficiency of the de-excitation process before the collision with the barrier; the population of $n = 0$ scattered molecules never rises above 5% even in this late barrier system.

To test the degree of vibrational adiabaticity, one run for the late barrier was performed at an initial translational energy of 100 meV with the H_2 in the $n = 1$ state. The results confirmed that the transmitted flux remained vibrationally excited with over 90% of the dissociation products diffusing in the $n = 1$ state.

The concept of 'vibrational adiabaticity' has important consequences for interpreting experimental scattering data. Recent results[77] for the scattering of H_2 from a Cu(110) surface have quite clearly shown strong evidence in favour of a dissociation mechanism involving vibrationally excited molecules. Figure 16 shows a sketch of results for the dissociation probability plotted as a function of the kinetic energy normal to the surface.[77] As can be seen from the curve, the sticking probability for a pure beam of hydrogen molecules becomes finite at ~ 180 meV and rises exponentially thereafter. This data has been confirmed by Anger, Winkler, and Rendulic.[78] For a given vibrational distribution (e.g. $T \sim 1150$ K), the results for a seeded beam are also shown in Figure 16. The sticking coefficient now shows almost

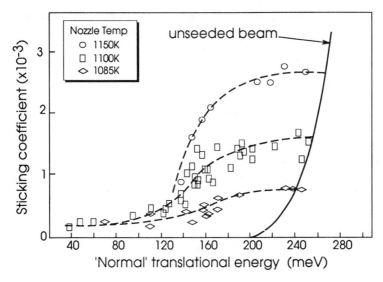

Figure 16 *The experimental results for the sticking of a pure beam of H_2 molecules (thick solid line). In this experiment, translational and vibrational energy change together, making it impossible to separate their individual effects on dissociation.[77] The calculated Boltzmann population of the $n = 1$ state is more than sufficient to account for the observed sticking.[79] To investigate the efficiency of vibrational energy, the nozzle is held at a fixed temperature, thereby maintaining the Boltzmann population. The translational energy is then decreased by anti-seeding with He. The independence of the sticking coefficient for normal translational energies greater than 160 meV, can be accounted for by assuming a saturation in the $n = 1$ dissociation probability (see Figure 15b)*

no dependence on the translational energy, implying that the vibrational energy plays a very important role in the dissociation process. The Boltzmann population of the $n = 1$ state is at least twice as large as the sticking coefficient for all energies, and is more than sufficient to account for the sticking data.[79] The two other curves in the figure, for vibrational temperatures of 1100 and 1085 K, show identical behaviour.

The minimal energy dependence of the seeded beam sticking coefficients in the range ~ 180–250 meV can be accounted for by drawing a parallel to the data obtained for the late barrier simulation, Figure 14b, where the $n = 1$ curve practically levels out before the $n = 0$ curve becomes finite. If this were the case, then an estimate of the barrier height perceived by the molecules in the $n = 1$ state can be made by measuring the translational energy at which ~ 50% of the saturation sticking coefficient is reached. For all three curves this is roughly 150 meV. The question then arises as to the true height of the barrier in the PES, assuming that the dissociation is energy limited by the bond breaking process. As has been seen even for a late barrier not all of the vibrational energy is available for barrier traversal, in fact from estimates based on the late barrier, approximately 60% of the available vibrational energy is released as the vast majority of dissociated products populate the $n = 1$ state. Therefore assuming dissociation to be vibrationally adiabatic, an estimated barrier height for the H_2/Cu results[77] is ~ 540 meV (obtained by adding the E_{trans} corresponding to 50% saturation ~ 150 meV to the vibrational energy release of 392 meV). This activation energy is far below the estimated 920 meV quoted by Hayden and Lamont[77] who assumed that the entire vibrational energy was available for surmounting the barrier (Figure 16). For this lower barrier, the dissociation of ground state H_2 molecules would still be negligible for the range of energies reported $E^i_{trans} < 260$ meV[77] but may have been finite by the experiments of Anger et al.[78] where the upper limit was $E^i_{trans} \sim 475$ meV.

Coupling the Reaction Co-ordinate to Rotations. To treat the collision of a hydrogen molecule with a static surface, six degrees of freedom should be considered. Of these six dimensions, three are internal degrees of freedom; a

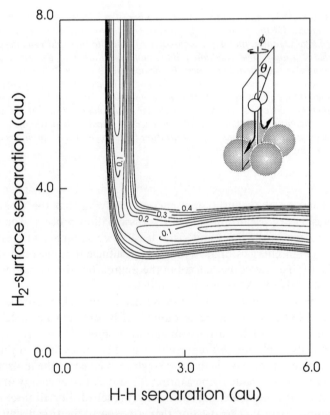

Figure 17 *A Z–x cut of the PES used in the three-dimensional study.[80] It represents a H_2 molecule coming down to a surface over the centre site. The activation barrier has a value of 175 meV. Energies are in eV and are measured with respect to a molecule infinitely far from the surface*

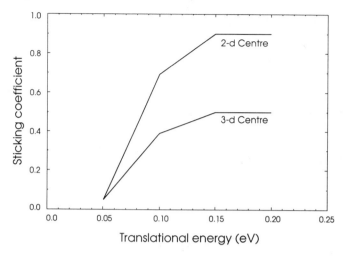

Figure 18 *The sticking coefficient as a function of translational energy of the molecule. Shown are the results of the two-dimensional, fixed orientation calculation and the three-dimensional calculation where the molecule is allowed to rotate in a plane perpendicular to the surface (cartwheeling). Results show that by including the rotational degree of freedom, the dissociation probability saturates at a value significantly less than unity.[80] This is because those molecules aligned with their axis perpendicular to the surface have an unfavourable dissociation geometry and are scattered into the gas phase*

polar (θ) and an azimuthal (ϕ) angle, a vibrational co-ordinate (x), and three centre-of-mass translations (X, Y, Z). Due to computational limitations only four of these six dimensions may be treated explicitly using exact quantum dynamics. To increase the dimensionality of the problem, Nielsen et al.[80] included one of the rotational degrees of freedom in addition to the (Z, x) PES discussed above. Both hydrogen atoms were constrained to lie in a plane perpendicular to the metal surface above a centre site in the surface resulting in a PES, $V(Z, x, \theta)$. The elbow plot ($\theta = \pi/2$) corresponding to this particular geometry is shown in Figure 17 where the height of the activation barrier was 175 meV. While this does not imply that azimuthal effects will be negligible or even uninteresting, there will be stronger dynamical coupling to the dissociation and rotational excitation probability for variations in the polar angle θ which was included in the calculation.[80] This may be expected as dissociation cannot occur for a molecule constrained to have its axis normal to the surface ($\theta = 0$) and impulsive collisions will predominantly excite cartwheel rotations with $\mathbf{J}$ parallel to the surface.[81]

The initial state was chosen to be that of a beam of molecules incident normally to the surface in the ground rotational and vibrational states. The accessible scattering channels are; (i) dissociation into the atomic well on the surface or, (ii) reflection from the surface with or without rotational excitation. No vibrational excitation can occur for the scattered flux as the

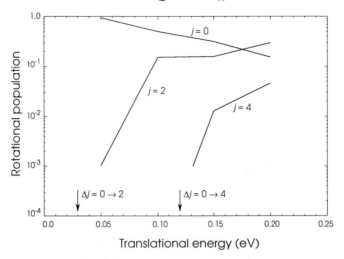

Figure 19 *The rotational excitation probabilities for the three-dimensional calculation described in this section.[80] This occurs because of the stereospecificity in the dissociation process. Those molecules with their molecular axis parallel to the surface normal do not dissociate and as a consequence, the scattered flux has lost a particular angular fraction [see equation (43)]. Translated into a final state distribution, this implies rotational excitation. Alternatively one can think of this as arising from a strongly anisotropic gas–surface potential. The thresholds for the first two rotational excitations are indicated by arrows*

translational energies considered lie well below the 514 meV threshold for the $v = 1 \leftarrow v = 0$ transition in gas-phase H_2.

The results for the dissociation probability are given in Figure 18 and clearly show that the inclusion of rotational motion having **J** parallel to the surface results in a saturation value for the sticking coefficient, $S_0 \sim 0.5$. This is to be contrasted with the results from the two-dimensional calculation shown in the same figure where θ has been fixed to be $\pi/2$. For this case, as was seen in the previous section, the dissociation probability rises smoothly from 0 to 0.9. To probe further into the dynamics, Figure 19 shows the asymptotic rotational state distributions for the scattered flux as a function of translational energy. At very low energies all of the molecules emerge in the $j = 0$ state, purely from conservation of energy as the thresholds for rotational excitation for the $0 \rightarrow 2$ and $0 \rightarrow 4$ transitions lie at 30 meV and 120 meV respectively. At these low translational energies it has been shown that 'mechanical' collision with a corrugated wall is not capable of exciting such a degree of rotational excitation in the H_2 molecule.[82] The only possibility left is that in the course of the reaction a selectivity occurs in the dissociation which, in turn, accounts for the perturbation in the state distributions.

To interpret this effect, Holloway and Jackson have proposed the following simple model.[83] Suppose that the initial rotational state distribution is isotropic, with $j = 0$ labelling the two-dimensional quantum number. They assumed that only those molecules having their molecular bond axis within a specific range of angles $\pm a$, were able to dissociatively chemisorb. This is in accord both with simple intuition and total energy calculations.[84] Preferential dissociation will therefore occur for those molecules having $\theta = \pi/2 \pm a$ where a will depend markedly upon the substrate in question and the initial translational energy. For a system with an activation barrier to dissociation it might be expected that at the (classical) threshold a would be very small, increasing as the barrier energy is surpassed. The scattered flux of molecules will therefore be depleted of those molecules lying flat, and this selectivity will result in a rotational distribution with an unusually large occupation of higher j states. This distribution will, of course, be modified by any attractive steering forces and impulsive collisions but, as discussed above, for the H_2 potential these are expected to be small.

To quantify this behaviour, Holloway and Jackson proposed an ansatz for the scattered wavefunction whose angular distribution,[83] originally isotropic, has two Gaussian shaped portions removed around $\theta = \pi/2$ and $3\pi/2$:

$$\psi_f(\theta) = \frac{1}{\sqrt{(2\pi)}} \left[1 - A \left(\exp\left\{ -\left(\frac{\theta - \pi/2}{a}\right)^2 \right\} \right. \right.$$
$$\left. \left. + \exp\left\{ -\left(\frac{\theta - 3\pi/2}{a}\right)^2 \right\} \right) \right] \quad (43)$$

where A is the amplitude of the Gaussian function. Normalization was chosen such that the integral of $\psi_f^* \psi_f$ over the interval $0-2\pi$ is $(1 - S_0)$. For $a < 90°$, S_0 is approximately:[85]

$$S \approx \frac{Aa}{\sqrt{(2\pi)}} [2\sqrt{2} - A] \quad (44)$$

Assuming that the angular width is small enough such that the removed fractions do not overlap, the resulting probability distributions of angular momenta in the scattered beam was found by projecting equation (43) onto the rotational wavefunctions $e^{ij\theta}/\sqrt{(2\pi)}$:

$$|P(0)|^2 = \left[1 - \frac{Aa}{\sqrt{(\pi)}} \right]^2 \quad (45)$$

and

$$|P(j)|^2 = \frac{(Aa)^2}{2\pi} \exp\{-j^2 a^2/2[1 + \cos(j\pi)]\} \quad (j \neq 0) \quad (46)$$

Figure 20 shows plots for the rotational distributions, $|P(j)|^2$, as a function of a for the cases of $A = 0.5$ and $A = 1.0$. These results show quite clearly that as the dissociative adsorption channel opens and a increases, the scattered

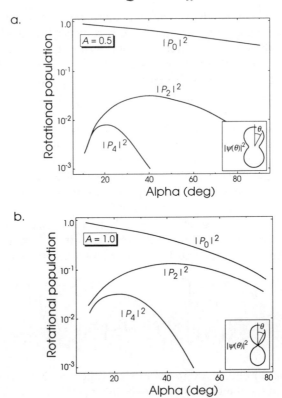

Figure 20 *Variation of the rotational state populations with the width of the Gaussian function used in the ansatz equation (43).[83] Two values for the amplitude have been chosen for illustration, (a) A = 0.5 and (b) A = 1.0. Within the insets are shown the probability distributions for the two cases with α = 45°*

fraction exhibits significant rotational excitation—a conclusion borne out in the scattering calculations shown in Figure 20. The $|P(j)|^2$ tend to bend over at the higher values of a and this is simply a consequence of the $\psi_f^* \psi_f$ reverting to an isotropic distribution as more particles dissociate.

It is clear from the figure that as the acceptance angle increases, the amount of excitation of the $j = 2$ state in particular is radically altered. The reason for this is that since the PES has θ periodic in π, the resulting wavefunction will adopt this functionality and consist predominantly of $j = 0$ and $j = 2$ states. In addition the higher the dissociation probability for a given angle, altered using A, the greater is the excitation of all the higher states as seen when comparing Figures 20a and b.

Other Diatomics

Thus far there has only been mentioned the dissociation of hydrogen and its isotopes; to what extent do conclusions drawn carry over to heavier diatom

Table 2 *Potential parameters for the N_2/Fe system given in equation (35)*[53]

$$\text{Morse} \begin{cases} V_{N_2} \\ W = 9.91 \\ a_M = 1.45 \\ V_0 = 18.0 \\ \lambda = 1.0 \end{cases} \quad \begin{array}{l} V_{2N} \\ V_1 = 0.5 \\ A_x = 0.6 \\ A_z = 0.6 \\ x_0 = 2.8 \\ Z_0 = 2.5 \\ V_2 = 250.0 \\ \lambda_m = 1.45 \end{array}$$

Coupling parameter: $\varkappa = 1.5$

systems? To answer this question it is instructive to return to the dissociation model presented earlier in this section, where the surface was allowed to recoil. In this calculation, the authors constructed a PES to represent the scattering of N_2 from a Fe surface and the results for the dissociation probability are presented in Figure 21.[53] Unlike the case of H_2 scattering, the results for the static *vs.* dynamic barrier are totally different with an observed downshift of ~ 0.3 eV for the latter. As before, the occupation of the surface atom vibrational states is approximately a Poisson distribution centred about the classical energy loss.

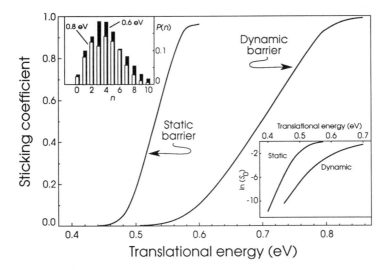

Figure 21 *Sticking coefficient for the N_2/Fe system.[53] Potential parameters are in Table 2. Results are given for a static and a dynamic barrier which is capable of recoiling. Unlike the case for H_2 (Figure 12) there is a considerable upshift for the dynamic barrier. The inset top left shows the occupations of the surface oscillator states following the scattering event. The black and white histograms refer to incident energies of 0.6 and 0.8 eV, embracing the region where back-scattering was, respectively, the majority and minority event. The inset bottom right show results plotted on a log scale*

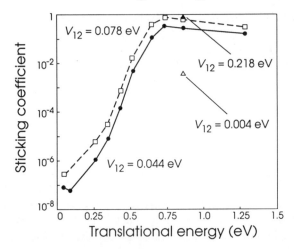

Figure 22 *Results for the sticking coefficient as a function of translational energy for the N_2/Re system calculated by Haase et al.[86] The coupling strength V_{12} which is a measure of the strength of the electronic non-adiabaticity has been varied parametrically*

Kosloff and co-workers have discussed the dissociative adsorption of N_2 on Re surfaces, again using a time dependent formalism but have explicitly addressed the effects of electronic non-adiabaticity and have included curve-crossing dynamics and multiple potential surfaces.[86] In this study the effects of surface recoil have not been included. The two diabatic surfaces in question correspond to the N_2/Re interaction and the 2N/Re PES. The restricted (Z, x) geometries have again been employed but the Hamiltonian was taken to be of the matrix form:

$$\mathcal{H} = \begin{bmatrix} \mathcal{H}_1 & V_{12} \\ V_{12} & \mathcal{H}_2 \end{bmatrix} \quad (47)$$

where the diagonal terms are simply those discussed in Section 3 and V_{12} is the interaction potential describing the coupling between the surfaces. In the language of Section 2, V_{12} is intimately related to E_a, the width of the antibonding resonance at the curve crossing. The initial wavepacket was given by a product state similar to equation (19) but excluding ϕ_0; initially it is prepared on the N_2/Re diabatic PES. Without going into specific details concerning the PES, it may be classified as an early barrier type, with an activation energy of 0.76 eV. The value of V_{12} at the curve crossing point was taken as a parameter, and in Figure 22 the dissociation probability is plotted as a function of the initial translational energy for $0.0044 \text{ eV} < V_{12} < 0.22 \text{ eV}$. Although the current wisdom on level widths outside surfaces is in a state of flux,[87] for molecular species this range of values is rather on the low side. The results show the now familiar increase of S_0 with kinetic energy. What is interesting is the variation with V_{12} which, for given translational energy, exhibits a linear dependence. This may be directly traced to the behaviour expressed in equation (10). At low translational energies the process is adiabatic and the wavepacket passes through the curve crossing region smoothly and does not experience the upper surface. As the energy increases, an increasing fraction stays on the diabatic curve and is reflected back into the gas phase. This work also demonstrated that even for molecules as heavy as N_2, there is still the possibility of a significant isotope effect. This effect will be enhanced if the barrier is located toward the exit channel, since in this case the effective mass which is tunnelling will be the reduced mass rather than the molecular mass. This observation was first made on the basis of a semi-classical study by Holloway et al. for the dissociation of H_2.[73,88]

The first time that a tunnelling model for dissociative adsorption on metal surfaces was proposed, was to offer an explanation for the large isotope effect observed for the CH_4–CD_4/W system.[89] Winters argued that a possible reaction path is where the symmetric v_1 mode of a methane molecule is excited to a higher vibrational level near the surface which allows the H atom to tunnel to an adjacent adsorption site. An attempt to quantify this was made on the basis of a one-dimensional rectangular barrier.

The large isotope effect has subsequently been observed in other CH_4/surface systems[90] but at this time there does not exist any detailed microscopic model to explain the data. An attempt to selectively pump the internal modes of the methane in order to enhance the reactivity for an adsorbed species however, produced a null result.[91,92]

6 Future Directions

In this chapter we have examined the various ways in which quantum phenomena associated with molecular internal degrees of freedom manifest themselves in scattering experiments. For both inelastic scattering and reactive scattering it has been demonstrated that, for a number of cases, a classical treatment of the dynamics simply will not suffice. For the case of a diatomic approaching a rigid surface, a full treatment of the problem would entail a six-dimensional quantum calculation. At this point in time the record stands at five,[82] where the vibrational co-ordinate was neglected in a study of H_2 scattering from a LiF surface. Unquestionably within the near future, computational memory and speed will enable a six-dimensional calculation to be performed. Will this bring new levels of understanding? Although this will be a computational *fait accompli*, in all likelihood the answer will probably be no. On the basis of the current methodology we have already begun to develop a good intuition as to the relevant and irrelevant co-ordinates for a number of scattering scenarios, particularly for reactive encounters. Also on the basis of previous experience, even when this calculation comes along, it will no doubt be necessary to interpret the results from a limited dimensionality standpoint anyway.

The coupling of the multi-dimensional wavepacket to the surface atomic motion is certainly more of an open question and a challenge. For a scattering problem, there exist mean field treatments in which the full density matrix for the particle–phonon system is replaced by a reduced matrix, which behaves like a wavefunction.[93] The phonons appear as a modification to the potential, controlled by time-dependent complex amplitudes quantifying their creation and annihilation. While this method has produced a series of nice results for the inelastic scattering of rare gases, when it comes to either trapping or dissociation the method appears to have some fundamental flaws. In particular for encounters where low energy–low momentum phonons are important, it would be necessary to have many grid points in the parallel direction and a very large sized mesh. Even if it were possible to overcome such technicalities, there is still a problem since in order to scatter from specific phonons with non-zero parallel momentum, we have gone beyond the concept of a mean field. In Jackson's method[93] what happens is that by introducing a phonon driving term that has not been averaged, the probability of creating or annihilating a phonon at all times is identically zero. An alternative approach, by Cerjan and Kosloff,[24] has its merits but from the outset it appears to be rather cumbersome to implement.

All of these dynamical problems hinge upon the fact that there is a potential energy surface which in some way is a good representation of the physical problem to be studied. Certainly over the last ten years there have not appeared in the literature copious numbers of total energy calculations for solitary molecules (or even atoms) outside semi-infinite surfaces. To the contrary, the way that surface dynamics has progressed, has been alluded to in various sections in this chapter. For the NO scattering, an attempt has been made to invert experimental data. Unfortunately since the data base is so restricted, the most commonly used PES[47] gets the binding properties of the system hopelessly wrong.[94] For the H_2/Cu system, as mentioned previously, there has been an *ab initio* calculation for H_2 interacting with a Cu_2 dimer.[9] The (elbow) PES obtained has been a remarkably versatile animal. In the original calculation the height of the activation barrier was 1.4 eV and the most important topological features were that there existed no molecular resonance (*cf.* Figure 9) and that the saddle point was located at a H—H separation that was larger than the gas-phase equilibrium value. The first parametrization of this PES for dynamical purposes took a more conservative view and modified E_{act} to be 1 eV.[22] Subsequent to this, the value of E_{act} was reduced even further in the next round of calculations and a value of 0.72 eV was used.[53] What real differences in the dynamics occur because of this somewhat arbitrary manipulation of an activation barrier? The important parameter when considering activated systems is the ratio of the barrier height to the vibrational quanta of the molecule, E_{act}/ω. For all of the various modifications to the H_2/Cu PES, this value has always been greater than unity which means that the overall dynamics will remain fairly similar with only a numerical shift occurring in some of the calculated quantities. There have been other calculated PESs for the H_2/Cu system that have been used for dynamical studies. In particular the calculation of Madhaven and

Whitten,[8] which gave total energies for a selection of high symmetry points, has been interpolated by Billing and co-workers and used in a mixed classical–quantum treatment.[95] Additionally, the effective medium H_2/Cu PES calculated by Nørskov[96] has provided a useful parametrization for a system where the ratio $E_{act}/\omega \ll 1$. With the recent advances in embedding methods,[12] it is hoped that the time is now right to see an increase in our understanding of the electronic properties of dissociative chemisorption and as a consequence, new potential surfaces.

Finally there is an open question as to what are the future directions for surface dynamics. In particular, what problems will look attractive and, moreover, lend themselves to theoretical investigation once the dissociation of diatomics is 'solved'? As discussed in Section 5, polyatomics suggest themselves (particularly CH_4 dissociation) since there is so much data already available. With fifteen degrees of freedom to investigate, this is a formidable problem and it will take much effort to decide which are the important co-ordinates in the hydrogen abstraction. It is probably too simplistic to consider the problem as a methyl radical bound to a hydrogen, and treat the system as a dimer since undoubtedly some of the internal modes will couple in a very non-trivial way. Obtaining even a zeroth order approximation to the PES will not be easy but will be necessary if a dynamical study is to have any credibility. Another track is to consider Eley–Rideal reactions. This looks a little more promising since it may be possible to use existing knowledge to go a few steps further. Consider the hypothetical reaction of a gas–phase hydrogen atom plucking an adsorbed hydrogen from the surface of a metal.[75] The problem is still that of a dimer but the initially prepared state is different to that of the scattering experiments discussed in Section 5. For more complex reactions [$e.g.$ $AB_{gas} + C_{ads} \rightarrow A_{ads} + BC_{gas}$] again it may be possible to profit from an elementary understanding of the AB surface and the BC surface interactions in order to investigate the exchange mechanism.

In conclusion this work contains a review of the application of time dependent quantum wavepacket techniques to the scattering of molecules from surfaces. Over the past five years, significant progress has been made in understanding the connection between potential energy surface topologies and the ensuing dynamics and it is anticipated that with the increase in experimental activity the next five years will be an exciting period for surface dynamics.

Acknowledgement. I thank current and previous members of my research group for providing the stimulating atmosphere in which much of this work was done. Particular thanks go to Mick Hand for helpful suggestions and George Darling for reading the manuscript.

References

1. 'Interaction of Atoms and Molecules with Solid Surfaces', ed. V. Bortolani, N. H. March, and M. P. Tosi, Plenum Press, New York, 1990.
2. H. Eyring, *J. Chem. Phys.*, 1935, **3**, 107.

3. R. Car and M. Parrinello, *Phys. Rev. Lett.*, 1985, **55**, 2471.
4. J. Harris, *Phys. Scr.*, 1987, **36**, 156.
5. L. Wilzén, F. Althoff, S. Andersson, and M. Persson, *Phys. Rev. B* in Press.
6. G. C. Corey and D. LeMoine, *Chem. Phys. Lett.*, 1989, **160**, 324.
7. J. Kimman, C. T. Rettner, D. J. Auerbach, J. A. Barker, and J. C. Tully, *Phys. Rev. Lett.*, 1986, **57**, 2053.
8. P. Madhaven and J. L. Whitten, *J. Chem. Phys.*, 1982, **77**, 2673.
9. J. Harris and S. Andersson, *Phys. Rev. Lett.*, 1985, **55**, 1583.
10. M. Weinert and J. W. Davenport, *Phys. Rev. Lett.*, 1985, **54**, 1547.
11. C. Umrigar and J. W. Wilkins, *Phys. Rev. Lett.*, 1985, **54**, 1551.
12. P. Feibelman, *Phys. Rev. Lett.*, 1989, **63**, 2488.
13. T. F. O'Malley, *At. Mol. Phys.*, 1971, **7**, 223.
14. J. W. Gadzuk, *Comments At. Mol. Phys.*, 1985, **16**, 219.
15. E. E. Nikitin, 'Theory of Elementary Atomic and Molecular Processes in Gases', Clarendon, Oxford, 1974.
16. J. v. Neumann and E. P. Wigner, *Phys. Z.*, 1929, **30**, 467.
17. J. K. Nørskov, A. Houmøller, P. K. Johansson, and B. I. Lundqvist, *Phys. Rev. Lett.*, 1981, **46**, 257.
18. E. Bauer, E. R. Fisher, and F. R. Gilmore, *J. Chem. Phys.*, 1969, **51**, 4173.
19. F. T. Smith, *Phys. Rev.*, 1969, **179**, 111.
20. S. Holloway, *J. Vac. Sci. Technol.*, 1987, **A5**, 476.
21. B. Jackson and H. Metiu, *J. Chem. Phys.*, 1986, **86**, 1026.
22. J. Harris, S. Holloway, T. Raman, and K. Yang, *J. Chem. Phys.*, 1988, **89**, 4427.
23. D. Halstead and S. Holloway, *J. Chem. Phys.*, 1990, **93**, 2859.
24. C. Cerjan and R. Kosloff, *Phys. Rev. B*, 1986, **34**, 3832.
25. B. Jackson, *J. Chem. Phys.*, 1987, **88**, 1383.
26. G. Drolshagen and E. J. Heller, *Chem. Phys. Lett.*, 1984, **104**, 129.
27. R. B. Gerber, R. Kosloff, and M. Berman, *Comput. Phys. Rep.*, 1986, **5**, 61.
28. R. Kosloff, *J. Phys. Chem.*, 1988, **92**, 2087.
29. V. Mohan and N. Sathyamurthy, *Comput. Phys. Rep.*, 1988, **7**, 213.
30. C. Leforestier, R. Bisseling, C. Cerjan, M. D. Feit, R. Freisner, A. Guldberg, A. Hammerlich, G. Jolicard, W. Karrlin, H.-D. Meyer, O. Roncero, and R. Kosloff, *J. Comput. Phys.*, in Press.
31. Throughout this work atomic units will be used.
32. D. Kosloff and R. Kosloff, *J. Comput. Phys.*, 1983, **52**, 35.
33. J. A. Fleck, J. R. Morris, and M. D. Feit, *Appl. Phys.*, 1976, **10**, 129.
34. H. Tal-Ezer and R. Kosloff, *J. Chem. Phys.*, 1984, **81**, 3967.
35. R. Kosloff and D. Kosloff, *J. Comput. Phys.*, 1986, **63**, 363.
36. C. Leforestier, *Chem. Phys.*, 1984, **87**, 241.
37. M. R. Hand and S. Holloway, *J. Chem. Phys.*, 1989, **91**, 7209.
38. S. Andersson, L. Wilzén, M. Persson, and J. Harris, *Phys. Rev. B.*, 1989, **40**, 8146.
39. J. Harris, S. Andersson, C. Holmberg, and P. Nordlander, *Phys. Scr.*, 1986, **T13**, 155.
40. P. Nordlander, C. Holmberg, and J. Harris, *Surf. Sci.*, 1985, **152**, 702.
41. J. Harris and P. Feibelman, *Surf. Sci.*, 1982, **115**, 133.
42. G. M. McClelland, G. D. Kubiak, H. G. Rennagel, and R. N. Zare, *Phys. Rev. Lett.*, 1981, **46**, 831.
43. A. W. Kleyn, A. C. Luntz, and D. J. Auerbach, *Phys. Rev. Lett.*, 1981, **47**, 1169.

44. C. T. Rettner, J. Kimman, F. Fabre, D. J. Auerbach, J. A. Barker, and J. C. Tully, *J. Vac. Sci. Technol.*, 1986, **A5**, 508.
45. A. Mödl, T. Gritsch, F. Budde, T. J. Chuang, and G. Ertl, *Phys. Rev. Lett.*, 1986, **57**, 384.
46. J. A. Barker, A. W. Kleyn, and D. J. Auerbach, *Chem. Phys. Lett.*, 1983, **97**, 9.
47. H. Voges and R. Schinke, *Chem. Phys. Lett.*, 1983, **100**, 245.
48. R. Schinke and R. B. Gerber, *J. Chem. Phys.*, 1985, **82**, 1567.
49. W. Brenig, H. Kasai, and H. Müller, *Surf. Sci.*, 1985, **161**, 608.
50. T. Brunner, R. Brako, and W. Brenig, *Phys. Rev. A*, 1987, **35**, 5266.
51. M. R. Hand, X. Y. Chang, and S. Holloway, *Chem. Phys.*, 1990, **147**, 351.
52. W. Brenig, *Z. Phys.*, 1979, **B36**, 81.
53. M. R. Hand and J. Harris, *J. Chem. Phys.*, 1990, **92**, 7610.
54. A. W. Kleyn, *Comments At. Mol. Phys.*, 1987, **19**, 133.
55. M. S. Child, 'Molecular Collision Theory', Academic Press, London, 1974.
56. H. J. Korsch and R. Schinke, *J. Chem. Phys.*, 1980, **73**, 1222.
57. $4\mu(E^i_{\text{trans}} + W_{\text{av}})/(1 + \mu^2)$, where W_{av} is the laterally averaged well depth and μ is the ratio m/M.
58. The expression for $P(E)$ in the collision of a particle with a harmonic oscillator, where E is the final oscillator energy, has a $1/\sin\phi$ dependence, where ϕ is the phase of the oscillator on collision (taking $y = 0$, p_y positive at $t = 0$). This then gives rise to a singularity at $\phi = 0$ and $\phi = \pi$, which correspond to minimum and maximum oscillator excitation, respectively. Integration over the rotational states in the three-dimensional case results in a smearing out of the singularities, although they are still visible. Of course, for the real surface case such an effect would be absent.
59. G. O. Sitz, Ph.D., Stanford University, 1987.
60. S. Holloway and J. W. Gadzuk, *Surf. Sci.*, 1985, **152/153**, 838.
61. S. Holloway and J. W. Gadzuk, *J. Chem. Phys.*, 1985, **82**, 5203.
62. J. W. Gadzuk and S. Holloway, *Chem. Phys. Lett.*, 1985, **114**, 314.
63. J. W. Gadzuk and S. Holloway, *J. Chem. Phys.*, 1985, **82**, 5203.
64. J. W. Gadzuk and S. Holloway, *Phys. Scr.*, 1985, **32**, 413.
65. J. W. Gadzuk and S. Holloway, *Phys. Rev. B*, 1986, **33**, 4298.
66. C. T. Rettner, F. Fabre, J. Kimman, and D. J. Auerbach, *Phys. Rev. Lett.*, 1985, **55**, 1904.
67. D. Newns, *Surf. Sci.*, 1985, **171**, 600.
68. H. Kasai and A. Okiji, *Surf. Sci.*, 1990, **225**, L33.
69. M. R. Hand and S. Holloway, *Surf. Sci.*, 1989, **211/212**, 940.
70. M. R. Hand and S. Holloway, *J. Phys.: Condens. Matter*, 1989, **1**, SB27.
71. A. Bringer and J. Harris, *J.Chem.Phys.*, 1989, **91**, 7693.
72. J. C. Polanyi. *Acc. Chem. Res.*, 1972, **5**, 161.
73. S. Holloway, A. Hodgson, and D. Halstead, *Chem. Phys. Lett.*, 1988, **147**, 425.
74. S. Latham, J. F. McNutt, and R. E. Wyatt, *J. Chem. Phys.*, 1978, **69**, 3746.
75. W. Brenig and H. Kasai, *Surf. Sci.*, 1989, **213**, 179.
76. R. B. Bernstein, 'Chemical Dynamics via Molecular Beam and Laser Techniques', Oxford University Press, Oxford, 1982.
77. B. E. Hayden and C. L. A. Lamont, *Phys. Rev. Lett.*, 1989, **63**, 1823.
78. G. Anger, A. Winkler and K. D. Rendulic, *Surf. Sci.*, 1989, **220**, 1.
79. J. Harris, *Surf. Sci.*, 1989, **221**, 335.
80. U. Nielsen, D. Halstead, S. Holloway, and J. K. Nørskov, *J. Chem. Phys.*, 1990, **93**, 2859.

81. G. O. Sitz, A. C. Kummel, and R. N. Zare, *J. Vac. Sci. Technol.*, 1987, **A5**, 513.
82. R. C. Mowrey, Y. Sun, and D. J. Kouri, *J. Chem. Phys.*, 1989, **91**, 6519.
83. S. Holloway and B. Jackson, *Chem. Phys. Lett.*, 1990, **172**, 40.
84. P. K. Johansson, *Surf. Sci.*, 1981, **104**, 510.
85. This obtains for $0° < a < 120°$ for $A = 0.5$ and $0° < a < 78.5°$ for $A = 1.0$.
86. G. Haase, M. Asscher, and R. Kosloff, *J. Chem. Phys.*, 1989, **90**, 3346.
87. P. Nordlander and J. C. Tully, *Surf. Sci.*, 1989, **211/212**, 207.
88. S. Holloway, D. Halstead, and A. Hodgson, *J. Elect. Spec.*, 1987, **45**, 207.
89. H. F. Winters, *J. Chem. Phys.*, 1976, **64**, 3495.
90. C. T. Rettner, H. E. Pfnür, and D. J. Auerbach, *Phys. Rev. Lett.*, 1985, **54**, 2716.
91. J. T. Yates, J. J. Zinck, S. Sheard, and W. H. Weinberg, *J. Chem. Phys.*, 1979, **70**, 2266.
92. S. G. Brass, D. A. Reed, and G. Ehrlich, *J. Chem. Phys.*, 1979, **70**, 5244.
93. B. Jackson, *J. Chem. Phys.*, 1988, **89**, 2473.
94. R. J. Behm and C. R. Brundle, *J. Vac. Sci. Technol.*, 1984, **A2**, 1040.
95. G. D. Billing, (to be published).
96. J. K. Nørskov, *J. Chem. Phys.*, 1989, **90**, 7461.
97. J. E. Lennard-Jones, *Trans. Faraday Soc.*, 1932, **28**, 333.

CHAPTER 4

The Dynamics of Hydrogen Adsorption and Desorption on Copper Surfaces

B. E. HAYDEN

1 Introduction

It may appear unusual at first that amongst the more general titles which constitute this volume on the dynamics of gas surface collisions, a contribution is included which concerns itself exclusively with the dynamics of the adsorption and desorption of a particular system, namely hydrogen on copper. The reason that the dynamical behaviour of this particular adsorption system is worthy of close inspection is that it has become the *classic example of activated dissociative adsorption of a molecule at a surface*. In addition, the barrier to dissociation can be accessed directly from the gas phase and, conversely, the barrier directly influences the final state of the desorbing molecules following recombination at the surface. From the dynamical point of view, therefore, the important part of the potential leading to reaction (catalytic dissociation) is accessible both experimentally and theoretically. Adsorption systems which exhibit dissociation of a fully accommodated molecule (precursor), where the thermal bath provides the activation energy, are much less accessible dynamically. The result is that the hydrogen/copper system has been the subject of extensive experimental investigation and theoretical treatment spanning some 150 years, and has become a model system for the development of a realistic potential energy hypersurface for a reactive surface collision. It is also likely that, as in the past, the system will continue to be a springboard for the development of ideas concerning surface reaction dynamics.

It was the absence of hydrogen adsorption on polycrystalline copper[1] which gave rise to a suggestion of a kinetic activation barrier to dissociative adsorption of the molecule. Hydrogen (once dissociated) has a high solubility in the bulk lattice,[2] and the adsorbed hydrogen atom at the surface is energetically more favourable.[3] One method of circumventing the apparent barrier to dissociation and producing the atomically adsorbed state is to use a hot filament source[4,5] which pre-dissociates (or vibrationally excites) the molecule. The apparent barrier also manifests itself in the low sticking probability of the molecule on single crystal surfaces.[6] Isothermal data obtained on copper films[7] yielded an activation energy to dissociative adsorption of between 300 and 400 meV. Some insight into the nature of this barrier was revealed in permeation experiments[8] in which the angular distribution of desorbing hydrogen molecules following surface recombination on polycrystalline copper was measured. The angular distributions revealed a strong anisotropy, peaked towards the surface normal. This was interpreted in terms of an activation barrier in a one-dimensional potential[9] following the analysis of van Willigen.[10] Desorption from single crystal surfaces also exhibited peak angular distributions[11] which, following the same analysis,[10] yielded barrier heights of 250 meV for Cu(111), 210 meV for Cu(100), and 84 meV for Cu(110). If, indeed, the final translational state of the desorbing hydrogen was a result of such an activation barrier, microscopic reversibility would predict that the energy required for dissociative adsorption could be provided in the form of translational energy of the impinging molecule.

It was precisely such an experiment, establishing an angular and translational energy dependence of hydrogen dissociation on copper single crystals, which most clearly demonstrated the existence of an activation barrier in the entrance channel to adsorption. A series of detailed molecular beam scattering experiments[12] for the reaction of H_2 with adsorbed deuterium atoms on Cu(110) and Cu(100) demonstrated a normal kinetic energy ($E_\perp$) dependence of the initial sticking probability (S_0). A series of 'S' shaped curves in $S_0(E_\perp)$ were interpreted in terms of a distribution of one dimensional barriers.[9,10] The average value of the barriers obtained were 120 meV on Cu(110) and 200 meV on Cu(100). While these results appeared to be consistent with the desorption results[11] within the framework of the one-dimensional potential, the barriers were too low to account for the very low sticking probability of thermal molecular sources[6] and the insensitivity of the diffracted H_2 component to energy in this range.[13] A measurement of the translational energy of desorbing hydrogen indicated a narrow distribution at 600–700 meV on Cu(100).[14] This result indicated a barrier of energy significantly higher than that obtained in the beam experiments.[12]

The barrier predicted theoretically for the hydrogen/copper system was in any case high (when compared to the hydrogen/transition metal systems), and a reflection of the 'Pauli Repulsive' interaction.[15] Values range from 0.7 to 1.8 eV.[16-19] In addition, a measurement of the vibrational and rotational states of desorbing hydrogen from Cu(110) and Cu(111) indicated a hot vibrational distribution.[20] This was the first clear indication that the transition state to dissociation involved an extended intermolecular hydrogen bond. As a result the copper/hydrogen adsorption system has been modelled in a large number of trajectory calculations based on a two-dimensional potential energy hypersurface in which the inter-proton separation is the important new dimension.[21–26]

Most recently it has been shown in sticking measurements using molecular beams[27,28] that the nominal 'translational' energy requirement for overcoming the barrier is indeed rather higher than previously reported.[12] It has also been shown that vibrational energy is effective in promoting dissociation,[29,30] and the two-dimensional potential energy surfaces based on a relatively high barrier now appear consistent with the adsorption[29,30] and desorption measurements.[20] However there remains controversy over these measurements.[31] We remain far from a complete understanding of the dynamics of the hydrogen/copper adsorption system. The catalogue of measurements outlined above has as much to offer in indicating where we should be going both experimentally and theoretically in the study of this system, and in others exhibiting direct dissociation. I will endeavour in this chapter to describe and interconnect many of the ideas and measurements developed during the hydrogen/copper saga to date, highlight their strengths and limitations, and emphasize what they have taught us about the way to proceed in the future.

2 Adsorbed Hydrogen on Copper

Dissociatively Adsorbed Hydrogen

Bulk hydrides of copper are formed endothermically and include phases in which hydrogen is incorporated in octahedral or tetrahedral co-ordination. The heat of solution for hydrogen in copper is relatively large at 570 meV.[32] Study of the adsorption of molecular hydrogen onto copper surfaces has a long history.[33] Its importance lies in the use of copper (usually alloyed with other transition metals) in hydrogenation catalysis and methanol synthesis,[34] and in the field of energy generation.[35] The additional motivation is now its importance as a prototype for dissociative chemisorption. Adsorption studies revealed that at very low temperatures (*ca.* 77 K) hydrogen is not chemisorbed on copper, while at higher temperatures (180–350 K) it is dissociatively chemisorbed with an appreciable activation barrier for the process.[7,36–41] A value of 300–400 meV has been obtained from isothermal data on polycrystalline films,[7] although values as high as 800 meV have been reported.[36] The chemisorption of hydrogen has since been studied on single crystal surfaces using a variety of techniques including photoemission,[6,42] vibrational spectroscopy,[5,43] low energy electron diffraction (LEED),[4,5,41–43] and ion[44] and helium[4] scattering. The low sticking probability of the molecule has been overcome in such studies by pre-dissociation (or vibrational excitation) of the hydrogen using a hot tungsten filament, since the atom (or vibrationally hot molecule) has a high sticking probability. Angle resolved UV photoelectron spectroscopy (ARUPS) measurements have revealed both a strong perturbation on the copper d-band on Cu(111)[6], and a weak interaction on Cu(110) which was reported to undergo little structural relaxation during adsorption.[42] LEED and vibrational studies on Cu(111)[43] have been interpreted in terms of hydrogen in ordered overlayers adsorbed in two-fold sites. A similar study on Cu(110) indicated a surface reconstruction induced by hydrogen. The hydrogen atoms are co-ordinated in pseudo four-fold or three-fold sites depending on the surface phase.[5] Ion scattering on the same surface indicates a substantial hydrogen induced reconstruction[44] involving a missing row structure, and He scattering data has even given rise to the suggestion that a sub-surface reconstruction takes place on Cu(110).[4] The temperature at which recombinative desorption takes place for the close packed crystal faces lies in the range 300–350 K[6,27] but extraction of adsorption heats is complicated by the existence of the activation barrier. In the case of Cu(111) it appears that a reliable *activation barrier to desorption* can be extracted.[27] A value of *ca.* 850 meV is obtained, which only slightly decreases with increasing hydrogen coverage. It is worthwhile noting that in the series of detailed desorption studies[27] the Cu(111) and Cu(100) surfaces exhibit classical second-order behaviour, while the hydrogen desorption from Cu(110) appears to be first order. The origin of this difference probably lies in the surface reconstruction of the Cu(110) surface which takes place at the lowest hydrogen coverages.[5] The concerted lifting of the reconstruction and recombinative desorption may be responsible for the unusual desorption behaviour on Cu(110).[27] This finding is important from the dynamical point of view since it implies that the desorption–permeation experiments may involve desorbing molecules coupled to the lifting of the reconstruction. In contrast, the dissociative adsorption at low temperatures[27–29] probably takes place on an unreconstructed surface since rearrangement of the copper atoms in the surface is likely to be slow with respect to the time associated with hydrogen dissociation. It will be seen that dynamical arguments involving sub-surface phases have previously been made[14] to account for excess translational energy following desorption. While it is likely that the explanation for the high translational energy[14] lies in a high barrier associated with the surface layer (Section 4), one should keep in mind that arguments involving detailed balance, microscopic reversibility, and comparison of measurements at different surface temperatures require careful consideration in circumstances where surface reconstruction takes place.

Effective medium theory predicts that the surface and sub-surface sites for hydrogen are considerably more stable than in the bulk, and this is connected with substantial relaxation of the copper lattice.[3] On Cu(110) the surface site is about 300 meV more stable than the most stable sub-surface site. *Ab initio* calculations[19] on a thirty-eight atom cluster also indicate that dissociated hydrogen binds exothermically to Cu(100) in several surface bonding sites with binding energies in the range 550–925 meV.

The Apparent Activation Barrier to Dissociation

As far as the atomically adsorbed ground state is concerned, there is both an experimental and theoretical consensus that the chemisorption on copper is rather stable thermodynamically and may give rise to surface reconstruction. What of the potential away from this state that will be important in the dissociative adsorption? One of the simplifications as far as the dynamics is concerned is that there is *no evidence of an intermediate molecular chemisorbed state of hydrogen on copper* which may be a precursor to the dissociative state. In cases where the precursor to dissociation is a stable chemisorbed molecule, energy for activated dissociation can be supplied from the thermal bath of the surface, and in such cases it appears difficult to access the barrier directly from the gas phase.[45] The effect of increasing translational energy in the impinging molecule in such cases will be simply to reduce the probability of trapping in the precursor (accommodation becomes more difficult and inelastic scattering more probable) and hence the dissociative sticking probability S_0 is reduced.[46] The formation of a less well bound molecular intermediate, which can also involve some charge transfer, may give rise to a resonant intermediate to dissociative adsorption. This leads to a more difficult dynamical problem since some scrambling of energy partitioned in the impinging gas-phase molecule can take place before the barrier is accessed. Such systems are usually characterized by dissociation probabilities dependent on energy partitioned in both the gas-phase molecule and the surface.[47-50] In addition, the gas-phase translational energy dependence appears to scale with total rather than normal energy. In the case of hydrogen on copper, only a weakly bound physisorption state exists, and there is no evidence for a precursor mechanism to dissociation. It is this weak interaction potential that the molecule experiences first on approach towards the surface prior to dissociation (*i.e.* in the entrance channel). It happens that this state has been extensively characterized in molecular sticking experiments at low temperatures[51,52] and has been the subject of theoretical treatment.[53,54] Hydrogen adsorbs molecularly at surface temperatures $T_s < 15$ K. The molecule exhibits resonant sticking associated with both coupled rotational excitation,[51] and via a resonant process involving quasi-bound states in the physisorption well.[52] Both measurements have allowed a detailed potential for the physisorbed state to be deduced, both in the direction of the surface normal (z) and the intermolecular bond length (r) which appears to be the same as the gas phase molecule. The bound states supported by the H_2/Cu potential have also been measured as selective adsorption resonances in diffraction experiments.[55] The potential in the z direction seems to be well fitted by a 9–3 potential with a well depth of 22 meV with $z = 2.8$ Å. Time-of-flight measurements of rotational Feshbach resonances[56] have been used to estimate a well depth of 22.2 meV at $z = 2.71$ Å.

The transition from the physisorption to the dissociative chemisorption has also been the subject of theoretical treatment because of the apparent

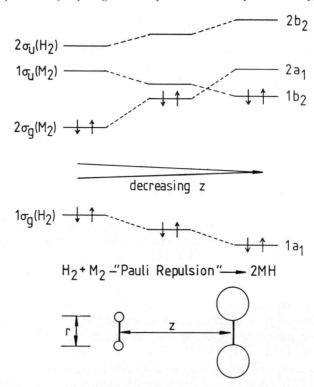

Figure 1 *The orbital energy level diagram for the cluster configuration[15] illustrating a symmetry forbidden chemical reaction and the origin of the activation barrier to dissociation through 'Pauli Repulsion'. The subscripts* u *and* g *refer to even and odd orbitals with respect to the symmetry plane. A calculation of this cluster is used to develop the 2D-PES (Figure 14). The switch in electronic configuration corresponds to the 'seam' in the PES*

activation barrier. The *barrier height itself is significantly higher on copper than on the transition metals*, and the origin of the barrier on the noble metals can be understood in terms of Pauli Repulsion between two closed shell systems.[15,23] Model calculations on Jellium surfaces[16,57] demonstrate that dissociation results from the partial filling of the $H_2(1\sigma_u)$ antibonding orbital. The interaction between the metal and hydrogen levels can be understood in more detail by considering the orbital hybridization in a copper cluster,[15,23] shown schematically in Figure 1. As the molecule approaches the 'surface' (a co-linear approach of H_2 to the Cu_2 dimer), the antibonding resonance* broadens and shifts downwards because of its interaction with the copper levels. However, this takes place at distances

---

* As the molecule approaches the surface, there is a strong mixing of the $M_2(\sigma_u)$ and $H_2(\sigma_u)$ levels. It therefore has the character of both metal and hydrogen antibonding orbitals. This resonance is the orbital moving down in energy and being occupied (Figure 1). This is equivalent to the partial filling of the $H_2(\sigma_u)$ level in the Jellium[16] picture. A further discussion can be found elsewhere.[17]

$z < 3$ a.u., and prior to this the interaction displays Pauli Repulsion characteristic of the closed shells. This results in an increase in the potential energy in the entrance channel which can vary from *ca.* 0.2 eV for H_2 on Na to more than 1 eV on Al,[16] and even larger on the noble metals.[17-19] In the case of hydrogen on copper, cluster calculations[15,17-19] yield a barrier height in the region 1.3 eV while the Jellium calculations[16] place it somewhat lower at about 700 meV. A two-dimensional PES has been generated (see Figure 14) on the basis of a Kohn–Sham local density calculation of the hydrogen Cu_2 dimer interaction.[17] The situation on the noble metals contrasts with the transition metals where hydrogen can dissociate with no apparent activation barrier. It is suggested[15] that the important function of the *d*-band in such systems is to act as 'sinks' for the *s* electrons, the metal *s* electrons reverting to the *d*-band rather than penetrating the $H_2(\sigma_g)$ core.

It is the potential energy hyper-surfaces generated from the information outlined above on which both classical and quantum mechanical trajectory calculations are based. The multi-dimensionality of the system clearly provides a major challenge, but significant progress can be made by restricting the dimensions to important degrees of freedom highlighted by the calculations of the potentials or by the results of experiment.

3 Angular Distributions in Desorption

Non-equilibrium Distributions

Molecules desorbing from a surface T_s have generally been regarded as being in quasi-equilibrium at the point of desorption,[58] and the desorbing flux is expected to be characterized by equilibrated energy partition; a Boltzmann energy distribution of internal states, and a Maxwellian velocity distribution. The angular distribution for such a desorbing flux is expected to be diffuse (or Knudsen) and follow a $\cos \phi$ function, where ϕ is the angle of inspection measured from the surface normal. There is, however, now a considerable body of experimental evidence which shows that energy distributions in desorbing fluxes generally deviate from such energy partitioning,[59] and simple arguments show that even under conditions of thermodynamic equilibrium these laws need not, and in general should not, be followed.[60] The flux of desorbing species can be strongly peaked towards the surface normal in distributions typically described by $\cos^n \phi$ where, for example, $n = 8$ for hydrogen on copper.[14] The desorbing flux of molecules can be translationally [14] or vibrationally[20] hot, rotationally cool,[20,61] and even exhibit a preferred molecular orientation,[62] or have an asymmetric rotational alignment.[63] None of these features comfortably fit into the quasi-equilibrium description. Rather, *state resolved experiments of desorbing molecules provide a dynamical probe of the potential energy surface in the region of the desorption transition state.*

In the case where two atoms combine at the surface to associatively desorb, the desorbing molecule retains information pertaining to its final interactions with the surface at the point at which the new molecular bond was formed. A simple picture based on transition state concepts is reasonable for the Cu–H_2 system[56] because the potential energy surface (PES) divides naturally into product and reactant regions. These are separated by a 'seam' in a two-dimensional PES (2D-PES) at which it is reasonable to picture a thermal distribution of desorbing trajectories. The seam passes through the saddle point or col of the transition state, and the energy falls off rapidly on either side of the seam (Figure 14). For the one-dimensional potential[9] which will be described first, the equivalent of the seam is the point on the 1D-PES associated with the activated transition state.

Permeation and Desorption

In the study of desorption, there are two ways to supply molecules to the surface prior to desorption; by permeation from the rear of the sample, or by adsorption from the gas phase. The latter method usually involves adsorption followed by temperature programmed desorption (TPD) which, because of the low fluxes, is repeated until the required statistics are obtained. Alternatively, a molecular beam can be used to supply the adsorbate, but one must ensure that the detected flux is due to desorption rather than, for example, elastic or diffuse scattering of the impinging beam (see below). The supply by permeation radically simplifies the experimental problems[59] since the desorbing flux can be high and continuous (unlike TPD), and one can be sure that all molecules observed are desorbing from the surface. Perhaps the major complication has been the question of whether the desorbing species desorbs with energy partitioned according to some surface PES, or a PES associated with the bulk material.[14,59,64]

Supply by permeation is limited, however, to a small number of systems. It can, nevertheless, be achieved for hydrogen through most metals, and was applied first by van Willigen[10] in a large number of measurements. Detection of the desorbing flux is usually achieved using an ion gauge or mass spectrometer because of the simplicity and sensitivity of these techniques. The position of the detector with respect to the surface normal is easily varied, and the detection can be combined with TOF techniques to obtain velocity distributons. The large fluxes have also allowed resonance enhanced multi-photon ionization schemes to be used to obtain rotational and vibrational state distributions.

Early studies[8,10,65-67] had shown that the angular distributions of hydrogen molecules desorbed from polycrystalline metal surfaces exhibited strongly peaked $\cos^n\phi$ distributions. Generally such peaked distributions were obtained from surfaces contaminated with adsorbed impurities such as carbon or sulphur,[8,66] and n was found to approach unity for the clean surfaces. *Copper appeared to be an exceptional case, however, since a non-diffuse distribution had been observed from the clean polycrystalline surface.*[8] This observation led to a more detailed study of the angular distributions of desorbing molecules from the low index planes of copper.[11] All of the clean

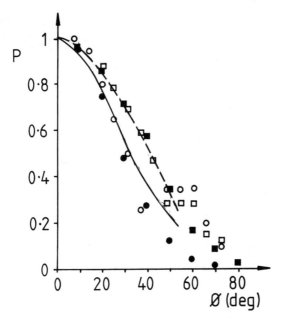

Figure 2 *The angular distributions of* HD *(open points) desorbing following the* $H_2 + D_{ads}$ *reaction on* Cu(110) *(□) and* Cu(100) *(○).*[12] $T_s = 850$ K. *Desorption of* H_2 *following permeation (filled points) with* $T_s = 1100$ K[11] *are also included for these crystal faces. The curves are the results of the detailed balance treatment*[68] *based on the* $S_0(E_\perp)$ *dependence*[12] *for* Cu(100) *(solid) and* Cu(110) *(dashed)*

copper surfaces studied exhibited non-diffuse distributions in the desorbing flux. With $T_s = 1100$ K, they were characterized by $\cos^n \phi$ functions[11] with $n = 2.5$ for Cu(110), $n = 5$ for Cu(100), and $n = 6$ for Cu(111). The results for Cu(110) and Cu(100) are included in Figure 2.

A One-dimensional Potential

The peaked distributions were interpreted[11] as arising directly from an activation barrier for the microscopically reversed adsorption process, following the analysis of van Willigen.[10] This model is based on the suggestion of Lennard-Jones[9] that the molecular and atomic potential cross above the asymptotic dissociation energy of the gas phase H_2 molecule, the one-dimensional potential thereby leading to an activation barrier for dissociative adsorption. A schematic of the one-dimensional potential is shown in Figure 3. The van Willigen model assumes that the adsorbed molecules are in thermal equilibrium with the surface, and the activation barrier is single valued over the entire surface. This equilibrium condition allows the angular distribution of desorbing molecules to be calculated from the angular dependent dissociative sticking probability using the principle of detailed balance. The model also assumes that the sticking probability is unity for

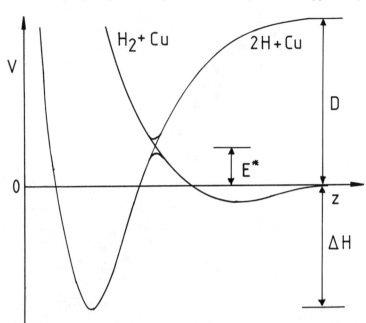

Figure 3 *A schematic of the one-dimensional potential* [V(R)] *following the suggestion of Lennard-Jones.*[9] *The crossing of the molecular and dissociative potentials above the energy zero gives rise to the activation barrier* E*. D *is the energy of dissociation for* H_2, *and* ΔH *is the energy of dissociative chemisorption*

molecules with sufficient translational energy to adsorb. The resultant velocity distribution of desorbing molecules is then predicted to be Boltzmann but with an additional velocity component (v) of all particles in the direction of the surface normal derived from the energy of the barrier:

$$v = (2E^*/M)^{1/2} \qquad (1)$$

where M is the molecular mass and E^* is the barrier height. The angular distribution of emitted particles is given by:[10]

$$N(\phi) = N_0[(E^* + kT_s \cos^2 \phi)/((E^* + kT_s) \cos \phi)] \exp(-E^* \tan^2 \phi/kT_s) \qquad (2)$$

The barriers required to explain the permeation results were calculated using equation (2) to be $E^* = 250$ meV for Cu(111), $E^* = 210$ meV for Cu(100), and $E^* = 84$ meV for Cu(110). It was clear that the activation barrier was sensitive to surface structure, with the more open packed face exhibiting the lowest barrier.

This unique desorption behaviour of hydrogen from clean copper[11] motivated the first molecular beam study[12] designed to probe the barrier directly during dissociative adsorption (Section 4). The same study also enabled an additional measurement of the angular dependence of the desorbing flux from Cu(110), Cu(100), and Cu(310). In this case the molecules were supplied to the surface from the gas phase in the form of a modulated molecular beam of hydrogen. In order to ensure that the detected molecules were exclusively from the associative desorption process (rather than the total scattered flux) the desorption of HD was monitored following the reaction of hydrogen with pre-adsorbed D atoms. This experiment allowed the measurement of the angular distribution over a slightly wider range of T_s than accessible in the permeation experiment.[11] The distributions for Cu(110) and Cu(100)[12] are also shown in Figure 2 and were the same as those found following permeation. The van Willigen model, however, predicts a dependence of the angular distribution of the desorbing flux on surface temperature T_s [equation (2)]: Such an effect was not observed. A possible explanation given at the time for this was that surface roughness causes additional broadening which obscures the predicted dependence. It turns out that a *distribution of one-dimensional barriers can also mask this dependence.*

4 Translational Energy Partitioning

Dissociation Promoted by Translational Energy

With a view to probing the activation barrier to adsorption by studying the adsorption process itself (at least indirectly), a series of detailed molecular beam experiments[12] were made on Cu(100), Cu(110), and Cu(310). The dissociative sticking probability was not measured directly, but inferred from the probability of HD production in the Langmuir–Hinshelwood reaction of molecular hydrogen (supplied from a supersonic nozzle source) with a constant (low) coverage of adsorbed deuterium atoms. The reaction rate was assumed to be limited by the dissociative sticking probability of the molecular hydrogen (S_0). In this case the HD product flux was proportional to S_0, and was measured as a function of the translational energy (E) of the incident H_2 flux (in the range 67–500 meV) and incident angle (ϕ) with respect to the surface normal. The translational energy of the beam was increased by heating the nozzle to high temperatures. Since the detector was free to rotate in the plane of scattering, the angular distributions of the desorbing HD as well as the scattered H_2 beam could be measured.

Arguably the most important results from these measurements were a *series of 'S' shape curves for the sticking probability as a function of the normal component of the incident translational energy* $E_\perp = E \cos^2 \phi$. These curves for Cu(110) and Cu(100) are shown in Figure 4, and clearly demonstrate that there is a barrier to dissociative adsorption which is overcome with translational energy in the impinging hydrogen molecule. Further, it is the *normal component of energy which is effective in promoting dissociation*. The component of energy normal to the surface is given by $E_\perp = E \cos^2 \phi$ where E is the translational energy in the beam. Scattering behaviour which scales with $E_\perp$ rather than E is said to exhibit normal energy scaling.

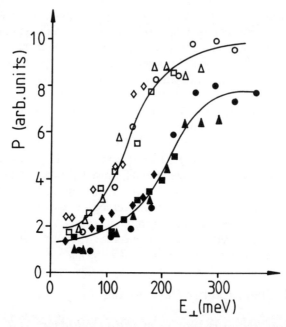

Figure 4 *The probability* P *for the reaction of* H_2 *with adsorbed* D *atoms to produce HD on Cu(110) and Cu(100)*[12] *as a function* $E_\perp = E \cos^2 \phi$. *Measurements were made with* $\phi = 25°$ (○), $\phi = 35°$ (△), $\phi = 45°$ (□) *and* $\phi = 50°$ (◇). $T_s = 850$ K

These results could again be interpreted in terms of the one-dimensional potential. However, rather than invoking the existence of a single barrier, which would result in a rather sharp translational onset of sticking, the broad onset observed required the existence of a *distribution of barriers with a spread of energies E**. The average activation barrier was obtained from the steepest part of the sticking curve (Figure 4) and estimated to be $E^* = 126$ meV for Cu(110) and $E^* = 210$ meV for Cu(100). This was in rather good agreement with the values obtained from the angular distributions of the desorbing molecules where the principles of detailed balance were assumed (Section 3). The adsorption and desorption results were compared in more detail[68] using the principles of detailed balance. Rather than rely on the single valued barrier [equations (2), and (5)], a distribution of barriers was assumed, a distribution given by the sticking measurements themselves. The proportion of molecules which will go on to adsorb from a Maxwellian distribution $f(v, \phi)$ is obtained by weighting the mono-energetic sticking probabilities $S_0(v, \phi)$ (obtained[12] from $S_0(E_\perp)$ in Figure 4) at each angle ϕ with the Maxwell–Boltzmann distribution $f_M(v)$ at T_s:

$$f(v, \phi) = f_M(v) S_0(v, \phi) \quad (3)$$

An example of these functions is plotted in Figure 5 for Cu(100).[11] The equilibrium rate of adsorption per unit area (R_{ads}) is obtained by integrating

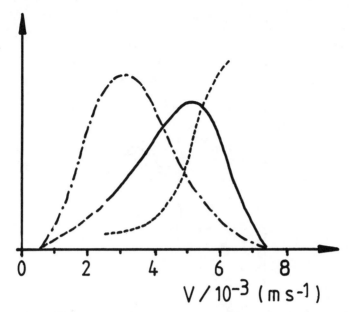

Figure 5 Schematic[68] to illustrate the folding of the mono-energetic sticking probability $S_0(v, \phi)$ (-----) with a Maxwellian $f_M(v)$ at T_s (-·--·--·--) to obtain the proportion of molecules $f(v, \phi)$ (———) which go on to adsorb [equation (3)]. This is related to the desorbing flux at T_s within the detailed balance framework

$f(v, \phi)$ over velocity and weighting by $\cos \phi$. (The magnitude of flux impinging on a unit surface area is proportional to $\cos \phi$). This gives:

$$R_{\text{ads}}(\phi) = \cos \phi \int v f_M(v) S_0(v, \phi) \mathrm{d}v \qquad (4)$$

The principle of detailed balance requires that $R_{\text{ads}}(\phi) = R_{\text{des}}(\phi)$ for equal velocity and solid angle increments. The predicted desorption distributions based on such a folding of the sticking data[12] are shown in Figure 2 together with the desorption results. Not only is the angular distribution in good agreement, the introduction of the distribution of one-dimensional barriers also predicts the observed independence of $R_{\text{des}}(\phi)$ over the range of T_s studied.[12]

While the desorption[1] and sticking[12] results appeared to be consistent, there were, however, signs from the beam study that the one-dimensional potential was an oversimplification. A measurement was made of the sticking of D_2 and H_2 on Cu(110), and the results are shown in Figure 6. The data indicate that for any particular normal translational energy, S_0 for D_2 was higher than H_2 by a factor of 1.3–1.4. It was pointed out at the time that this ratio was equal to the square root of the relative molecular masses and was perhaps associated with the time the molecule spent near to the surface

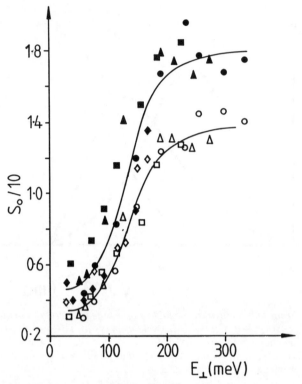

Figure 6 *The dissociative adsorption probability S_0 of H_2 (open points) and D_2 (filled points) extracted from the reaction data.*[12] $T_s = 850$ K. *Measurements were made with* $\phi = 25°$ (○), $\phi = 35°$ (△), $\phi = 45°$ (□) *and* $\phi = 50°$ (◇). $T_s = 650$ K.

where dissociation could take place (also proportional to this value). Another possibility was the density of available adatom vibrational states, which also varies with the square root of the mass, although no suggestion of how this particular vibrational co-ordinate could be involved in the reaction path was made. An alternative explanation for the observed isotope effect will be discussed in Section 5 in the light of more recent sticking measurements.

Excess Translational Energy following Recombination

The limitations of the one-dimensional potential were highlighted in a permeation experiment in which not only were the angular distributions measured, but also the distribution of translational energies for the desorbing molecules using time-of-flight (TOF) techniques.[14] The most probable value of the *translational energy* E of the molecule expected using the van Willigen[10] model should be given by:[69]

$$E(\phi) = 2kT_s + kT_s\{E^{*2}/(E^* + kT_s \cos^2 \phi)kT_s \cos^2 \phi\} \qquad (5)$$

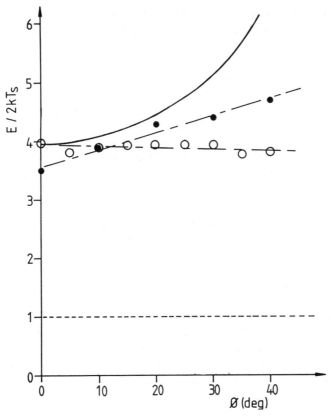

Figure 7 *The translational energy (E) of D_2 desorbing from Cu(100) at T_s = 1000 K as a function of angle (ϕ) measured following permeation[14] (open points). The fine dashed curve would be the prediction for no activation barrier in adsorption. The solid line is the prediction of $E(\phi)$ based on a single valued barrier in a one-dimensional potential [equation (5)]. The calculation of $E(\phi)$ within a detailed balance framework from the sticking data of H_2 on Cu(100)[27] is included in the filled points*

The measured angular distributions were somewhat sharper ($\cos^8 \phi$) than observed previously, but more striking was the measurement of a very narrow energy distribution of desorbing molecules with a mean energy of 700 meV. The one-dimensional barrier which would be required to yield such a distribution would be $E^* = 600$ meV [equation (5)], and the corresponding angular distribution would be $\cos^{13} \phi$ [equation (2)]. Adhering to the one-dimensional model,[14] the higher apparent barrier observed in desorption[14] than accessed during adsorption[12] was explained by assuming that the permeating atoms recombined in a sub-surface site. Desorption then occurred over a 'bulk' one-dimensional barrier (which was suggested to be higher than that of the surface) without any equilibration in the surface layer. A similar model was used to explain the fast component of desorbing hydrogen from palladium.[59] Support for this picture was drawn from the observation that the measured angular and translational distributions were insensitive to the differences in surface geometry between Cu(111) and Cu(100). More recent sticking measurements,[27,28] however, no longer make it necessary to invoke the sub-surface argument, even adhering to the one-dimensional picture. An additional breakdown of the one-dimensional potential, however, was the *observed independence of translational energy with desorption angle*,[14] not predicted by equation (5). The observed[14] and predicted [equation (5)] velocity dependence is shown in Figure 7. No satisfactory explanation was given at the time, although more recently such behaviour has been shown to be explained once again in terms of a distribution of one-dimensional barriers[27] or alternatively within the framework of a 2D-PES.[70]

Direct Measurements of Dissociative Sticking

The barrier assigned to the 1D-PES[11,12] was clearly too small to reconcile the later permeation results.[14] It would also predict a room temperature sticking probability of *ca.* 10^{-4} which was clearly not observed experimentally from thermal sources of molecular hydrogen.[6] Hydrogen diffracted at high translational energy also gave no indication of a dissociation barrier around 126 meV on Cu(110),[13] with scattering dominated by the repulsive wall of the entire unit cell over the energy range studied (63–280 meV). The reaction of hydrogen (as a function of $E_\perp$) with adsorbed oxygen on Cu(110) to produce water[71] was similar to the early beam study[12] in that it provided an indirect measurement of the hydrogen dissociation probability on copper. The rate of water production as a function of $E_\perp$ from a constant coverage

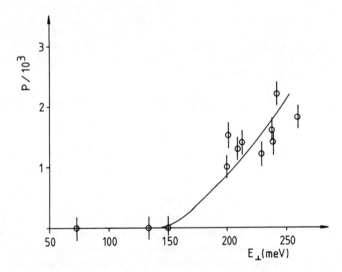

Figure 8 *The probability of* H_2 *reacting with oxygen at constant coverage* (0.5 ML) *to produce water on* Cu(110) *as a function of* $E_\perp$.[71] $T_s = 650$ K

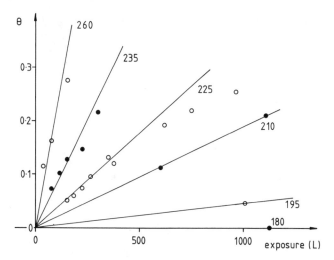

Figure 9 *The coverage of adsorbed H_2 as a function of beam exposure for a series of $E_\perp$.*[28] $T_s = 140$ K

of oxygen is shown in Figure 8. The results were obtained with the beam impinging normal to the surface ($\phi = 0$). The reaction is shown to proceed following Langmuir–Hinshelwood kinetics and is limited by the rate of hydrogen dissociation on the clean (oxygen free) areas of Cu(110) surface. The reaction probability (Figure 8) provides an upper limit to S_0 on the oxygen free part of the surface since there is always a finite probability for hydrogen recombination. Comparison of the results of Figure 8 with Figure 4 clearly indicates a different translational energy requirement for dissociative adsorption. No reaction at $E_\perp < 150$ meV was observed, and the maximum reaction probability for producing water per incident hydrogen molecule (the upper limit to the dissociative sticking probability in a Langmuir–Hinshelwood mechanism) was 2×10^{-3} at $E_\perp = 240$ meV. This was therefore another indication that the dissociation barrier was indeed higher than previously believed.

What was required was a direct measurement of S_0 as a function of $E_\perp$ rather than relying on a more indirect measure through a surface reaction.[12,71] This was achieved[27,28] by cooling the sample to low temperature, exposing the surface to a supersonic beam source of hydrogen for finite time, and subsequent thermal desorption of the fraction sticking to the surface. This is an alternative to the method introduced by King and Wells[72] which is more suitable for systems exhibiting high sticking probabilities. Hydrogen coverage as a function of exposure time allows extrapolation of an initial sticking probability, and absolute values obtained by normalizing the desorption curves to that of a saturated monolayer. Curves obtained in this way as a function of translational energy are shown in Figure 9. The results from the two independent groups[27,28] for Cu(110) were in agreement, and showed an onset for dissociative sticking at $E_\perp = 180$ meV, and a value of

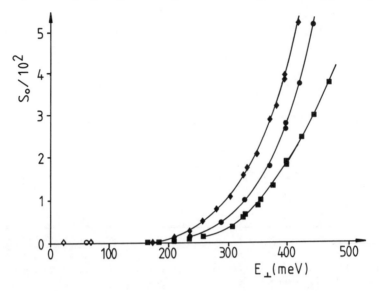

Figure 10 *The initial sticking probability S_0 of H_2 as a function of $E_\perp$ on Cu(111) (◆), Cu(100) (●) and Cu(110) (■).*[27] $T_s = 100$ K

$S_0 = 2 \times 10^{-3}$ at 250 meV. This was in good agreement with the observation of a dissociation limited water reaction (Figure 8). Rather extensive sticking measurements on the three low index planes of copper were carried out by one of the groups,[27] and the results are shown in Figure 10. The *apparent barrier for Cu(110) and Cu(100) is significantly higher than obtained previously*. The functional shape of $S_0(E_\perp)$ is now exponential rather than 'S' shaped, although presumably at even greater energies the curve would level off, even within the framework of a distribution of one-dimensional barriers.[12,27] Extraction of an average barrier[12] directly from the maximum in $dS_0/dE_\perp$ is difficult, but for Cu(110) $E^* > 450$ meV. The results of Figure 10 also highlight an important difference with the previous beam study[12] in the *order of the apparent barrier heights for the different crystal faces* (Figures 4 and 10), with the highest packed surface now yielding the lowest barrier, and the more open Cu(110) face requiring the highest translational energy for dissociation. There was also[27] reported to be *no observable isotope effect on $S_0(E_\perp)$*. Rather than explore the differences between this and the previous[12] beam study in more detail, an extensive analysis using detailed balance was carried out[27] to compare these new results with the velocity distribution in the permeation experiment.[12] Assuming a single one-dimensional barrier, the value $E^* = 600$ meV obtained (Section 4) from the desorption result,[12] is clearly more in line with the new sticking data.[27,28] Assuming a distribution of one-dimensional barriers, the authors[28] make use of the sticking curves themselves to predict the sticking of a Maxwellian source, and through detailed balance relate this to the desorption at $T_s = 1000$ K. The same approach (Section 3) was used to compare previous data.[11,12] An important consequence of the treatment was that not only is the velocity of desorbing molecules at $\phi = 0$ predicted, but as a consequence of the barrier distribution, the sharp angular dependence $E(\phi)$ predicted by the single one-dimensional barrier [equation (5)] is relaxed, and a slower increase of the function is predicted. This result is shown together with the data in Figure 7.

5 Vibrational and Rotational Energy Partitioning

Hot Vibrational and Cold Rotational Distributions

Although van Willigen[10] did not treat the intramolecular degrees of freedom, the hypothesis of thermal equilibrium of molecules prior to desorption implies that the rotational and vibrational populations are those of a Boltzmann gas at the surface temperature:

$$N_J = N_0(2J + 1)\exp(-E_J/kT_s) \quad (6)$$

$$N_v = N_0 \exp(-E_v/kT_s) \quad (7)$$

where E_J and E_v are the rotational and vibrational level energies. The non equilibrium behaviour in the translational co-ordinate outlined above prompted a detailed study of the quantum state distributions in the remaining degrees of freedom of the desorbing hydrogen. Using the technique of permeation, the ro-vibrational distributions of H_2 and D_2 recombinatively desorbing from Cu(110) and Cu(111) were measured.[20] This was achieved using the technique of resonance-enhanced multi-photon ionization (REMPI) combined with TOF mass spectrometry. The rotational distributions were found to be non-Boltzmann [equation (6)] with mean rotational energies of 80–90% of T_s. The rotational temperatures were also found to be the same for H_2 and D_2, and for the desorption from Cu(110) and Cu(111). In addition, the *ortho* and *para* nuclear spin modifications of both H_2 and D_2 were statistically populated. The implication was that rotational energy does not play an important role in the desorption and thus, by microscopic reversibility, in the adsorption process.

The result for the rotational distributions was in sharp contrast to the observed vibrational populations[20] which are summarized in Table 1. The ratio $N(v = 1)/N(v = 0)$ was found to be as much as one hundred times greater than that predicted by equation (7). This important result was the first clear indication that the hydrogen vibrational co-ordinate was important in the dynamics of hydrogen desorption, and presumably in adsorption. The transition state, however, could not be that assumed in the one-dimensional potential, but had to involve some extension of the H—H bond. A detailed balance comparison[20] with the early sticking data[12] assumed that the forward and reverse rates were equal ($R_{des} = R_{ads}$), but now at a perturbed transition state. Within such a framework, the result that, for example, $N(v = 1)$ is about fifty times greater than for $N(v = 0)$ during desorption at

Table 1 *The measured ratio of vibrationally hot molecules following permeation[20] together with a detailed balance prediction[68] based on the sticking measurements.[29]*

Surface	Molecule	$N(v=1)/N(v=0)$	
		observed[20]	predicted[68]
Cu(110)	H_2	0.052	0.025–0.28
Cu(110)	D_2	0.24	0.07–0.31
Cu(111)	H_2	0.084	0.033
Cu(111)	D_2	0.35	0.068

equilibrium at T_s [for H_2 on Cu(110)], implies that $S_0(v=1)$ should similarly be enhanced over $S_0(v=0)$ in the adsorption process. This assumption neglects possible correlations between vibration, ϕ, and $E_\perp$ which would be incorporated in the trajectory calculations and detailed balance treatment that followed. Assuming a Boltzmann population of vibrational states at the nozzle temperatures T_n used in the beam experiments,[12] the data obtained at $\phi = 50°$ could not be predicted from the data at $\phi = 25°$ assuming such an enhanced sticking for $H_2(v=1)$. This led to the conclusion[20] that translational and vibrational dynamics for adsorption and desorption were dissimilar. In the event the authors resorted to the model of Comsa and David[14] involving a bulk desorption process (Section 4) despite the fact that the vibrational,[20] unlike the translational,[14] results were different for the two faces studied. A mechanism involving temporary negative ion formation[73] was suggested to account for the excess vibrational energy. A recombinatively formed hydrogen begins to leave the surface on an H_2^- potential and at some point loses an electron through resonant tunnelling to one of the unoccupied copper levels above the Fermi level. The relative positions of the H_2^- and H_2 potentials determine the extent of vibrational excitation on the neutral molecule. The additional electron density in the $H_2(1\sigma_u)$ of H_2^- would lead to the required bond perturbation. There is no evidence experimentally or theoretically, however, for the existence of a negative ion state for molecular hydrogen on copper [Section (2)], and it will be seen that it is unnecessary to invoke such a resonance state to account for vibrational excitation.

Vibrationally Enhanced Dissociation

It has already been pointed out that in the light of the more recent sticking data[27,28] (which indicated a higher barrier than previously believed[12]), the adsorption and desorption[14] data could be reconciled within a detailed balance framework for translation. There appeared to be no direct evidence for a vibrational channel to dissociative adsorption.[28] This was indeed at odds with the desorption results.[20] In addition it was pointed out that the dissociative sticking observed[27] could be accounted for by the statistically

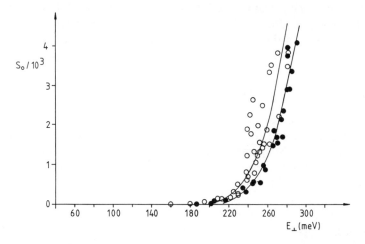

Figure 11 *The initial sticking probability* S_0 *of* H_2 (○) *and* D_2 (●) *in pure beams as a function of* $E_\perp$ *on* Cu(110).[30] $T_s = 140$ K

significant populations of vibrationally hot molecules in the beams.[27] This was because the increased translational energy was achieved by heating the nozzle. The role of vibrational energy was discounted[27] on the basis that there was no measurable isotope effect. The desorption result[20] had spawned a number of classical and quantum mechanical trajectory calculations based on a late barrier in a 2D-PES (Section 5) which accounted for the vibrational excitation in desorption, and predicted a vibrational coupling in adsorption. In particular the results of such calculations demonstrate how the translational requirement for dissociative adsorption can be significantly lower than the theoretical barrier height since vibrational energy conversion can take place over a barrier if it resides somewhat in the exit channel. A consequence of this is that the translational energy requirement for e.g. $H_2(v = 1)$ is significantly lower than for $H_2(v = 0)$. Similarly, the sticking of H_2 should be enhanced over D_2 at any particular normal translational energy $E_\perp$ for any particular vibrational state (including the ground state) because of the difference in vibrational quanta. No sticking result had revealed the predicted isotope effect. Indeed the opposite was observed (Figure 6) on Cu(110)!

More recent sticking measurements[30] on Cu(110) clearly reveal a difference in sticking of the two isotopes for any translational energy (or T_s). The results are shown in Figure 11 and reveal an enhanced sticking of H_2 over D_2 at any $E_\perp$. Such a result can be clearly understood within the framework of the 2D-PES. The inability to observe such an effect previously[28] led to the suggestion[29] that the smeared distributions of the different quantum states excited in the hot nozzle may screen the isotope effect. This problem makes interpretation of the pure beam experiments difficult, particularly since it appears that changing the nozzle heating method influenced the sticking measurement.[30] This result itself was taken as an indication that vibrationally partitioned energy in the beam was influencing the dissociation. Taken together with the isotope effect (Figure 11), it indeed appeared that vibrational energy could be channelled into dissociative adsorption.

Vibrational and Translational Coupling

The pure beam experiment clearly has the disadvantage that both translational and vibrational energy are imparted to the molecules by heating the nozzle. Energy is also partitioned to some extent into rotation, but the desorption result indicated that this will be rather ineffective in promoting dissociation. In an attempt to separate the translational and vibrational contributions to sticking, a series of cold seeding experiments was carried out[29,30] for beams at constant T_N. In this way, the vibrational state populations of hydrogen and deuterium could be kept constant, and the translational energy and incident angle ϕ varied. The results for such experiments carried out for H_2 and D_2 are shown in Figure 12 and 13 which also include the pure beam sticking curves. These results are obtained with the beam impinging along the surface normal. In the case of H_2, the beam has been cold-seeded (or anti-seeded) with either Ne or He, and in the case of D_2, Ne has been used. The translational energy was measured during seeding using TOF techniques. A set of points at constant T_N (vibrational populations held constant) shows that S_0 starts off on the pure beam curve as expected: As $E_\perp$ is reduced by seeding, the function $[S_0(E_\perp)]_v$ (constant vibrational population) does not follow the pure beam curve. Co-incidence of the curves would have been expected if the primary effect of heating the nozzle in the pure beam experiment was simply to increase $E_\perp$ and in so doing promote the dissociation with purely translational energy. In contrast, only a very slow decrease in S_0 is observed when $E_\perp$ is changed independently. This behaviour was interpreted[29,30] in terms of vibrational energy, in addition to translational energy, at the higher T_N which enables the molecules to overcome the apparent activation barrier to dissociation on Cu(110). In fact, in the case of H_2, it was concluded that the sticking was dominated by the dissociation of $H_2(v = 1)$ in the beam. Assuming a Boltzmann population of $H_2(v = 1)$ in the beam at T_N, there was always a sufficient population of the vibrationally excited state to account for the sticking. The result was the first clear indication that vibrational energy could be channelled into overcoming the activation barrier to H_2 dissociation on copper.[29] The functional difference between $S_0(E_\perp)$ for the pure beam and the seeded curve $[S_0(E_\perp)_v]$ for D_2 (Figure 13) is further evidence for the role of vibrational energy.[30]

The cold seeding curves (Figures 12 and 13) also reveal that there is a *translational requirement for the sticking of the vibrationally excited states.* Indeed, the sharp decay in S_0 (130 < $E_\perp$(meV) < 180) for H_2 for all curves at various T_N was associated with the *translational onset for the sticking of $H_2(v = 1)$*, an analysis made within the framework of the 2D-PES. It can be seen[74] that such an analysis predicts that the translational onset for $D_2(v = 1)$

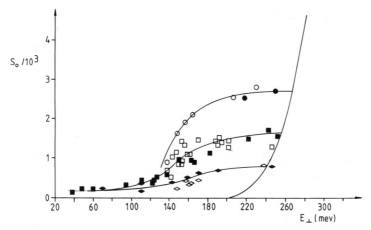

Figure 12 The initial sticking probability S_0 of H_2 in the pure beam (solid curve taken from Figure 11) is plotted together with the results of the cold seeding experiment[30] which yield $[S_0(E_\perp)]_v$. Hydrogen is seeded in He (open points) and Ne (filled points) and results are shown for three nozzle temperatures: $T_N = 1150$ K ($\bigcirc$), $T_N = 1100$ K ($\square$), and $T_N = 1085$ K ($\diamond$). $T_s = 140$ K

would be higher than for $H_2(v = 1)$ since less energy is available in the vibrational quantum. In the case of D_2, however, sticking is observed at $E_\perp < 130$ meV (Figure 13). If only $D_2(v = 0)$ and $D_2(v = 1)$ were present in the beam, one would have expected D_2 sticking to reduce to zero for $E_\perp < 130$ meV. There is, however, a small concentration of $D_2(v = 2)$ in the beam which would in fact be expected to have an even lower translational onset than $H_2(v = 1)$. The sticking of D_2 at lowest values of $E_\perp$ was therefore

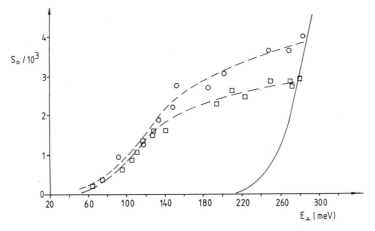

Figure 13 The initial sticking probability S_0 of D_2 in the pure beam (solid curve taken from Figure 11) is plotted together with the results of the cold seeding experiment[30] which yield $[S_0(E_\perp)]_v$. Deuterium is seeded in Ne and results are shown for two nozzle temperatures: $T_N = 1220$ K ($\bigcirc$), $T_N = 1180$ K ($\square$). $T_s = 140$ K

ascribed to $D_2(v=2)$ which was estimated to have a translational onset of 60 meV.[30] At the higher values of $E_\perp$, $D_2(v=1)$ in addition to $D_2(v=2)$ contributes to sticking because its translational onset is being approached (it was estimated to be *ca.* 260 meV), and there is a high concentration of $D_2(v=1)$ in the beam.

The seeding curves (Figures 12 and 13) and the difference in the sticking of the pure beam isotopes (Figure 11) on Cu(110) are at odds with the original conclusions of Anger *et al.*[27] who discounted any contribution of vibrational energy to the sticking. A more recent comparison[31] by the same group of the sticking of H_2 and D_2 on Cu(111), however, does reveal a similar enhancement of sticking for H_2 to that revealed for Cu(110) in a pure beam experiment. In addition an anti-seeding experiment was carried out.[31] The seeded curve followed the pure beam curve at highest $E_\perp$, but deviated significantly at low $E_\perp$. The functional shape of $[S_0(E_\perp)]_v$ is quite different from that observed on Cu(110) (Figure 12 and 13), and the results were interpreted as showing that the pure beam experiment is primarily dominated by the sticking of $H_2(v=0)$. The deviation at low $E_\perp$ is, nevertheless, ascribed to the sticking of $H_2(v=1)$ *which indeed appears therefore to have a lower translational energy requirement than $H_2(v=0)$ on Cu(111)*. This observation is therefore consistent with the idea that vibrational energy can indeed be channelled into the dissociation.

A Two-dimensional Potential Energy Surface

The results of the desorption[20] and sticking measurements[29-31] demonstrate the limitations of the one-dimensional potential. An important co-ordinate in addition to z (the molecule surface separation) will be r (the interatomic separation). Indeed following the very first beam study in which an isotope effect was observed, an attempt to improve the one-dimensional potential to account for such observations was made by carrying out a series of classical trajectory calculations on a purely empirical, but multi-dimensional, PES.[75] A modification of the three-body LEPS surface was chosen to describe the interaction of H_2 and D_2 with Cu(100) and Cu(110). Despite continued modification of the potential with increasing degrees of freedom, the trajectory calculations were rather unsuccessful in reproducing the experimental results. In particular, in none of the calculations could the isotope effect favouring the sticking of D_2 over H_2 be predicted. Indeed the opposite to the observed effect was predicted as arising from the conversion of vibrational energy in the region of the barrier.

Following the pioneering work of Polanyi,[76] it is well known that vibrational energy can be effectively utilized in promoting a gas-phase chemical reaction. This will particularly be the case when the PES displays a 'late' barrier, for example when the saddle point corresponding to a transition state, corresponds to a region of extension in the intramolecular bond being broken. In such a case, the classification adapted for homogeneous gas phase reactions is an exit channel barrier. The reaction Li + FH → LiF + H pro-

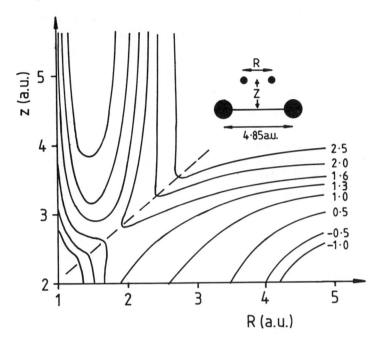

Figure 14 *The 2D-PES for the Cu_2H_2 cluster (Figure 1) derived using a Kohn–Shan energy scheme and local density approximation.[15,17] The energy of the contours are in eV. The dashed line represents the 'seam' which divides the product and reactant regions[15]*

vides an extreme example[77] since this is found to proceed (in a restricted geometry) at all translational energies, providing the energy partitioned in vibration is sufficient to overcome the activation barrier. The model of resonance mediated dissociation invoked to account for the desorption results is a surface related example: The scrambling of translational and vibrational degrees of freedom in the intermediate state allows energy partitioned in both degrees of freedom to be channelled into overcoming the barrier.

There have been a number of detailed 2D-PES in the z and r co-ordinates specifically developed to account for the role of vibrational energy during adsorption and desorption in the hydrogen/copper system. Construction of the PES itself has been based on a series of detailed electronic structure calculations.[15–19] The molecule appears to retain its gas phase properties relatively close into the surface, and the weak van der Waals interaction is generally ignored. There is clearly no evidence for an intermediate resonance state. The molecule feels the effect of the Pauli repulsion on closer approach and experiences a significant energy cost in distending the H—H bond. The associated activation barrier is calculated to be *ca.* 1 eV, a value often taken as a compromise between the Jellium and cluster calculations.[23] An example of such a 2D-PES is shown in Figure 14 based on the model 'surface' interaction of a co-linear hydrogen approaching a Cu_2 dimer.[15] The 'seam' which passes through the col at the classical transition state divides the PES into reactant and product regions, since it corresponds to the electronic 'switch' illustrated in Figure 1. The sharp seam generated by the cluster calculation is smeared out for dissociation on an extended metal surface because of the broadness (10 eV) of the $M_2(\sigma_u)$ resonance involved in the electronic 'switch' (Figure 1). The 2D-PES is characterized by a transition state somewhat in the exit channel.

Both classical and quantum mechanical trajectory calculations carried out over such surfaces have been successful in predicting the vibrational excitation in desorption, and the enhanced sticking in adsorption. The desorption result can be predicted by assuming detailed balance holds for molecules at the col of the 2D-PES, where it is reasonable to assume a thermal distribution of desorbing trajectories.[23] Encouragingly there seems to be some convergence as to the height of the activation barrier E necessary to account for the results. In the case of desorption,[20] a quantum mechanical treatment using a square barrier of height $E = 700$ meV was found to predict the observed[14] vibrational distributions (Table 1). The sticking results for both $H_2(v = 1)$[29,30] and $D_2(v = 2)$[30] are found to be consistent with the predicted, ca. 60%, vibrational energy conversion at a barrier $E = 600$ meV.[78] Such calculations[74] not only demonstrate vibrational adiabaticity characteristic of a late barrier, but also emphasize the ability of hydrogen to tunnel through a barrier of finite height and width. Classical trajectory calculations are also able to predict the observed translational onset[30] of both $H_2(v = 1)$[26,79] and $D_2(v = 2)$[79] using a barrier height $E = 720$ meV. These calculations clearly cannot account for tunnelling but demonstrate the same vibrational adiabaticity (also termed 'bootstrapping'[26,79]) evident in the quantum mechanical calculations where vibration to translational energy conversion is facilitated already in the entrance channel. This is illustrated graphically in two trajectories plotted in Figure 15. The underlying potential is that used above.[26,79] Results are shown in the case of $D_2(v = 2)$ at incident translational energies $E_\perp = 50$ meV and $E_\perp = 63$ meV (just above and below the translational onset).[30] At $E_\perp = 50$ meV, already significant conversion of vibrational to translational energy is evident. The classical turning point on this 'in and out' trajectory at $z = 4.1$ a.u. is about 150 meV. This coupling is a result of the curvature of the potential energy contours at the extreme of the vibrational motion. The molecule in this case does not reach the point of 'no return' and re-conversion of vibrational to translational energy takes place on the outward journey. This conversion illustrates the mechanism for vibrational excitation for molecules leaving from a thermal bath at the seam in desorption.[23] At $E_\perp = 63$ meV the molecule reaches the point of no return and reaction takes place. It appears therefore that the decision whether there is sufficient energy for dissociation is made relatively early in the trajectory along the z co-ordinate, and depends critically on the shape of the PES in the entrance channel.

Two 'rounds' of detailed balance calculations have been described (Sec-

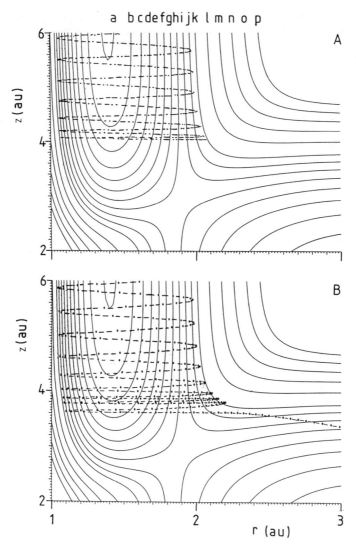

Figure 15 *The classical trajectories over a 2D-PES[26] illustrating the effect of vibrational adiabaticity (or 'bootstrapping') in the entrance channel. The trajectories are for $D_2(v = 2)$ over a potential where the saddle point energy is at 720 meV. An initial value $E_\perp = 50$ meV leads to an 'in and out' trajectory (A), while for $E_\perp = 63$ meV dissociation takes place (B). The potential energy contours (in eV) are a = 0, b = 0.1, c = 0.2, d = 0.3, e = 0.4, f = 0.5, g = 0.6, h = 0.7, i = 0.8, j = 0.9, k = 1.0, l = 1.2, m = 1.4, n = 1.6, o = 1.8, p = 2.0*

tions 3 and 4) to relate the desorption and adsorption data. It is clearly appropriate to ask whether the treatment can again be extended to correlate the translational–vibrational coupling observed in adsorption[29–31] to the vibrational distributions in desorption.[20] In a first approximation such a calculation involves defining a translational onset, and an associated width of translational barriers, for each of the vibrational states. This is an advance on the earlier detailed balance treatment[20] which effectively assumed an identical translational onset and width for both $H_2(v = 0)$ and $H_2(v = 1)$, and simply a constant enhancement of $S_0(E_\perp)$ for $H_2(v = 1)$ over $H_2(v = 0)$. The translational onset and width can now be extracted from the sticking data,[27–30] and folded (Section 3) with a Maxwellian velocity distribution, with the vibrational quanta weighted by a Boltzmann distribution, at the surface temperature T_s. A calculation along these lines[70] used a fit to the normal and variable incidence sticking data[27,29] to calculate the vibrational distributions in desorption. The results included in Table 1 (lower predicted values) show an underestimate of the observed vibrational excitation. Recognizing the danger in attempting to cross correlate any two sets of data within a detailed balanced framework, the authors also attempt to include seeded sticking data for Cu(110)[29] to fix the contribution of H_2 ($v = 1$) and normal incidence data[27] to establish the contribution of H_2 ($v = 0$).This results in a different set of translational thresholds which in turn lead to the upper values predicted (Table 1) for the vibrational population ratios. The uncertainty in the values of the translational threshold, and the sensitivity of the result to the values chosen, make it difficult to make any detailed conclusions at this stage on the basis of detailed balance calculations. These calculations[70] nevertheless also highlight the deviation one would expect from a normal energy scaling in the pure beam experiment. Close inspection of the sticking data[28] indeed reveals a systematic deviation from such a scaling on all the surfaces investigated. This observation was made previously[28] on data for Cu(110) and was suggested at the time to be an indication of a vibrational contribution to activation.

A detailed balance argument or trajectory calculation based on the 2D-PES[23] also predicts that since the translational requirement for the dissociation of $H_2(v = 1)$ will differ from $H_2(v = 0)$, the translational energy of the two vibrational states will be different in desorption. The detailed balance prediction[70] based on the sticking data[29,30] predicts a velocity distribution with two resolvable components in the TOF. Only a single distribution was observed[14] in the TOF on Cu(100) and Cu(111). On the other hand, it was reported[70] that the insensitivity of the translational energy on ϕ (Figure 7) can be accounted for by the inclusion of the vibrational contribution.

6 Conclusions

Detailed Balance and the Multi-dimensional Potential Energy Surface

There is clearly scope for implementing a comparison of desorption and adsorption data within the framework of detailed balance. Despite the initial discrepancies between adsorption and desorption, and the possibility of structural re-arrangement and the desorption of sub-surface species, it appears that adsorption and desorption are microscopically reversible. *The proviso for implementation of detailed balance must be some understanding of the important degrees of freedom and their inter-relationship (coupling)*, particularly as the number of quantum states accessible experimentally increases. There is now some indication of the coupling between translation and vibration, thus allowing a better application of detailed balance[70] with a wider data base than originally implemented.[20] As for the angular dependence of S_0 for the various vibrational states, it is conceivable that a detailed balance treatment of the translationally partitioned energy in a 'pseudo one-dimensional potential' may be appropriate, taking each vibrational quantum state individually. The details of the coupling in the 2D-PES would be included in such a treatment in the translational onsets for $H_2(v = n)$ etc. extracted from the sticking data. The angular dependence of sticking probability $S_0(\phi)$ has been found to be sharply peaked near the onset for sticking.[27,28,30] In the case of Cu(110) for example, $S_0(\phi)$ closely follows a $\cos^{15}\phi$ function at the sticking onset in the pure beam experiment. It appears also that this distribution broadens[27] as S_0 increases at higher $E_\perp$. This can be understood within the framework of the one-dimensional potential with a distribution of barriers; at higher values of E, there is a sufficient normal energy component $E_\perp$ at even grazing incident angles ϕ to facilitate dissociation over the lowest of the barriers, leading to concomitant broadening of the distribution $S_0(\phi)$ on the other hand a similar trend of broadening in $S_0(\phi)$ as E increases is predicted assuming a vibrational contribution to dissociation in the 2D-PES[26] since the excited vibrational states also have a translational energy requirement. Measurement of $[S_0(\phi)]_v$ at the translational onset of $H_2(v = 1)$ using the seeded beam[30] also produced a sharp $\cos^{15}\phi$ sticking distribution, but the limited available range of $E_\perp$ prevented further analysis.

Trajectory calculations for simple surface reactions will undoubtedly play an important role in surface dynamics as has been evidenced in gas-phase scattering with the development of realistic potential energy surfaces. Both quantum mechanical[74] and classical[26,79] trajectory calculations appear to be rather successful in predicting the dynamical behaviour of the direct dissociation of H_2 on copper. The success of the 2D-PES demonstrates the importance of vibrational and translational degrees of freedom. The PES in these limited co-ordinates enables an understanding of how molecules with low translational energies dissociate on a surface with a predictably high activation barrier. It is also clear that additional degrees of freedom will be important. In particular it is necessary to consider the impact parameter of H_2 over the surface unit cell since there have been differences established for the sticking probability as a function of translation and vibration as a function of crystal face. Similar effects are apparent in desorption. There is also some consensus that the system will be best described by a distribution of barrier heights. It is likely that such a distribution will mainly be derived from differences in the barrier height over the unit cell. A distribution of

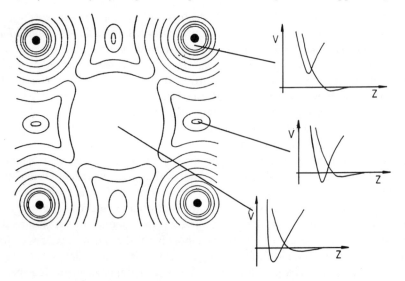

Figure 16 *A contour plot[12] of a* Cu(100) *unit cell where the lines represent a constant value of the dissociation barrier* E^*. *The solid points represent the copper nuclei. The insets illustrate how the one-dimensional potential varies from sites of strong chemisorption potential (high co-ordination) to weak chemisorption (e.g. above a copper atom) with a concomitant change in the barrier* E^*

one-dimensional barriers was suggested in the original beam study[12] and is illustrated in Figure 16. The regions of the cell where the atomic potential well is deepest (strong chemisorption) yield the lowest barriers E^* in the one-dimensional potential. Since these sites are also of highest co-ordination (Section 2), the deep chemisorption well also corresponds to the softest repulsive wall in the molecular potential in the unit cell. This is the potential which gives rise to the corrugation in the elastic scattering potential, and concomitant diffraction. The softening of the repulsive wall at the high co-ordination sites also acts to enhance dissociation by lowering the barrier. The effect of increasing $E_\perp$ is to allow access to the dissociation barrier over an increasing area of the unit cell (the distribution of barriers). The 'opening up' of the accessible barriers in the unit cell with $E_\perp$ will clearly have implications for the elastically scattered component since the proportion of the unit cell repulsive molecular potential being probed over the unit cell is also changing. It has been demonstrated in a series of wavepacket calculations of the H_2 diffracted intensities[80] that a study of the elastic component as a function of $E_\perp$ should allow a mapping of the PES in the unit cell co-ordinates (reactive potential topology). In the case of anisotropic unit cell structures [*e.g.* Cu(110)] there may also be shadowing effects in angular dependencies measured along orthogonal azimuths.[30]

There are an increasing number of trajectory calculations becoming available. Very recent calculations[81] on a multi-dimensional PES derived from *ab initio* cluster calculations[19] consider the additional contributions of impact parameter over the unit cell, rotational alignment, the motion of the copper atoms (non-rigid surface), and interaction with electron-hole pairs. A surprising result is the favourable dissociation predicted for hydrogen molecules with their axis perpendicular to the surface. There is also the suggestion that the normal energy scaling and angular distribution can be explained by damping via electron-hole pair formation. A useful alternative approach to multi-dimensionality can be found elsewhere[82] based on a potential derived using the effective medium approximation, with the increasing dimensionality introduced to include all four molecular degrees of freedom. An important conclusion concerning the rotational degree of freedom is that the final rotational state distribution of the scattered fraction will provide valuable information regarding the orientation of the molecules at the transition state.[82,83] The effect of coupling of translational energy to the metal lattice giving rise to any recoil effects in surface dissociation[84] appears to be unimportant in the case of hydrogen, although it may play a role for molecules of higher mass.

Rather than simply the quantitative prediction (of what is often a limited set) of detailed experiments, it is surely the phenomenological revelation such calculations bring with each new degree of freedom* which will provide the greatest asset to the experimentalist.

An Experimental Prospective

There are discrepancies in the experimental data obtained to date on the hydrogen/copper system which remain to be explained. Surfaces can be contaminated or exhibit defects, both of which can profoundly alter dissociation dynamics. It is all too easy, however, to reject the data that fails to conform to a particular analysis on such grounds. This is particularly true when it is clear that the dynamical behaviour of even this simple system is quite complex. The difference between the early beam data[12] and that obtained more recently[27,28] is a striking case. It has correctly been argued[27] that there are dangers in using the hydrogen/deuterium reaction as a measure of S_0 since at high turnover the rate limitation ceases to be dissociation. The curves of Figure 4 therefore should not in reality be 'S' shaped but should continue to increase. Nevertheless, even within the limits of experimental error (estimated to be 50%[12]), the values of S_0 are significantly larger at lower $E_\perp$ than obtained elsewhere. In the light of the seeded sticking data,[29,30] however, it has been suggested[30] that the early measurements may have been dominated by significant concentrations of vibrationally hot molecules in the beam. The curve for Cu(110) (Figure 4) has a translational onset near to the value obtained for $H_2(v = 1)$ on the same surface (Figure 12). This hypothesis also allows an explanation for the observed reverse isotope effect (Figure 6) which proved a stumbling block in the early theoretical treatments.[75] It can be seen that the hot nozzle produces sig-

---

* As an experimentalist, the introduction of increased degrees of freedom (and therefore variables) can prove to be counter productive. For an assessment of the technical merits of the trajectory calculation and the potential, the reader is referred elsewhere in this volume.[74]

nificant concentrations of $D_2(v = 2)$ in addition to $D_2(v = 1)$. $D_2(v = 2)$ has been shown to have a lower translational onset than $H_2(v = 1)$ which dominates the sticking under similar nozzle conditions. The region of $E_\perp$ where the enhanced sticking of D_2 is observed (Figure 6) corresponds to the translational energies where $D_2(v = 2)$ will give rise to enhanced sticking in the pure D_2 beam (Figure 12 and 13). It has also been pointed out[30] that both sets of hydrogen measurements (Figure 6 and 12) are characterized by residual sticking at low $E_\perp$.

Since vibrational energy does appear to play a role in the dissociation process, this brings into perspective the difficulties in assessing vibrational populations in hot nozzles. Whether a pure beam experiment will be dominated at any particular T_N by ground state or vibrationally hot molecules will clearly be a reflection of the efficiency of vibrational excitation (and de-excitation) in the nozzle expansion. This in turn is a function of the expansion conditions and perhaps the nozzle itself. Additionally, the height of the barrier and the extent of coupling will also influence which limiting sticking condition (vibrational or translational energy) is reached first as a nozzle is heated. It is therefore feasible that the pure beam experiments are sensitive to different quantum states to different extents. These differences may well also be responsible for the different order of barriers across the different crystal faces. In any event, the selection of discrete vibrational states using hot nozzles, even with the well spaced vibrational quanta of H_2, are quite crude. In the long run only state selective sticking experiments will provide the necessary detail to convincingly map the multi-dimensional PES of even a simple reaction such as the direct dissociation of hydrogen on copper.

References

1. O. Beek, A. E. Smith, and A. Wheeler, *Proc. R. Soc. London, A*, 1940, **177**, 62.
2. T. Springer in 'Hydrogen in Metals I; Basic Properties', ed. G. Alefeld and J. Volkl, Springer–Verlag, Berlin, 1978, and references therein.
3. K. W. Jacobsen and J. K. Nørskov, *Phys. Rev. Lett.*, 1987, **59**, 2764.
4. K. H. Rieder and W. Stocker, *Phys. Rev. Lett.*, 1986, **57**, 2548.
5. B. E. Hayden, D. Lackey, and J. Schott, *Surf. Sci.*, 1990, **239**, 119.
6. F. Greuter and E. W. Plummer, *Solid State Commun.*, 1983, **48**, 37.
7. C. S. Alexander and J. Pritchard, *J. Chem. Soc. Faraday Trans.*, 1972, **68**, 202.
8. T. L. Bradley and R. E. Stickney, *Surf. Sci.*, 1973, **38**, 313.
9. J. E. Lennard–Jones, *Trans. Faraday Soc.*, 1932, **28**, 333.
10. W. van Willigen, *Phys. Lett.*, 1968, **28A**, 80.
11. M. Balooch and R. E. Stickney, *Surf. Sci.*, 1974, **44**, 310.
12. M. Balooch, M. J. Cardillo, D. R. Miller, and R. E. Stickney, *Surf. Sci.*, 1974, **46**, 358.
13. J. Lapujoulade, Y. le Cruer, M. Lefort, Y. Lejay, and G. Maurel, *Surf. Sci.*, 1981, **103**, L85.
    J. Lapujoulade and J. Perreau, *Phys. Scr.*, 1983, **T4**, 138.
14. G. Comsa and R. David, *Surf. Sci.*, 1982, **117**, 77.

15. J. Harris, S. Andersson, C. Holmberg, and P. Nordlander, *Phys. Scr.*, 1986, **T13**, 160.
16. P. K. Johansson, *Surf. Sci.*, 1981, **104**, 510.
17. J. Harris and S. Andersson, *Phys. Rev. Lett.*, 1985, **55**, 1583.
18. P. E. M. Seigbahn, M. R. A. Blomberg, and C. W. Bauschlicher, *J. Chem. Phys.*, 1984, **81**, 1373.
19. P. Madhavan and J. L. Whitten, *J. Chem. Phys.*, 1982, **77**, 2673.
20. G. D. Kubiak, G. O. Sitz, and R. N. Zare, *J. Chem. Phys.*, 1984, **81**, 6397.
21. S. Holloway, A. Hodgson, and D. Halstead, *J. Electron Spectrosc.*, 1987, **45**, 207.
22. S. Holloway, A. Hodgson, and D. Halstead, *Chem. Phys. Lett.*, 1988, **147**, 425.
23. J. Harris, S. Holloway, T. S. Rahman, and K. Yang, *J. Chem. Phys.*, 1988, **89**, 4427.
24. M. R. Hand and S. Holloway, *J. Chem. Phys.*, 1989, **91**, 7209.
25. W. Brenig and H. Kasai, *Surf. Sci.*, 1989, **213**, 170.
26. J. Harris, *Surf. Sci.*, 1989, **221**, 335.
27. G. Anger, A. Winkler, and K. D. Rendulic, *Surf. Sci.*, 1989, **220**, 1.
28. B. E. Hayden and C. L. A. Lamont, *Chem. Phys. Lett.*, 1989, **160**, 331.
29. B. E. Hayden and C. L. A. Lamont, *Phys. Rev. Lett.*, 1989, **63**, 1823.
30. B. E. Hayden and C. L. A. Lamont, *Surf. Sci.*, 1991, **243**, 31.
31. H. F. Burger, M. Leisch, A. Winkler, and K. D. Rendulic, *Surf. Sci. Lett.*, in press.
32. R. B. McLellan and C. G. Harkins, *Mater. Sci. Eng.*, 1975, **18**, 5.
33. M. Dumas, *Ann. Chim. Phys.*, 1984, **III 8**, 189.
    W. Hampe, *Z. Anal. Chem.*, 1874, **13**, 352.
    V. G. Rienacker and B. Sarry, *Z. Inorg. Chem.*, 1948, **257**, 41.
34. C. C. Bond, 'Catalysis by Metals', Academic, New York, 1962, p. 149.
    H. H. Kung, *Catal. Rev. Sci. Eng.*, 1980, **22**, 235.
35. See references 8, 11, 12 and references therein.
36. T. Kwan, *Bull. Chem. Soc. Jpn.*, 1950, **23**, 73.
37. R. J. Mikovsky, M. Boudart, and H. S. Taylor, *J. Am. Chem. Soc.*, 1954, **76**, 3814.
38. D. D. Eley and D. R. Rossington in 'Chemisorption', ed. W. Garner, Butterworths, London, 1957, p. 137.
39. M. Kiyomiya, N. Momma, and I. Yasomori, *Bull. Chem. Soc. Jpn.*, 1974, **47**, 1852.
40. J. Pritchard, T. Catterick, and R. K. Gupta, *Surf. Sci.*, 1975, **53**, 1.
41. I. Kojima, M. Kiyomiya, and I. Yasomori, *Bull. Chem. Soc. Jpn.*, 1980, **53**, 2123.
42. A. P. Baddorf, I. W. Lyo, E. W. Plummer, and H. L. Davis, *J. Vac. Sci. Technol.*, 1987, **A5**, 782.
43. E. M. McCash, S. F. Parker, J. Pritchard, M. A. Chesters, *Surf. Sci.*, 1989, **215**, 363.
44. R. Spitzl, H. Niehus, B. Poelsema, and G. Comsa, *Surf. Sci.*, in press.
45. B. E. Hayden and D. C. Godfrey, *Surf. Sci.*, 1990, **232**, 24.
46. D. J. Auerbach and C. T. Rettner in 'Kinetics of Interface Reactions', ed. M. Grunze and H. J. Kreuzer, Springer Series in Surface Science vol. 8, Springer-Verlag, Berlin, 1987, p. 125.
47. A. C. Luntz, M. D. Williams, and D. S. Bethune, *J. Chem. Phys.*, 1988, **89**, 4381.

48. C. T. Rettner, H. E. Pfnur, and D. J. Auerbach, *J. Chem. Phys.*, 1986, **84**, 4163.
49. C. T. Rettner and H. Stein, *Phys. Rev. Lett.*, 1987, **59**, 2768.
50. A. C. Luntz and D. S. Bethune, *J. Chem. Phys.*, 1989, **90**, 1274.
51. S. Andersson, L. Wilzen, and J. Harris, *Phys. Rev. Lett.*, 1985, **55**, 2591.
52. S. Andersson, L. Wilzen, M. Persson, and J. Harris, *Phys. Rev. B*, 1989, **40**, 8146.
53. M. D. Stiles and J. W. Wilkins, *Phys. Rev. Lett.*, 1985, **54**, 595.
54. A. E. DePristo, C.-Y. Lee, and J. M. Hutson, *Surf. Sci.*, 1986, **169**, 451.
55. J. Lapujoulade and J. Perreau; *Phys. Scr.*, 1983, **T4**, 138.
56. U. Harten, J. P. Toennies, and Ch. Wöll, *J. Chem. Phys.*, 1986, **85**, 2249.
57. J. K. Nørskov, A. Houmoller, P. K. Johansson, and B. I. Lundquist, *Phys. Rev. Lett.*, 1981, **46**, 257.
58. M. A. Morris, M. Bowker, and D. A. King in 'Comprehensive Chemical Kinetics', Vol. 19, ed. C. Bamford and C. Tipper, Elsevier, New York, 1983.
59. G. Comsa and R. David, *Surf. Sci. Rep.*, 1985, **5**, 145.
60. G. Comsa, *J. Chem. Phys.*, 1968, **48**, 3235.
61. M. Asscher, W. L. Guthrie, T. H. Lin, and G. A. Somorjai, *J. Chem. Phys.*, 1983, **78**, 6992.
62. L. V. Novakoski and G. M. McClelland, *Phys. Rev. Lett.*, 1987, **59**, 1259.
63. D. C. Jacobs, K. W. Kolasinski, R. J. Madix, and R. N. Zare, *J. Chem. Phys.*, 1987, **87**, 5038.
64. G. Comsa and R. David, *Surf. Sci.*, 1980, **95**, L210.
65. A. E. Dabiri, T. J. Lee, and R. E. Stickney, *Surf. Sci.*, 1971, **26**, 522.
66. T. L. Bradley, A. E. Dabiri, and R. E. Stickney, *Surf. Sci.*, 1972, **29**, 590.
67. R. L. Palmer, J. N. Smith, H. Salzberg, and D. R. O'Keefe, *J. Chem. Phys.*, 1970, **53**, 1666.
68. M. J. Cardillo, M. Balooch, and R. E. Stickney, *Surf. Sci.*, 1975, **50**, 263.
69. G. Comsa and R. David, *Chem. Phys. Lett.*, 1977, **49**, 512.
70. H. Michelsen and D. J. Auerbach, *Phys. Rev. Lett.*, 1990, **65**, 2833.
71. B. E. Hayden and C. L. A. Lamont, *J. Phys. Condens. Matter*, 1989, **1**, SB33.
72. D. A. King and M. G. Wells, *Surf. Sci.*, 1971, **29**, 454.
73. J. W. Gadzuk, *J. Chem. Phys.*, 1983, **79**, 6341.
    J. W. Gadzuk and J. K. Nørskov, *J. Chem. Phys.*, 1984, **81**, 2828.
74. Refer to Chapter 3 by S. Holloway in this volume.
75. A. Gelb and M. Cardillo, *Surf. Sci.*, 1976, **59**, 128. *ibid.*, 1977, **64**, 197. *ibid.* 1978, **75**, 199.
76. J. C. Polanyi, *J. Chem. Phys.*, 1959, **31**, 1338.
77. I. Noordbatcha and N. Sathyamurthy, *Chem. Phys.*, 1983, **77**, 67.
78. D. Halstead and S. Holloway; *J. Chem. Phys.*, 1990, **93**, 2859.
79. J. Harris, *J. Electron. Spectrosc*, 1990, **54/55**, 115.
80. M. Karikorpi, S. Holloway, N. Henriksen, and J. K. Nørskov, *Surf. Sci.*, 1987, **179**, L41.
81. M. Cacciatore, M. Capitelli, and G. D. Billing, *Surf. Sci.*, 1989, **217**, L391.
    M. Cacciatore and G. D. Billing, *Surf. Sci.*, 1990, **232**, 35.
82. U. Nielsen, D. Halstead, S. Holloway, and J. K. Nørskov, *J. Chem. Phys.*, 1990, **93**, 2859.
83. S. Holloway and B. Jackson, *Chem. Phys. Lett.*, 1990, **172**, 40.
84. M. Hand and J. Harris, *J. Chem. Phys.*, 1990, **92**, 7610.

CHAPTER 5

Kinetics of Surface Reactions

W. HENRY WEINBERG

1 Introduction

The importance of achieving a detailed understanding of the rates and mechanisms of surface reactions cannot be overstated. Quite apart from the satisfaction obtained from confronting and overcoming the formidable scientific and intellectual challenges, the technological applications and implications associated with this problem are simply enormous. Just a few of these technological applications are the following: heterogeneous catalysis, crystal growth, thin film growth (including, for example, evaporation, sputtering, molecular beam epitaxy, and chemical vapour deposition), tribology and lubrication (*i.e.*, chemical modification of surfaces), corrosion inhibition (*i.e.*, chemical passivation of surfaces), and polymer adhesion at surfaces via chemical reaction of functional groups located on the polymer.

A very impressive array of experimental techniques has been developed over the course of the past few decades that can be applied to the problem of surface reactivity. These include a variety of diffraction,[1-4] scattering[5-11] and direct imaging techniques[12-14] for surface structural determination, vibrational[15-18] and electronic[19-22] spectroscopies for a characterization of surface reactants and reaction intermediates, and molecular beam scattering and mass spectrometric techniques[5,23,24] for elucidating the dynamics and kinetics of surface reactions. Rather recent theoretical advances have also been impressive. Reliable electronic structural calculations have been carried out on model systems in order to characterize surface reaction intermediates and in order to obtain limited information concerning the multi-dimensional potential energy surfaces that govern their behaviour.[25-39] Moreover, qualitatively reasonable potentials have been employed in connection with Monte Carlo and molecular dynamics simulations of surface rate processes.[40-47]

In this chapter, an attempt will be made to systematize much of the existing experimental data within the context of a few general 'rules' or concepts. In other words we shall attempt to answer the question, 'How should one think about surface reactivity?' This question will be addressed both from the point of view of macroscopic kinetics (thermally averaged rate coefficients) and microscopic phenomena (the dynamics of energy exchange and accommodation, and *elementary* surface reaction rate coefficients). The organization of this chapter is as follows. In Section 2, we group surface reactions, such as chemisorption or bimolecular catalytic reactions, into two classes, which we define as 'direct' and 'trapping-mediated', and we argue intuitively that trapping-mediated reactions should, in general, be more commonly observed experimentally. In Section 3, we present some elementary results of transition state theory that support our point of view that trapping-mediated reactions should be preferred over direct reactions. We also explain when direct reactions would be expected to become important. In Section 4, we discuss the related concepts of trapping and accommodation, an understanding of which is necessary in order to appreciate fully the meaning and mechanism of a trapping-mediated surface reaction. In Section 5, we present a detailed discussion of the initial rate of dissociative chemisorption of ethane on the Ir(110)–(1 × 2) surface. Quantitative rate parameters are deduced for the trapping-mediated reaction. We demonstrate that 'real world' surface chemistry involving this system would be dominated by the trapping-mediated reaction channel, but we also present data showing when the direct reaction channel would become important. The latter involves translationally hot molecular beams of ethane. Finally, in Section 6, we present a brief summary which contains both conclusions as well as a preview of expected future developments in this exciting and important area of chemistry.

2 Direct *vs.* Trapping-mediated Surface Reactions

A direct reaction in this context is defined as one which occurs at the surface in a single collision from the gas phase with a time scale typically quite less than 10^{-12} s. Two examples of such a reaction are direct dissociative chemisorption, *e.g.* $A_2(g) \rightarrow 2A(a)$, and the so-called Eley–Rideal surface reaction mechanism, *e.g.* $A(a) + AB(g) \rightarrow A_2B(g)$.* In this type of reaction, the most relevant temperature is that of the gaseous reactant. Since the reactant is not trapped at the surface, it cannot accommodate to the surface temperature.

The other, quite different, type of reaction is a trapping-mediated surface reaction. In this case there is a 'real' (bound) intermediate in the reaction, which is usually termed a 'precursor'. As discussed in Section 4, the lifetime of the precursor is usually sufficiently long compared to a vibrational period that it accommodates to the surface temperature. Although the elementary

---

* Here, and hereafter, (a) denotes an adsorbed species, and (g) denotes a gas-phase species. Whenever (a) is replaced by (p), this implies physical adsorption; and whenever (a) is replaced by (c), this implies chemisorption.

surface reaction rate coefficients will be functions of surface temperature only, the *measured* reaction rate could be a function of both the gas and the surface temperature. This is because the trapping probability into the precursor state, a dynamic effect, is a function of the gas temperature. Although it need not be the case, the lifetime of the precursor is often quite short compared to the reciprocal of the impingement rate of the reactant. When this condition is fulfilled, the concentration of precursors on the surface will be very low, and a pseudo steady-state approximation (the time derivative of the surface concentration of the precursor is zero) may be employed when formulating rate expressions (*vide infra*).

The most general case of trapping-mediated dissociative chemisorption may be written as follows:

$$A_2(g) \rightleftharpoons A_2(p) \rightleftharpoons A_2(c) \rightarrow 2A(c). \tag{1}$$

Here, the physically adsorbed A_2 molecule is a precursor to the molecularly chemisorbed state, which, in turn, is a precursor to the dissociatively chemisorbed products of the reaction. Projections of a possible potential energy surface describing this reaction are shown in Figure 1. In the top panel, Figure 1a, the potential energy along the reaction co-ordinate ρ is shown; and the potential minima corresponding to the physically adsorbed A_2, the molecularly chemisorbed A_2, and the dissociatively chemisorbed 2A are noted explicitly. The two barriers, both of which lie below the 'vacuum zero' of energy (the gas-phase A_2 molecule infinitely far from the surface and at rest), have associated with them two transition states which are denoted by † and ‡ in Figure 1. To the left of the minimum corresponding to dissociative chemisorption in Figure 1a, there are two branches of the potential, one of which is denoted by $\perp$ and the other of which is denoted by $//$. The $\perp$ branch corresponds to motion perpendicular to and into the surface, which is repulsive due to the Pauli Principle. The $//$ branch corresponds to motion parallel to the surface, which is periodic on a single crystalline surface and describes adatom hopping on the surface. A contour plot which captures the important features of the potential energy surface is shown in Figure 1b. The intramolecular (A—A) spacing is plotted on the ordinate, and the spacing between the surface and the centre-of-mass of the A_2 molecule is plotted on the abscissa. Contours of constant potential are indicated schematically. Note that the three potential minima and the two barriers (transition states) are labelled in the same way as they are in the one-dimensional representation of Figure 1a. The final dissociatively chemisorbed state corresponds to two A adatoms typically separated by one lattice constant on the surface, *i.e.*, a typical separation of 2.5–3 Å, or more. Since this is a large separation compared to most molecular bond lengths, one would, in general, expect 'late' barriers to dissociative chemisorption, *i.e.*, barriers that are well into the exit channel of the reaction, as is indicated by the ‡ in the contour plot of Figure 1b.

It should be intuitively clear from Figure 1 that trapping-mediated disso-

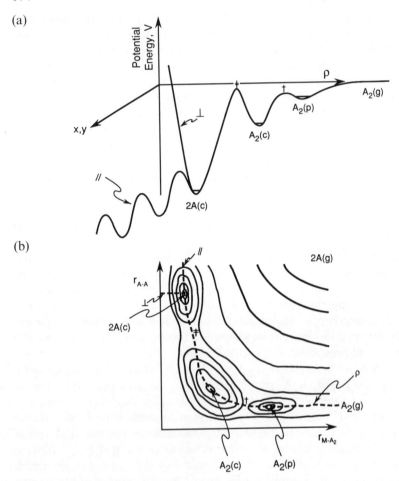

Figure 1 (a) *One-dimensional potential energy diagram (along the reaction co-ordinate, ρ) for the hypothetical surface reaction* $A_2(g) \rightleftharpoons A_2(p) \rightleftharpoons A_2(c) \rightarrow 2A(c)$ (b) *Potential energy surface in the form of a contour diagram for the reaction in Figure 1(a). The reaction co-ordinate, ρ, is shown as a dashed line. Note that the* ⊥ *and* // *labelled curves represent motions of the dissociated surface species* [i.e. $A(c)$] *perpendicular and parallel (x,y directions) to the surface, respectively, and that the transition states are denoted by the* † *and* ‡ *symbols*

ciative chemisorption is far more probable than direct dissociative chemisorption except in unusual cases. In order for the A_2 molecule to chemisorb dissociatively in a direct fashion, it must have sufficient kinetic energy to scatter inelastically from the repulsive part of the potential surface and redirect sufficient energy along the reaction co-ordinate (the A—A intramolecular stretching mode in this case) in order for the reaction to occur. Note that the other inelastic channels in this scattering event are dissipative and do not lead to reaction. These dissipative channels include phonon and electron-hole pair excitation in the solid, and other intramolecular energy exchange within the A_2 molecule, *e.g.* rotational degrees of freedom and translational degrees of freedom parallel to the surface (see Chapter 1). The probability of such a direct dissociation process may be relatively low even when the kinetic energy exceeds the barrier associated with the lowest energy path. However, species that become trapped on the surface have numerous opportunities to sample and surmount this barrier. In many cases the trapped species may dissociate with high probability. This intuitive expectation of the dominance of trapping-mediated surface reactions is rendered quantitative in Section 3, and an example [the dissociative chemisorption of ethane on the Ir(110)–(1 × 2) surface] that illustrates the point beautifully is presented in Section 5.

The other example of a direct surface reaction that we mentioned at the beginning of this section is the Eley–Rideal mechanism. This is to be contrasted with the Langmuir–Hinshelwood mechanism, which implies a reaction between two adsorbed species, and which clearly fits into our definition of a trapping-mediated reaction. It is probably fair to state that the Langmuir–Hinshelwood reaction mechanism is the more correct way to describe most surface reactions, which is another way of saying that trapping-mediated reactions are usually more important than direct reactions. In the next section some simple theoretical arguments will be presented that put this conclusion on a firmer foundation.

3 Transition State Theory of Surface Reaction Rates

In this section the rates of Langmuir–Hinshelwood and Eley–Rideal surface reactions will be formulated within the context of an elementary, equilibrium statistical mechanical treatment of transition state theory. This will allow a quantitative assessment of the relative importance of these two mechanisms: the trapping-mediated Langmuir–Hinshelwood reaction *vs.* the direct Eley–Rideal reaction.

Langmuir–Hinshelwood Reaction Mechanism

Consider the following surface reaction which obeys the Langmuir–Hinshelwood mechanism:

$$S\text{—}A + S\text{—}BA \rightleftharpoons (S \ldots A \ldots B \ldots A)^{\ddagger} \rightarrow A_2B(g) \qquad (2)$$

where S denotes a surface binding site. The one-dimensional potential energy diagram along the reaction co-ordinate ρ is shown in Figure 2 for this activated and (assumed) exothermic reaction. The rate of the reaction is just the rate (usually written as a flux) at which the reactants cross (irreversibly) a hypothetical surface at the saddle point of the potential energy surface (the barrier in Figure 2) which separates the reactants from the products. Hence, it follows that:

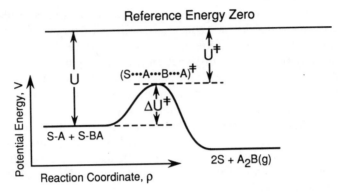

Figure 2 *One dimensional potential energy diagram (along the reaction co-ordinate, ρ) for the hypothetical Langmuir–Hinshelwood reaction, S—A + S—BA → (S ... A ... B ... A)‡ → S + A₂B(g), where S denotes a surface binding site*

$$R^{L-H} = \frac{C^{\ddagger}_{ABA}}{d\rho} \int_0^{\infty} \dot{\rho} f(\dot{\rho}) d\dot{\rho} = \frac{C^{\ddagger}_{ABA}}{d\rho} \langle \dot{\rho} \rangle \quad (3)$$

where $C^{\ddagger}_{ABA}$ is the two-dimensional concentration of 'activated complexes' at the transition state on the potential energy surface, $d\rho$ is a differential length along the reaction co-ordinate at the transition state, $\dot{\rho}$ is the centre-of-mass velocity $(d\rho/dt)$ of reacting species along the reaction co-ordinate, and $f(\dot{\rho})$ is a normalized velocity distribution function.

The concentration of activated complexes $C^{\ddagger}_{ABA}$ may be evaluated from the equilibrium reaction that is assumed in equation (2) via the equilibrium coefficient of this reaction $K(T)$, *i.e.*:

$$K(T) = \left(\frac{C^{\ddagger}_{ABA}}{C_A C_{BA}}\right)\left(\frac{1}{\beta}\right) = \frac{(q^{\ddagger}_{ABA})' q^{\ddagger}_{\rho}}{q_A q_{BA}} \quad (4)$$

where C_A and C_{BA} are the two-dimensional concentrations of the reactants, β is the area of an adsorption site on the surface, q is a single-particle canonical ensemble partition function, $q^{\ddagger}_{\rho}$ is the single *translational* degree of freedom of the ABA complex along the reaction co-ordinate, and the prime on $(q^{\ddagger}_{ABA})'$ reminds us that the one degree of freedom along the reaction co-ordinate $(q^{\ddagger}_{\rho})$ has been removed from this partition function, *i.e.*, there are $3n - 1$ (not $3n$) degrees of freedom in $(q^{\ddagger}_{ABA})'$, where n is the number of atoms per activated complex. For example, if A and B are taken to be atoms in this example, then there are eight degrees of freedom in $(q^{\ddagger}_{ABA})'$. The degree of freedom contained in $q^{\ddagger}_{\rho}$ can very loosely be thought of as what would be the frustrated translational degree of freedom perpendicular to the surface of the ABA‡ complex were it bonded to the surface in our usual picture of adsorption.

Writing equation (4) as:

$$C_{ABA}^{\ddagger} = \left[\frac{(q_{ABA}^{\ddagger})' q_{\dot{\rho}}^{\ddagger}}{q_A q_{BA}}\right] C_A C_{BA} \beta \tag{5}$$

and substituting equation (5) into equation (3) allows us to write for the rate:

$$R^{L-H} = \left(\frac{C_A C_{BA}}{d\rho}\right)\left[\frac{(q_{ABA}^{\ddagger})' q_{\dot{\rho}}^{\ddagger}}{q_A q_{BA}}\right] \langle \dot{\rho} \rangle \beta \tag{6}$$

In order to evaluate $\langle \dot{\rho} \rangle$, we perform the averaging with the properly normalized Maxwell–Boltzmann distribution function, namely:

$$f(\dot{\rho}) d\dot{\rho} = \left(\frac{M}{2\pi k_B T_s}\right)^{1/2} e^{-M\dot{\rho}^2/2k_B T_s} d\dot{\rho} \tag{7}$$

where M is the mass of the ABA complex, and T_s is the surface temperature. Using this distribution function, we find that:

$$\langle \dot{\rho} \rangle = \left(\frac{k_B T_s}{2\pi M}\right)^{1/2} \tag{8}$$

and the rate expression of equation (6) becomes:

$$R^{L-H} = \frac{C_A C_{BA}}{d\rho} \left[\frac{(q_{ABA}^{\ddagger})' q_{\dot{\rho}}^{\ddagger}}{q_A q_{BA}}\right] \left(\frac{k_B T_s}{2\pi M}\right)^{1/2} \beta \tag{9}$$

In order to evaluate $q_{\dot{\rho}}^{\ddagger}$, we make use of the classical phase integral for the one-dimensional translation along ρ, namely:

$$q_{\dot{\rho}}^{\ddagger} = \frac{1}{h} \int_{p_\rho = -\infty}^{p_\rho = +\infty} e^{-p_\rho^2/2Mk_B T_s} dp_\rho d\rho \tag{10}$$

which gives:

$$q_{\dot{\rho}}^{\ddagger} = \left(\frac{2\pi M k_B T_s}{h^2}\right)^{1/2} d\rho \equiv \frac{d\rho}{\Lambda} \tag{11}$$

where Λ is an effective de Broglie wavelength. Substituting equation (11) into equation (9) allows us to write the rate of reaction as

$$R^{L-H} = \frac{k_B T_s}{h} \left[\frac{(q_{ABA}^{\ddagger})'}{q_A q_{BA}}\right] \beta \, C_A C_{BA} \tag{12}$$

If, as is essentially always the case, the reactants occupy specific sites, then this localization implies that each of the partition functions that appears in equation (12) is a vibrational partition function. This allows us to write:

$$(q^{\ddagger}_{ABA})' = \left(\prod_{i=1}^{8} q^{\ddagger}_{v,ABA,i}\right) e^{-U^{\ddagger}/k_B T_s} \tag{13}$$

and

$$q_A q_{BA} = \left(\prod_{j=1}^{3} q_{v,A,j}\right)\left(\prod_{k=1}^{6} q_{v,BA,k}\right) e^{-U/k_B T_s} \tag{14}$$

where we have assumed that the vibrational degrees of freedom are separable, and the electronic ground states are non-degenerate. The 'Boltzmann factors' that result from the electronic partition functions contain the potentials $U^{\ddagger}$ and U which are defined in Figure 2. Within the harmonic approximation, the individual vibrational partition functions are given by:

$$q_{v,i} = \frac{e^{-hv_i/2k_B T_s}}{1 - e^{-hv_i/k_B T_s}} \tag{15}$$

We define a new vibrational partition function $\hat{q}_{v,i}$ as follows:

$$\hat{q}_{v,i} \equiv (1 - e^{-hv_i/k_B T_s})^{-1} \tag{16}$$

and we include the zero-point energy contribution explicitly in the Boltzmann factor, *i.e.*:

$$U^{\ddagger}_0 \equiv U^{\ddagger} + \frac{h}{2}\sum_{i=1}^{8} v^{\ddagger}_i \tag{17}$$

and

$$U_0 \equiv U + \frac{h}{2}\sum_{j=1}^{3} v_j + \frac{h}{2}\sum_{k=1}^{6} v_k \tag{18}$$

Combining equations (13)–(18) into equation (12) gives:

$$R^{L-H} = \left(\frac{k_B T_s}{h}\right)\left[\frac{\prod_{i=1}^{8} \hat{q}^{\ddagger}_{v,ABA,i}}{\left(\prod_{j=1}^{3} \hat{q}_{v,A,j}\right)\left(\prod_{k=1}^{6} \hat{q}_{v,BA,k}\right)}\right] \beta e^{-\Delta U^{\ddagger}_0/k_B T_s} C_A C_{BA} \tag{19}$$

where $\Delta U_0^{\ddagger} \equiv U^{\ddagger}_0 - U_0 > 0$. We might, for completeness, multiply the right-hand side of equation (19) by $\langle \kappa \rangle$ which takes into account the possibility of barrier recrossing by the product molecule. In other words, $\langle \kappa \rangle$ is the probability that once the product molecule has been formed by crossing the barrier, it remains a product molecule; this is the so-called dynamical correction to transition state theory.[48] We can write the reaction rate (in units of flux, recall) in terms of a Langmuir–Hinshelwood reaction rate coefficient as follows:

$$R^{L-H} = k^{L-H} C_A C_{BA} \tag{20}$$

where

$$k^{L-H} \equiv k^{(0)L-H} e^{-\Delta U_0^{\ddagger}/k_B T_s} \tag{21}$$

and $k^{(0)L-H}$ is given by:

$$k^{(0)L-H} = \langle \kappa \rangle \frac{k_B T_s}{h} \Psi \beta \tag{22}$$

where Ψ is defined to be the bracketed ratio of partition functions in equation (19). Note that the dimensions of this second-order surface rate coefficient, given by equation (21), are area per unit time.

In order to obtain a 'physical feeling' for the magnitude of the pre-exponential factor of the Langmuir–Hinshelwood reaction rate coefficient, let us assume that $T_s = 400$ K, $\beta = 6.7 \times 10^{-16}$ cm² (a site concentration of 1.5×10^{15} cm^{-2}), and the *ratio* of the vibrational partition functions Ψ (with the factor involving the zero-point energy having been extracted) is approximately unity. In this case we find that

$$k^{(0)L-H} \cong 5.6 \times 10^{-3} \langle \kappa \rangle \text{ cm}^2 \text{ s}^{-1}, \tag{23}$$

and $\langle \kappa \rangle \leq 1$. Hence, a 'normal' value of the pre-exponential factor would be expected to be approximately 10^{-3} cm² s^{-1}. When 'unusual' values of the pre-exponential factor are observed experimentally, they are frequently rationalized in terms of various *ad hoc* assumptions concerning the localization or delocalization of the reactants and the transition state.* In actual fact, most experimentally observed pre-exponential factors which appear unusual are probably not unusual at all. Rather, they are generally a consequence of an artifact in the analysis of the data. The root of the problem lies in something known as the 'compensation effect', which has been discussed recently in terms of a temperature-dependent distribution of reacting surface configurations.[47,49]

Eley–Rideal Reaction Mechanism

We shall now derive an expression for the rate of an Eley–Rideal reaction in order to compare it with the results of equations (20)–(22) for the rate of a

---

* When unusually large pre-exponential factors are observed, the transition state is assumed to be delocalized, and the reactants are localized. When unusually small pre-exponential factors are observed, the transition state is assumed to be localized, whereas one or both reactants are delocalized. Occasionally, these arguments are couched in terms of 'entropy' of the transition state *vs.* the reactants. In either case it is recommended that they be viewed with scepticism.

Langmuir–Hinshelwood reaction. In this way, we shall be able to assess the relative contribution of each under different experimental conditions. The Eley–Rideal analogue of the reaction given in equation (2) may be written as:

$$\text{S—A} + \text{BA(g)} \rightleftharpoons (\text{S} \ldots \text{A} \ldots \text{B} \ldots \text{A})^{\ddagger} \rightarrow \text{S} + \text{A}_2\text{B(g)} \qquad (24)$$

Recall that this is a direct reaction since BA is not adsorbed on the surface. The rate of this Eley–Rideal reaction is defined to be (in flux units, as before):

$$R^{\text{E-R}} = k^{\text{E-R}} \left(\frac{\theta_A}{\beta}\right) C_{\text{BA,g}} \qquad (25)$$

where θ_A/β, the ratio of the fractional surface coverage of A to the site area, is the two-dimensional concentration of A, $C_{\text{BA,g}}$ is the three-dimensional (gas-phase) density of BA, and $k^{\text{E-R}}$ is the Eley–Rideal reaction rate coefficient (in units of volume divided by time), which may be written as:

$$k^{\text{E-R}} = k^{(0)\text{E-R}} e^{-\Delta U_0^{\ddagger}/k_B T_g} \qquad (26)$$

where T_g is the gas-phase temperature. Since it is not quite clear what is the appropriate temperature (T_s, T_g, or something intermediate between the two in non-isothermal systems) to use in connection with the partition function of the activated complex, in what follows we shall assume $T_s = T_g \equiv T$ when evaluating the pre-exponential factor of the Eley–Rideal reaction rate coefficient. We shall, however, use T_g in the Boltzmann factor for this direct chemical reaction, although this also is clearly an approximation.

We next introduce the probability of reaction $P_r(\theta_A)$, which is a function of the fractional surface coverage of A, as follows:

$$R^{\text{E-R}} = P_r(\theta_A) F_{\text{BA}} = P_r(\theta_A) \left[\frac{P_{\text{BA}}}{(2\pi M_{\text{BA}} k_B T)^{1/2}}\right] \qquad (27)$$

where F_{BA} is the impingement flux of BA onto the surface. Assuming that BA is an ideal gas and equating the two expressions for the reaction rate in equations (25) and (27) allows us to write the rate coefficient of the Eley–Rideal reaction as follows:

$$k^{\text{E-R}} = [P_r(\theta_A)] \left(\frac{\beta}{\theta_A}\right) \left[\frac{k_B T}{h q_{t,\text{BA}}}\right] \qquad (28)$$

where β, as before, is the area of an adsorption site of reactant A, and $q_{t,\text{BA}}$ is the canonical ensemble partition function of a single translational degree of freedom of a BA molecule, which in this case is given by:

$$q_{t,\text{BA}} = \frac{1}{\Lambda} = \frac{(2\pi M_{\text{BA}} k_B T)^{1/2}}{h} \qquad (29)$$

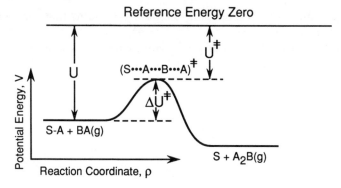

Figure 3 *One dimensional potential energy diagram (along the reaction co-ordinate, ρ) for the hypothetical Eley–Rideal reaction, $S{-}A + BA(g) \rightleftharpoons (S\ldots A \ldots B \ldots A)^{\ddagger} \rightarrow S + A_2B(g)$, where S denotes a surface binding site*

The rate of this Eley–Rideal reaction may also be written, in units of flux, as:

$$R^{E-R} = \frac{C^{\ddagger}_{\widehat{A}BA}}{d\rho} \langle \dot{\rho} \rangle \qquad (30)$$

in exactly the same way as the Langmuir–Hinshelwood rate was written in equation (3). Following the same derivation presented earlier in equations (4)–(11), we find that the Eley–Rideal reaction rate may be written as:

$$R^{E-R} = \langle \kappa \rangle \left(\frac{k_B T}{h}\right) \left[\frac{(q^{\ddagger}_{\widehat{A}BA})'}{q_A q_{BA}}\right] \left(\frac{\theta_A}{\beta}\right) C_{BA,g} \qquad (31)$$

where the notation is the same as that employed earlier, and $\langle \kappa \rangle$, the dynamical correction to transition state theory is included explicitly here. The one-dimensional potential energy diagram along the reaction co-ordinate for this reaction is shown in Figure 3. Note, by comparing equations (25) and (31), that the Eley–Rideal reaction rate coefficient is given by:

$$k^{E-R} = \langle \kappa \rangle \left(\frac{k_B T}{h}\right) \left[\frac{(q^{\ddagger}_{\widehat{A}BA})'}{q_A q_{BA}}\right] \qquad (32)$$

where q_{BA} in this case refers to the single-molecule canonical ensemble partition function of a gaseous BA molecule. If we assume that BA is an ideal gas, we can replace $C_{BA,g}$ by $P_{BA}/k_B T$; and we also extract one of the translational degrees of freedom from q_{BA} in equation (31), i.e., one factor of $q_{t,BA}$ given by equation (29). This allows us to rewrite equation (31) as follows:

$$R^{E-R} = \langle \kappa \rangle \left[\frac{(q^{\ddagger}_{\widehat{A}BA})'}{q_A q'_{BA}}\right] \left(\frac{\theta_A}{\beta}\right) F_{BA} \qquad (33)$$

where the primes on the partition functions remind us that one degree of freedom has been extracted.

As in the case of the Langmuir–Hinshelwood mechanism discussed earlier, we assume that both the reactant A and, of course, the transition state are localized. This allows us to write:

$$(q_{ABA}^\ddagger)' = \left(\prod_{i=1}^{8} q_{v,ABA,i}^\ddagger\right) e^{-U^\ddagger/k_B T_g} \qquad (34)$$

and

$$q_A q_{BA}' = \left(\prod_{j=1}^{3} q_{v,A,j}\right) (q_{t,BA}^2 q_{r,BA}^2 q_{v,BA}) e^{-U/k_B T_g} \qquad (35)$$

where the potentials associated with the electronic part of the partition function (the Boltzmann factors) are shown in Figure 3, and $q_{r,BA}^2$ is the partition function of the two rotational degrees of freedom. Extracting the zero-point energy from the various vibrational partition functions, as was done in connection with equations (15)–(18), allows us to rewrite the Eley–Rideal reaction rate as:

$$R^{E-R} = \langle \kappa \rangle \left[\frac{\prod_{i=1}^{8} \hat{q}_{v,ABA,i}^\ddagger}{\left(\prod_{j=1}^{3} \hat{q}_{v,A,j}\right)(q_{t,BA}^2 q_{r,BA}^2 \hat{q}_{v,BA})} \right] e^{-\Delta U_0^\ddagger/k_B T_g} \left(\frac{\theta_A}{\beta}\right) F_{BA} \qquad (36)$$

where $\Delta U_0^\ddagger$ is defined below equation (19), and the subscript zero implies that the zero-point energy is included in this potential.

In order to obtain a physical feeling for the pre-exponential factor of the rate coefficient implied by equation (36), which in this case is dimensionless, let us assume that the ratio of the vibrational partition functions is unity, $T = 400$ K, $\beta = 6.7 \times 10^{-16}$ cm^2, and BA is taken to be carbon monoxide which implies that $1/q_{t,BA}^2 \cong 2.7 \times 10^{-18}$ cm^2 and $q_{r,BA}^2 \cong 150$. With these assumptions, equation (36) becomes:

$$R^{E-R} \cong 2.7 \times 10^{-5} \langle \kappa \rangle e^{-\Delta U_0^\ddagger/k_B T_g} \theta_A F_{BA} \qquad (37)$$

It is important to note that equation (37) is of nearly the same form as that which would be derived for the initial rate of chemisorption of BA with a localized transition state, but with an additional factor of θ_A on the right-hand side. This, of course, is as it should be since the chemisorption reaction can occur at any unoccupied site on the surface, whereas the Eley–Rideal reaction can only occur at a site that is occupied by an A adatom.

The Eley–Rideal reaction rate coefficient, given by equation (32), may be written as:

$$k^{E-R} = \langle \kappa \rangle \left(\frac{k_B T}{h}\right) \left[\frac{\prod_{i=1}^{8} \hat{q}_{v,ABA,i}^{\ddagger}}{\left(\prod_{j=1}^{3} \hat{q}_{v,A,j}\right) (q_{t,BA}^2 q_{r,BA}^2 \hat{q}_{v,BA})} \right] e^{-\Delta U_0^{\ddagger}/k_B T_g} \quad (38)$$

If we make the same assumptions as in the previous paragraph concerning the temperature (400 K), the value of $1/q_{t,BA}$ (1.6×10^{-9} cm), the value of $q_{r,BA}^2$ (150), and the cancellation of the vibrational partition functions, equation (38) becomes:

$$k^{E-R} \cong 2.5 \times 10^{-16} \langle \kappa \rangle \, e^{-\Delta U_0^{\ddagger}/k_B T_g} \quad \text{cm}^3 \, \text{s}^{-1} \quad (39)$$

for the physically relevant case of a localized surface reactant and a localized (at the saddle point of the potential energy surface) transition state. The reaction probability contained in equation (28) may be written as:

$$P_r(\theta_A) = k^{E-R} \left(\frac{\theta_A}{\beta}\right) \left(\frac{h q_{t,BA}}{k_B T}\right) \quad (40)$$

and we note that while P_r is a function of θ_A, the ratio $P_r(\theta_A)/\theta_A$ is not a function of θ_A. Using the same assumptions concerning $q_{t,BA}$, β and T that were introduced earlier, and making use of the order-of-magnitude estimate for k^{E-R} given by equation (39), we find that:

$$\frac{P_r(\theta_A)}{\theta_A} \cong 2.7 \times 10^{-5} \langle \kappa \rangle \, e^{-\Delta U_0^{\ddagger}/k_B T_g} \quad (41)$$

which, of course, could have been written down directly from equation (37). If we ignore (*i.e.*, take to be unity) the θ_A in the denominator on the left-hand side of equation (41), for the reasons discussed below equation (37), equation (41) gives the initial probability of chemisorption of the BA molecule. Furthermore, this order-of-magnitude estimate, which applies to a localized transition state, is not just valid for direct chemisorption. It also applies to trapping-mediated chemisorption (with the gas temperature replaced by the surface temperature in the Boltzmann factor), and to both molecular and dissociative chemisorption. Indeed, depending on where the potential curves cross with respect to the reference energy (the vacuum zero), *i.e.*, the curves for physical adsorption and molecular chemisorption for the case of molecular chemisorption or for molecular adsorption and dissociative chemisorption for the case of dissociative chemisorption, $\Delta U_0^{\ddagger}$ could be either greater or less than zero; and $e^{-\Delta U_0^{\ddagger}/k_B T}$ could be either less or greater than unity. When the Boltzmann factor, $e^{-\Delta U_0^{\ddagger}/k_B T}$, is greater than unity, which corresponds to unactivated chemisorption with respect to the gas-phase energy zero, then it would appear that the probability of chemisorption, *cf.*, equation (41), could exceed unity. This is, of course, not the case. When the probability, calculated from an expression like equation (41), appears to exceed unity, then its value is in fact just unity, and the adsorption reaction is flux limited. In this case an equilibrium approximation, implicit in the derivation of the probability of chemisorption that is analogous to equation (41) for the Eley–Rideal reaction probability, simply breaks down.

Langmuir–Hinshelwood vs. Eley–Rideal Reaction Rates

It is possible to compare quantitatively the rates of the Eley–Rideal and Langmuir–Hinshelwood mechanisms by taking the ratio of equations (36) and (19). Noting that $C_i = \theta_i/\beta$, this gives:

$$\frac{R^{\text{E-R}}}{R^{\text{L-H}}} = \frac{\langle \kappa \rangle \left(\prod_{i=1}^{8} \hat{q}^{\ddagger}_{\text{v,ABA},i} \right) \left(\frac{\theta_A}{\beta} \right) F_{\text{BA}} \, e^{-\Delta U_0^{\ddagger}/k_B T}}{\left(\prod_{j=1}^{3} \hat{q}_{\text{v,A},j} \right) (q^2_{\text{t,BA}} q^2_{\text{r,BA}} \hat{q}_{\text{v,BA}})} \left/ \langle \kappa \rangle \left(\frac{k_B T_s}{h} \right) \left[\frac{\prod_{i=1}^{8} \hat{q}^{\ddagger}_{\text{v,ABA},i}}{\left(\prod_{j=1}^{3} \hat{q}_{\text{v,A},j} \right) \left(\prod_{k=1}^{6} \hat{q}_{\text{v,BA},k} \right)} \right] \left(\frac{\theta_A \theta_{\text{BA}}}{\beta} \right) e^{-\Delta U_0^{\ddagger}/k_B T} \right. \quad (42)$$

where we assume $T_s = T_g \equiv T$. If both the dynamical corrections to transition state theory, $\langle \kappa \rangle$, and the activation energies of the reaction, $\Delta U_0^{\ddagger}$, are the same for the Eley–Rideal and the Langmuir–Hinshelwood mechanisms, this expression reduces to:

$$\frac{R^{\text{E-R}}}{R^{\text{L-H}}} = \left(\frac{h}{k_B T} \right) \left(\frac{F_{\text{BA}}}{\theta_{\text{BA}}} \right) \left[\frac{\prod_{i=1}^{6} \hat{q}_{\text{v,BA},i}}{q^2_{\text{t,BA}} q^2_{\text{r,BA}} \hat{q}_{\text{v,BA}}} \right] \quad (43)$$

Next we make the same assumptions that we made earlier, namely, that $T = 400$ K and that BA is a carbon monoxide molecule, which implies that $q^{-2}_{\text{t,BA}} \cong 2.7 \times 10^{-18}$ cm^2 and $q^2_{\text{r,BA}} \cong 150$. Furthermore, if we assume that the ratio of vibrational partition functions in equation (43) is approximately unity, then we find that:

$$\frac{R^{\text{E-R}}}{R^{\text{L-H}}} \cong 2.2 \times 10^{-33} \left(\frac{F_{\text{BA}}}{\theta_{\text{BA}}} \right) \quad (44)$$

with the flux in units of cm^{-2} s^{-1}. If the flux is in units of site^{-1} s^{-1}, and assuming that $\beta^{-1} \cong 1.5 \times 10^{15}$ cm^{-2} is the site concentration, then equation (44) may be written as:

$$\frac{R^{\text{E-R}}}{R^{\text{L-H}}} \cong 3.2 \times 10^{-18} \left(\frac{F_{\text{BA}}}{\theta_{\text{BA}}} \right) \quad (45)$$

With the flux on a per site basis, a pressure of 10^{-6} Torr, for example, corresponds approximately to $F_{BA} \sim 1$ site^{-1} s^{-1}.

The implication of equations (44) and (45) would seem to be that the rate of the direct Eley–Rideal reaction is *always completely negligible* compared to that of the trapping-mediated Langmuir–Hinshelwood reaction. In fact, except in rather unusual cases, the Langmuir–Hinshelwood reaction rate dominates that of the Eley–Rideal reaction; but the domination is not quite so dramatic as that which seems to be implied by equations (44) and (45). Let us next try to understand intuitively why our calculated ratio R^{E-R}/R^{L-H} is so very small, and then we shall examine our derivation more carefully in order to decide when the Eley–Rideal rate could become important compared to the Langmuir–Hinshelwood rate.

When two reactants are adsorbed adjacent to one another, as in the Langmuir–Hinshelwood mechanism, they can attempt to react on the order of 10^{12}–10^{13} times each second. Think of this as either an asymmetric atom–surface stretching frequency or as a frustrated translational frequency of an adsorbed molecule parallel to the surface. In a direct reaction, such as the Eley–Rideal mechanism, this 'attempt frequency' is replaced by the impingement flux, which, for example, is *ca.* 1 site^{-1} s^{-1} at 10^{-6} Torr. Such high pressures would be required to overcome this limitation, that adsorption (perhaps even condensation) of the reactant would become extremely important, thus favouring the Langmuir–Hinshelwood mechanism. We shall return to this point later. Indeed, the situation is even worse than implied by this argument based on attempt frequencies, because there is a 'loss of entropy' of one of the reactants in forming the transition state in the Eley–Rideal mechanism. This loss of entropy is reflected, for example, by a prefactor of 2.7×10^{-5} (rather than unity) in equations (37) and (41). This argument can also be couched in terms of 'effective' or 'virtual' pressures of an adsorbed molecule. For example, saturation coverage ($\theta = 1$) on a lattice, the site concentration of which is 1.5×10^{15} cm^{-2}, corresponds to an effective three-dimensional pressure of an ideal gas at 400 K of approximately 3000 atm.!

We can write equation (45) in an alternate pedagogic form if we make the approximation that the adsorption–desorption equilibrium of the BA molecule is not perturbed significantly either by the Langmuir–Hinshelwood or Eley–Rideal reactions or by the presence of the A adatoms. Furthermore, if the surface temperature is sufficiently high that $\theta_{BA} \ll 1$, which is in the spirit of our trying to decide when the Eley–Rideal mechanism would be most competitive with the Langmuir–Hinshelwood mechanism, we may write:

$$\text{BA(g)} \underset{k_{d,BA}}{\overset{F_{BA}\xi_{BA}}{\rightleftharpoons}} \text{BA(a)} \tag{46}$$

where ξ_{BA} is the probability of trapping of the BA molecule into the chemisorbed state, and $k_{d,BA}$ is the rate coefficient of desorption of the adsorbed BA. In the pseudo-equilibrium physical situation we are imagining:

$$\frac{d\theta_{BA}}{dt} = F_{BA}\xi_{BA} - k_{d,BA}\theta_{BA} \cong 0 \tag{47}$$

which implies that:

$$\frac{F_{BA}}{\theta_{BA}} = \frac{k_{d,BA}}{\xi_{BA}} \tag{48}$$

The rate coefficient for desorption, $k_{d,BA}$, may be written in a Wigner–Polanyi form as:

$$k_{d,BA} = k_{d,BA}^{(0)} \, e^{-E_{d,BA}/k_B T_s} \tag{49}$$

where a normal value of $k_{d,BA}^{(0)}$, the pre-exponential factor of the first-order desorption rate coefficient, is approximately 10^{13} s^{-1}; $E_{d,BA}$ is the activation energy of desorption of the adsorbed BA molecule; and the temperature that appears in the Boltzmann factor is the surface temperature. Substituting equations (48) and (49) into equation (45) gives:

$$\frac{R^{E-R}}{R^{L-H}} \cong \frac{3.2 \times 10^{-5}}{\xi_{BA} \, e^{E_{d,BA}/k_B T_s}} \tag{50}$$

Within the context of equation (50), we can now inquire into when might the direct Eley–Rideal reaction become competitive with the trapping-mediated Langmuir–Hinshelwood reaction. Clearly, within the approximations inherent in equation (50), which have been delineated explicitly, the crossover from the Langmuir–Hinshelwood rate being greater to the Eley–Rideal rate being greater occurs at:

$$\xi_{BA} \, e^{E_{d,BA}/k_B T_s} \cong 3.2 \times 10^{-5} \tag{51}$$

and the Eley–Rideal rate becomes more significant at lower values of $\xi_{BA} \, e^{E_{d,BA}/k_B T_s}$.

When might we expect $\xi_{BA} \, e^{E_{d,BA}/k_B T_s}$ to be sufficiently small for the Eley–Rideal mechanism to become important? The answer to this question is when the AB molecule is bound very weakly to the surface and the surface temperature is very high (which tends to make $e^{E_{d,BA}/k_B T_s}$ smaller), *and* when both the surface and especially the gas temperature are sufficiently high to make ξ_{BA} very small indeed. It has been observed experimentally on numerous occasions that the trapping probability decreases slightly with increasing surface temperature and decreases precipitously with increasing gas temperature.[50–62] Hence, we tentatively conclude that very high gas temperatures would tend to emphasize the Eley–Rideal mechanism. Indeed, if the gaseous reactant were supplied from a supersonic molecular beam with not only a very high translational energy, but also with a very narrow translational energy distribution, then it is possible that the reaction might only follow the

Eley–Rideal mechanism. For example, we could picture the surface temperature being sufficiently low that the rate of the Langmuir–Hinshelwood reaction is negligible, whereas the gas 'temperature' could actually be greater than the activation energy of the reaction. Although this peculiar set of experimental conditions could lead to a greater *relative* rate of the Eley–Rideal mechanism compared to the Langmuir–Hinshelwood mechanism, we would nevertheless expect the *absolute* rate of the direct reaction to be quite low. Whether or not it would ever be anything more than a scientific curiosity remains an open question.

An equivalent way of considering this issue is to imagine that the barrier for reaction is sufficiently high with respect to the binding energy of the adsorbed reactant that the surface temperature would have to be so high in order to allow the surface reaction to occur that the lifetime of the adsorbed reactant would be too short both to accommodate to the surface temperature and, of more importance, to execute even one successful hop on the surface. A physical situation to which this argument may apply is the direct *vs.* the trapping-mediated dissociative chemisorption of methane on the Ni(111) surface.[63–65] We shall return to this important issue of accommodation (both perpendicular and parallel to the surface) in the next section. Next, however, we look more carefully and more critically at the approximations that led to equations (44), (45), and (50).

The most general expression for the ratio of the direct Eley–Rideal reaction rate to the trapping-mediated Langmuir–Hinshelwood reaction rate is given by equation (42). Various approximations were then adopted in order to simplify this expression to the forms found in equation (44), (45), and (50). We now examine these approximations in somewhat more detail. First, it was implicitly assumed that $\langle \kappa \rangle$, the dynamical correction to transition state theory, which takes into account the possibility of barrier recrossing, is equal for the two reaction mechanisms. Clearly, this need not be the case; but it is yet unclear just what typical relative values of $\langle \kappa \rangle$ are for the two different mechanisms (direct *vs.* trapping-mediated). Future theoretical calculations will be necessary to evaluate these dynamical corrections, and to determine whether any systematic trends exist concerning their ratio that appears in equation (42). Tentatively, there seems to be no reason to believe that this ratio is vastly different from unity.

Secondly, we assumed that the ratio of vibrational partition functions that appears in equation (43) is equal to unity, *i.e.*:

$$\left(\prod_{i=1}^{6} \hat{q}_{v,BA,i}\right) / \hat{q}_{v,BA} = 1 \qquad (52)$$

Let us evaluate this approximation by considering the special case of CO that is end-on bonded through the 5σ-orbital of the carbon atom to a late (*e.g.* Group VIII) transition metal surface. The six vibrational degrees of freedom of the chemisorbed CO consist of two frustrated translations parallel to the surface, two frustrated rotational degrees of freedom, one frustrated translation perpendicular to the surface, and one intramolecular vibration. Typical values for these frequencies for chemisorbed CO are the following: frustrated translations parallel to the surface, 50 cm^{-1}; frustrated rotational degrees of freedom, 350 cm^{-1}; frustrated translation perpendicular to the surface, 450 cm^{-1}; and internal vibrational degree of freedom, 2000 cm^{-1}.[66] The vibrational frequency of the free CO molecule which applies to the partition function that appears in the denominator of equation (52), is 2168 cm^{-1}. Using these frequencies and assuming, reasonably, a temperature of 400 K implies that this ratio of vibrational partition functions for the specific case of chemisorbed CO is on the order of 100 (rather than unity).

Thirdly, we have already discussed the fact that we assumed the surface temperature to be equal to the gas temperature, which was simply denoted by T in equations (42) and (43), for example. It is more correct to consider that the Boltzmann factor for the Eley–Rideal reaction [*e.g.* in the numerator of equation (42)] involves the gas temperature, whereas the Boltzmann factor for the Langmuir–Hinshelwood reaction [*e.g.* in the denominator of equation (42)] involves the surface temperature. Hence, if $T_g > T_s$ the Eley–Rideal reaction would be favoured relative to our order of magnitude estimates of equations (44), (45), and (50). Note, however, that the converse is also true. If $T_s > T_g$, then the Langmuir–Hinshelwood reaction would be favoured to an even greater extent than is suggested by equations (44), (45), and (50).

This argument concerning the possible difference in gas and surface temperatures, and the consequences thereof is an important one. Moreover, it leads naturally into a discussion of an even more profound implicit assumption that was made in going from equation (42) to equation (43), for example. This assumption is that the activation energy of reaction, $\Delta U_0^{\ddagger}$, is the same for both the Eley–Rideal and the Langmuir–Hinshelwood mechanisms. In fact, there is no particular reason to suppose that the activation barrier for forming the transition state for the Langmuir–Hinshelwood reaction of equation (2) is the same as that for forming the transition state for the Eley–Rideal reaction of equation (24). Although everything depends on the *details* of the electronic structure of the reactants (and the transition states), one might argue intuitively that frequently the barrier for the Eley–Rideal reaction would be lower than that for the Langmuir–Hinshelwood reaction. A simplistic way to think about this is to consider that some of the 'S—BA' bonding in equation (2) must be broken in order to form the transition state, whereas this is not the case in equation (24) for the Eley–Rideal reaction.* Likewise, one might intuitively expect the possible occurrence of two different transition states in the two mechanisms. For

---

* An opposing argument can also be put forward, however. We have implicitly assumed that the reaction cross-section for the Eley–Rideal mechanism is the area of a surface unit cell, *e.g.* via the introduction of the $1/\beta$ factor in equation (25). One might expect the corresponding impact parameter to be actually smaller, and one might expect also an (unaccounted for) dependence on the *orientation* of the impinging AB reactant. These are important issues that need to be addressed theoretically.

example, suppose the A_2B product molecule in equations (2) and (24) is a linear triatomic molecule. It would then be plausible that the transition state for the Eley–Rideal mechanism would also be linear (see Scheme I), whereas the transition state for the Langmuir–Hinshelwood mechanism might be expected to be bent (see Scheme II).

$$\begin{pmatrix} A \\ \vdots \\ B \\ \vdots \\ A \\ \vdots \\ S \end{pmatrix}^{\ddagger} \qquad \begin{pmatrix} & A \\ A\cdots B & \\ S & S \end{pmatrix}^{\ddagger}$$

I II

Let us apply these intuitive ideas to the specific example of CO oxidation on a late transition metal surface. Clearly, this is a special case of the $A + BA \rightarrow A_2B$ reaction that we have been discussing more generally above. In particular, let us consider the CO oxidation reaction on rhodium, realizing that the arguments will be qualitatively the same on all other late transition metal surfaces on which the CO is end-on bonded. The potential energy diagram along the reaction co-ordinate for the Langmuir–Hinshelwood reaction in the low-coverage limit is shown by the solid curve in Figure 4.[67] The dashed curve shows qualitatively the (unknown) potential along the reaction co-ordinate for the Eley–Rideal mechanism.* The potential energy diagram for the Langmuir–Hinshelwood mechanism was constructed as follows. The zero of energy (the reference energy) is taken to be a CO molecule and one-half an O_2 molecule in the gas phase infinitely far from the surface and at rest. The binding energy of the CO to the surface is simply equal to the heat of adsorption which, in this case, is 31.6 kcal mol^{-1} in the low-coverage limit.[68] Furthermore, it is known that, in the low-coverage limit, the heat of dissociative chemisorption of O_2 on Rh is ca. 85 kcal mol^{-1}.[69] Since dissociative adsorption of the O_2 involves cleavage of the O—O bond followed by the formation of two Rh—O bonds, we can employ a Born cycle to calculate the binding energy of the oxygen to the surface. In this case:

$$D(Rh\!=\!\!O) = [D(O\!=\!\!O) + h]/2 = (118.9 + 85)/2 = 102.0 \text{ kcal mol}^{-1} \quad (53)$$

where $D(O\!=\!\!O)$ is the bond dissociation energy of molecular oxygen, and h is the heat of dissociative chemisorption. Hence, with our chosen zero of energy, the reactants for the Langmuir–Hinshelwood mechanism [CO(a) + O(a) in Figure 4] lie at -74.1 kcal mol^{-1}, (i.e., $-31.6 - 102.0 + 118.9/2$ kcal mol^{-1}). Since the activation energy for the CO oxidation reaction via a Langmuir–Hinshelwood mechanism was measured to

---

* The reason this is unknown is because the Eley–Rideal mechanism has not been observed experimentally.

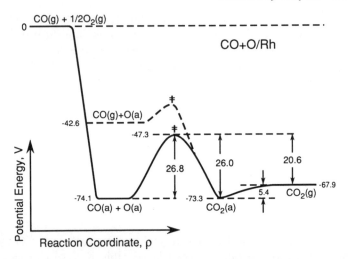

Figure 4 *One-dimensional potential energy diagram (along the reaction co-ordinate, ρ) illustrating the catalytic oxidation of CO on Rh. The solid curve represents the Langmuir–Hinshelwood mechanism [CO(a) + O(a) →CO$_2$(g)] for this reaction, and the dashed curve represents the Eley–Rideal mechanism [CO(g) + O(a) →CO$_2$(g)]. All energies are in kcal mol^{-1} in the limit of zero coverage, and ‡ denotes the transition state separating reactants from products*

be 26.8 kcal mol^{-1} in the low-coverage limit,[70] the transition state, which should be thought of as Scheme II above with A = oxygen and B = carbon, lies at −47.3 kcal mol^{-1}, as shown in Figure 4. We also note in passing that the energy of the CO$_2$ product molecule with this energy zero is −67.9 kcal mol^{-1}, and the CO$_2$ has a heat of physical adsorption of *ca.* 5.4 kcal mol^{-1}.[71] This implies that the low-coverage barrier for the dissociative chemisorption of CO$_2$ on Rh is 20.6 kcal mol^{-1} with respect to the free CO$_2$ energy zero, and it is 26.0 kcal mol^{-1} with respect to the bottom of the physically absorbed well in the case of trapping-mediated dissociative chemisorption.

We next consider what all this has to do with the Eley–Rideal mechanism of CO oxidation. Notice that the energy level of the reactants in the Eley–Rideal mechanism, an O adatom and a gas-phase CO molecule, is −42.6 kcal mol^{-1} (*i.e.,* −102.0 + 118.9/2 kcal mol^{-1}). This level lies above that of the transition state for the Langmuir–Hinshelwood mechanism, the latter of which was determined from experimental measurements! This quantifies our intuitive arguments made earlier concerning the fact that one might expect the activation energy for the Eley–Rideal mechanism to be different from that of the Langmuir–Hinshelwood mechanism. It is not likely that there is no activation energy at all to the Eley–Rideal mechanism of CO oxidation. Rather, it is much more likely that there is a *different* transition state (probably more similar to Scheme I than to Scheme II, above). This point of view is indicated schematically by the dashed line on the potential energy diagram of Figure 4.

Lamentably, we do not know the exact level of the transition state for the Eley–Rideal mechanism since neither has this mechanism been observed experimentally nor have accurate *ab initio* electronic structural calculations been carried out for this reaction. We are, however, in a position to reevaluate our approximations that led to equation (50) and to estimate the *maximum* plausible value of the ratio of R^{E-R} to R^{L-H}. In order to do this, we include the ratio of vibrational partition functions as 100 rather than as unity, as we previously assumed. Of more significance, however, we *assume there is no barrier at all* for the Eley–Rideal mechanism, whereas we use the measured value of 26.8 kcal mol^{-1} for the activation barrier of the Langmuir–Hinshelwood reaction. In this case, equation (50) becomes:

$$\frac{R^{E-R}}{R^{L-H}} = \frac{3.2 \times 10^{-5}(100)}{\xi_{CO}\, e^{E_{d,CO}/k_B T_s}\, e^{-\Delta U_0^{\ddagger L-H}/k_B T_s}} \tag{54}$$

If we assume thermal energies with a Maxwell–Boltzmann distribution of gas-phase momenta, then $\xi_{CO} \simeq 1$; and if we use $E_{d,CO} = 31.6$ kcal mol^{-1} and $\Delta U_0^{\ddagger L-H} = 26.8$ kcal mol^{-1} at a surface temperature of 400 K, where there is a measurable reaction rate, then we find:

$$\frac{R^{E-R}}{R^{L-H}} \cong 7.6 \times 10^{-6} \tag{55}$$

The order-of-magnitude estimate of equation (55) is extremely important. It tells us that any contribution of a direct reaction is almost always negligible, even under the most favorable circumstances when it is an unactivated reaction. In almost all technological applications, one can ignore the direct channel of the reaction compared to the trapping-mediated channel. Only when either one of the reactants is supplied in a supersonic molecular beam with extremely high translational energies or the surface lifetime of the trapped intermediate is extremely short would one expect a sufficiently small value of ξ in order for the direct channel to be observed. We shall explore what we mean by extremely short surface lifetime and some of its consequences in the next section.

Likewise, trapping-mediated chemisorption, including activated dissociative chemisorption, will almost always be far more important than direct chemisorption. In both cases, the form of the rate will involve a preexponential factor multiplied by a Boltzmann factor. The pre-exponential factor will be larger for the trapping-mediated case, however, due to the loss of entropy in going from the reactant to the transition state in the direct reaction. Moreover, the apparent activation energy will be greater for the direct reaction since the barrier almost always lies in the exit channel on the potential energy surface, *cf.* Figure 1. In this case also, direct (dissociative) chemisorption would only be expected either within the context of a translationally hot supersonic molecular beam experiment or when the trapping-mediated barrier is sufficiently large that the necessary *surface* temperature to surmount it precludes a sufficient lifetime of the trapped reactant.

Since essentially all of the 'real-world' surface chemistry is trapping-mediated rather than direct, we shall concentrate almost entirely on describing and quantifying this kind of surface reactivity in the remainder of this chapter.

4 Trapping and Accommodation

Trapping

Since trapping at a surface is the first step in all trapping-mediated reactions, it is important to have as complete a picture as is possible concerning this dynamical phenomenon. The reason we refer to trapping as a dynamical phenomenon is because it involves momentum and energy exchange between the gas-phase atom or molecule and the repulsive part of the gas–surface potential, *cf.* Figure 1b. Tentatively, we might expect that an atom or molecule is trapped at the surface if it exchanges sufficient energy in the gas–surface collision to have negative total energy (kinetic plus potential) after the collision: negative with respect to our usual zero of energy which is the gas molecule infinitely far from the surface and at rest. A schematic picture that helps clarify this discussion is shown in Figure 5, which depicts a gas with a three-dimensional Maxwell–Boltzmann velocity distribution interacting with a one-dimensional potential of (say) physical adsorption. Only the dissipative collision of a molecule with the most probable velocity is shown explicitly, but keep in mind that each molecule with its own particular velocity (kinetic energy) collides with the repulsive barrier and undergoes energy exchange.

There are a number of dissipative (inelastic) channels available in the gas–surface collision. If the impinging gas-phase particle is an atom, *e.g.* a rare gas atom, then kinetic energy can only be lost to the solid. Electronic excitations of the rare gas are considered to be inaccessible. In this case, the principal inelastic channels are phonon excitations and electron-hole pair creation, *i.e.* vibrational and electronic excitations. The detailed nature of the substrate, *e.g.* the Debye temperature and the densities of phonon and electronic states, will dictate the dynamics of the individual collisions. If the impinging gas-phase particle is a molecule, then there are additional dissipative channels that are available. For example, suppose the impinging gas molecule is an alkane (a saturated hydrocarbon). This choice is made because, in the absence of dissociative chemisorption, the alkanes are physically adsorbed on the surface with the same sort of van der Waals interactions that govern the rare gas atoms. In this case, there can be intramolecular energy transfer that converts translational energy to either rotational or vibrational energy. (The conversion of translational energy in a direction perpendicular to the surface to translational energy parallel to the surface will be discussed later.) For example, in the specific case of ethane

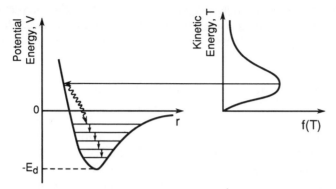

Figure 5 *Illustration of the trapping mechanism of gas-phase atoms or molecules [represented by the Maxwell–Boltzmann distribution of kinetic energies, f(T)] at a surface via inelastic interactions with the repulsive part of the surface potential (curve shown on the left)*

(C_2H_6), there are three rotational degrees of freedom and eighteen vibrational degrees of freedom.

Hence, we tentatively *define* the thermally-averaged trapping probability to be the *fraction* of molecules in the Maxwell–Boltzmann distribution of Figure 5 that lose sufficient energy to fall below the zero of energy of that figure. There are two different experimental measures of the trapping probability. First, if the surface temperature is sufficiently low that the lifetime of the trapped molecule on the surface is sufficiently long (depending on the experimental probe of choice), then the trapping probability would simply be the fraction of the impingement flux that adsorbs 'irreversibly'. The surface residence time is simply the reciprocal of the first-order (remember we are considering non-dissociative adsorption) rate coefficient of desorption, the latter of which may be written as:

$$k_d = k_d^{(0)} e^{-E_d/k_B T_s} \tag{56}$$

where E_d is the binding energy of the adsorbed molecule, *cf.* Figure 5, and $k_d^{(0)}$ is the pre-exponential factor of the desorption rate coefficient. Let us consider the specific case of the molecular adsorption of ethane on the reconstructed Ir(110)–(1 × 2) surface, a case that is discussed in detail in Section 5.[72] The values of k_d^0 and E_d have been measured and found to be 10^{13} s^{-1} and 7.7 kcal mol^{-1}, respectively. This implies that in order to effect a surface lifetime

$$\tau = (k_d^{(0)})^{-1} e^{+E_d/k_B T_s} \tag{57}$$

of at least ten seconds, a surface temperature of 117 K, or lower, is required. If the surface lifetime were one second, the corresponding surface temperature would be 126 K. A molecular beam reflectivity technique allows one to measure the *initial* trapping probability with surface lifetimes as short as 1–10 s.[73] In order to measure the trapping probability as a function of surface coverage, it is necessary, of course, to maintain irreversible adsorption over the course of the entire measurement, *e.g.* up to one monolayer coverage. Depending on the flux and the magnitude of the trapping probability, this will obviously imply a longer time scale.

The second way to measure experimentally the trapping probability is complementary to the first and is carried out at elevated surface temperatures at which the surface lifetime is negligibly small (or, more precisely, at which the surface concentration is negligibly small, and this is, consequently, only a measure of the 'initial' trapping probability, *i.e.* the probability of trapping on a clean surface). This method is typically used in connection with molecular beam scattering experiments in which the angular and/or velocity distributions of the 'desorbing' atoms or molecules are measured. These distributions are referred to as the 'trapping-desorption' component of the total 'desorbing' flux. The other component, which is not trapped at the surface, is known as the 'direct inelastic' flux. The ratio of the trapping-desorption component to the total flux is the trapping probability. Since an atom or molecule that is trapped at the surface desorbs with a cosine angular distribution and with a Maxwell–Boltzmann velocity distribution characterized by the surface temperature (remember we are *not* discussing activated adsorption here), both angular and velocity distribution measurements may be carried out in order to quantify the trapping probability.[50,51,74–82] Next, we shall illustrate how this protocol can be implemented in practice, and we arbitrarily choose the case of Ar trapping on the Pt(111) surface as an example. We note in passing that there were a number of early attempts to describe trapping in terms of what were known as cube models (or variations on that theme) in which momentum parallel to the surface was assumed to be conserved.[83–85] Today these models are of primarily historical interest, and have been replaced by far more sophisticated trajectory calculations in connection with the solution of a generalized Langevin equation, *e.g.* references 86 and 87.

In Ar scattering from the Pt(111) surface, the scattered flux can be separated easily into two components: a broad distribution which contains those Ar atoms that have accommodated on the surface, and a lobular peak near the angle of specular reflection, which contains those Ar atoms that have scattered inelastically but have not been trapped at the surface.[51] This observation is typical of almost all scattering measurements, and the trapping probability is just the ratio of the integrated desorbing flux contained in the trapping-desorption distribution to the total integrated scattered flux. Such a scattering distribution is shown in Figure 6 where the scattered intensity is plotted as a function of the scattering angle in Cartesian rather than polar co-ordinates.[88] In this example of Ar scattering from Pt(111), a somewhat unusual but not unprecedented[78,80] bimodal *velocity* distribution of the scattered Ar was observed as well. A typical distribution is shown in Figure 7, where the mass spectrometric intensity is plotted as a function of

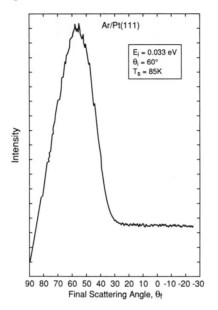

Figure 6 *Scattering distribution of Ar from the Pt(111) surface as a function of the final scattering angle, θ_f, at a surface temperature of 85 K*
(Reproduced with permission from reference 88)

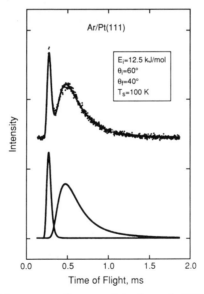

Figure 7 *Time-of-flight spectra of scattered Ar from the Pt(111) surface at a surface temperature of 100 K. The upper curve represents the experimental data, and the bottom curves represent the deconvolution of these data into a 'direct-inelastic' scattering component (sharp peak, high energy) and a 'trapping-desorption' scattering component (diffuse peak, lower energy)*
(Reproduced with permission from reference 88)

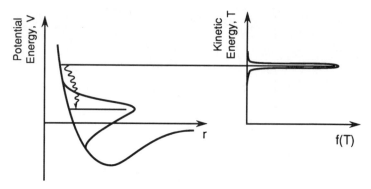

Figure 8 *Schematic diagram illustrating the interaction of a molecular beam of atoms or molecules [represented by the narrow Gaussian kinetic energy distribution, f(T)] with the repulsive part of the surface potential to produce the 'direct-inelastic' and 'trapping-desorption' scattering components observed in Figure 7*

the flight time. If the surface temperature is lower than the effective gas temperature, the trapping probability would be the ratio of the desorbing flux in the 'slow' time-of-flight peak to the total desorbing flux. The diagram shown in Figure 8 helps to clarify this discussion. A narrow (Gaussian) velocity distribution of Ar scatters inelastically from the cold Pt surface. This inelastic scattering both broadens the scattered velocity distribution of the Ar and reduces the mean velocity. Those atoms which have positive total energy scatter in the direct-inelastic lobular peak of Figure 6 with the 'fast' time-of-flight distribution peak of Figure 7. Likewise, those (trapped) atoms which have negative total energy are desorbed in a broad angular distribution and correspond to the 'slow' peak in the time-of-flight velocity distribution of Figure 7. Ar scattering from Pt(111) is discussed more fully below insofar as the energy-scaling relationship for trapping is concerned.

Accommodation vs. *Trapping*

The concepts of trapping and of accommodation are related but are by no means identical, and it is quite important to understand clearly what each implies and the differences between them. An energy accommodation coefficient, $a(T)$, is defined by:

$$a(T) \equiv (E_i - E_f)/(E_i - E_s) \quad (58)$$

where T refers to the gas temperature, E_i and E_f are the thermally averaged initial (before scattering) and final (after scattering) gas energies, and E_s is the thermal energy of the solid. Similar accommodation coefficients can be defined in terms of initial and final temperatures, again obtained from a proper thermal average. The person who has contributed more than anyone else to the measurement and to our understanding of accommodation

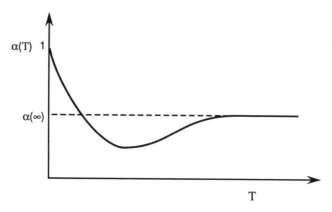

Figure 9 *The dependence of the energy accommodation coefficient, a(T) [as defined in equation (58)] on gas temperature, T*

coefficients is Lloyd Thomas, and the reader is referred to the original literature for a thorough discussion of this issue.[89–94]

The variation in the accommodation coefficient with (gas) temperature is shown schematically in Figure 9. At very low temperatures, the accommodation coefficient is large, due entirely to the fact that trapping dominates under these conditions. As the temperature increases, the accommodation coefficient decreases since the trapping probability is decreasing. The minimum in $a(T)$ is due to two opposing effects. As we shall discuss later, the trapping probability decreases rapidly with increasing gas temperature or, equivalently, gas kinetic energy (it also decreases much less rapidly with increasing surface temperature). However, energy exchange between the gas and the surface *increases* with increasing gas temperature, and the accommodation coefficient approaches the well known asymptotic limit of:[95]

$$a(\infty) = 2.4\, \mu^*/(1 + \mu^*)^2 \qquad (59)$$

as the gas temperature approaches infinity, and the trapping probability approaches zero. In equation (59), μ^* is the ratio of the mass of the gas atom or molecule to the (effective) mass of a surface atom.

Without thinking too deeply, one might reason that the accommodation coefficient will always be greater than the trapping probability (or equal to one another when both are unity). The logic here would be that each molecule that is trapped is accommodated completely, whereas there can be partial accommodation in the absence of trapping. This logic, however, is not quite correct for two related reasons, both of which have to do with the somewhat subtle but rather important concept of trapping and accommodation in a direction perpendicular to the surface *vs.* the two Cartesian directions parallel to the surface. Previously, we defined the trapping probability as the fraction of incident molecules whose total energy after scattering is negative with respect to the vacuum zero of energy, *cf.* Figure 5. This is a reasonable and a correct definition if the trapping probability is a universal function of the incident kinetic energy, *independent of the angle of incidence*. When this relationship is observed, the system is said to obey *total-energy scaling*. This situation would be expected when the impinging gas-phase particle experiences, in some sense, a three-dimensional interaction potential in which all three components of its linear momentum must be 'accommodated', *i.e.* there is rapid exchange between the parallel and perpendicular components of the momentum.

The opposite extreme is when the gas particle experiences an effective one-dimensional potential perpendicular to the surface. In this case only the perpendicular component of the linear momentum will have to undergo a dissipative collision in order for trapping to occur. When this type of gas–surface interaction occurs, the trapping probability is a universal function of the normal component of the incident momentum, *i.e.* it is a universal function of $E_i \cos^2 \theta_i$, where E_i is the incident kinetic energy, and θ_i is the angle of incidence with respect to the surface normal.* This relationship is known as *normal-energy scaling*. In view of this, one can easily imagine the possibility of an atom or a molecule being 'trapped' at a surface, but yet not having the parallel components of its momentum (energy) accommodated to the surface temperature. This presents the possibility that the trapping probability can actually be greater than the accommodation coefficient! Intuitively, one might expect *total-energy scaling* to be more likely on a geometrically rough surface, and conversely for normal-energy scaling to be more likely on a geometrically smooth surface. Next, we shall evaluate this naïve expectation as well as delve more deeply into the entire issue of energy scaling in trapping by considering some experimental molecular beam results for the trapping of Ar on the Pt(111) surface.[51,96-98]

Mullins, *et al.*,[51] employing molecular beam techniques, used the flux-weighted intensity of the trapping-desorption component of the scattering distribution as a measure of the trapping probability of Ar on Pt(111). Numerous kinetic energies were employed at three different angles of incidence, and the measurements were repeated at three different surface temperatures. The experimental results are shown in Figures 10–12 for surface temperatures of 80, 190, and 273 K, respectively. The trapping probability decreases strongly with increasing translational energy (gas 'temperature'), which should come as a surprise to no one. Notice also, however, that the trapping probability has a strong dependence on the angle of incidence, increasing with increasing angle of incidence. The important conclusions of this observation are that total-energy scaling does not apply to this system, and that the parallel components of the momenta do not have to be entirely dissipated in order for trapping to occur.

It is extremely interesting that for this system of Ar scattering from Pt(111) that *neither* total-energy *nor* normal-energy scaling is observed. Rather, the abscissae of Figures 10–12, which permit the data to be plotted on a

---

* Recall that the normal component of the momentum is proportional to $\cos\theta_i$, and the kinetic energy is proportional to the square of the momentum.

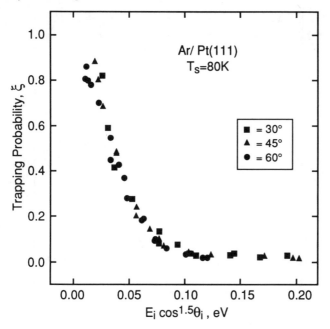

Figure 10 *Trapping probability of* Ar *on* Pt(111) *as a function of* $E_i \cos^{1.5}\theta_i$ *at a surface temperature of* 80 K
(Reproduced with permission from *Chem. Phys. Lett.*, 1989, **163**, 111)

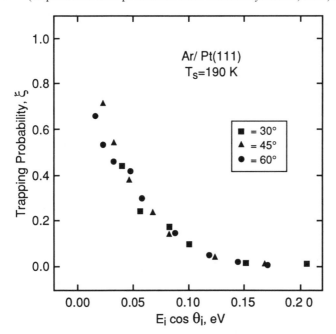

Figure 11 *Trapping probability of* Ar *on* Pt(111) *as a function of* $E_i \cos\theta_i$ *at a surface temperature of* 190 K
(Reproduced with permission from *Chem. Phys. Lett.*, 1989, **163**, 111)

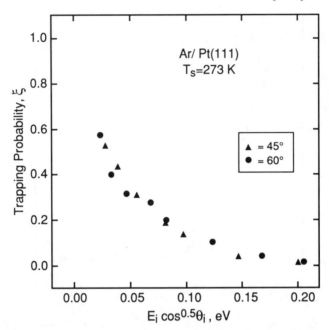

Figure 12 *Trapping probability of Ar on Pt(111) as a function of $E_i \cos^{0.5} \theta_i$ at a surface temperature of 273 K*
(Reproduced with permission from *Chem. Phys. Lett.*, 1989, **163**, 111)

universal curve, are of the form $E_i \cos^n \theta_i$, where n is neither zero nor two. Even more interesting is the fact that the scaling exponent n is a function of surface temperature, varying from *ca.* 1.5 at 80 K to *ca.* 0.5 at 273 K. This exponent is plotted as a function of surface temperature in Figure 13. Although there is no justification for extrapolating this line either to below 80 K or to above 273 K, it could be significant that normal-energy scaling is predicted in the low-surface temperature limit, and total-energy scaling is predicted for $T_s \geq 375$ K.

How should we qualitatively think about these trapping-dynamics results from a physical point of view? If normal-energy scaling is obeyed, only the component of momentum perpendicular to the surface must be dissipated 'instantaneously' in order for trapping to occur. This corresponds to slow rates of coupling between perpendicular and parallel components of the momentum, and trapping can occur in this case even if the total energy 'just after' the collision (within the order of 10^{-12} s) is positive. The parallel components of the momentum then are dissipated by energy exchange with the substrate. If this exchange rate is rapid compared to the rate of exchange between the parallel and perpendicular components, then the total energy of the particle becomes negative (*i.e.* it becomes trapped in this stricter sense), and normal-energy scaling is preserved. Let us now consider the opposite extreme, namely, that of total-energy scaling. In this case, the rate of exchange between the parallel and perpendicular components of the momen-

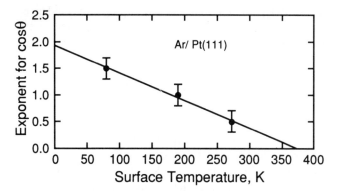

Figure 13 *Dependence of the exponent* n *in the scaling relation* $E_i \cos^n \theta_i$ *on the surface temperature. The solid line is a linear fit to the experimental data*
(Reproduced with permission from *Chem. Phys. Lett.*, 1989, **163**, 111)

tum is sufficiently rapid that all three components of the momentum must be dissipated 'during the collision' (with a concomitant negative total energy) in order for trapping to occur.* The data for Ar trapping on Pt(111) are quite interesting because they obey neither normal- nor total-energy scaling, and especially because the scaling relation that is observed is a function of surface temperature. This very likely implies that the rate of coupling between the parallel and perpendicular components of the momentum is comparable to the rate of dissipation of the parallel components of the momentum. It is reasonable to suppose that 'thermal roughness' (fluctuations) enhances the rate of coupling between the parallel and perpendicular components of the momentum. If this is true, then Figure 13 is easy to understand: In the limit of low-surface temperatures, parallel momentum exchange with the lattice dominates, and normal-energy scaling is observed; whereas in the limit of high-surface temperatures, exchange between the parallel and perpendicular components of the momentum dominates, and total-energy scaling is observed. Detailed calculations of the temperature-dependent coupling constants are needed to describe quantitively the interesting intermediate regime for which $0 < n < 2$.**

Another interesting feature to note in Figures 10–12 is the surface temperature dependence of the trapping probability. For example, at a low-

---

* We might expect the coupling constant of this exchange to be larger on a more corrugated surface.

** Very recent calculations and experiments suggest an alternative explanation for the increasing importance of parallel momentum as the surface temperature is raised. M. Head-Gordon, J. C. Tully, C. T. Rettner, C. B. Mullins and D. J. Auerbach, (*J. Chem. Phys.* 1991, **94**, 1516) argue that the residence time for atoms trapped at high surface temperature is too short for parallel momentum to be fully accommodated prior to desorption. Under these conditions, the trapping probability cannot be unambiguously defined. However, these authors contend that the trapping-desorption flux directed close to the surface normal will indeed fall with increasing parallel momentum—which is sufficient to account for the observed trends. The issue of the differential accommodation of momenta both parallel to and perpendicular to the surface is discussed later in this section.

incident kinetic energy of *ca.* 0.032 eV and an angle of incidence of 60°, the trapping probability decreases from *ca.* 0.8 at 80 K, to 0.65 at 190 K, to 0.6 at 273 K. This is a modest variation in the trapping probability with surface temperature compared to the variations that are observed with varying gas temperature. Indeed, a close inspection of Figures 10–12 reveals that at higher kinetic energies, there is a slight *increase* in the trapping probability with increasing surface temperature. The reason for this somewhat counter-intuitive result is due to the fact that whereas surface atoms that are moving in a direction toward the impinging Ar atoms will reduce the trapping probability of 'slow' incident species, at high energies, where trapping is less probable, collisions with surface atoms that are moving away from the incident Ar have the *net* effect of increasing the trapping probability. This effect was seen in connection with the ultrasimple 'cube' models of many years ago.[5,83,84]

This discussion of trapping and accommodation allows us to draw an important conclusion concerning an additional mechanism of surface diffusion and surface reactions. In particular, prior to dissipation of its parallel momentum, a trapped particle can execute free translational motion parallel to the surface. This is to be contrasted with the much slower motion associated with activated hopping from site to site. This type of 'ice skating motion' would especially be expected when the trapping probability scales with $E_i \cos^n \theta_i$ with $n > 0$. In this case the dissipation of parallel momentum is sufficiently slow compared to dissipation of normal momentum that it is manifested in measurements of the *average* trapping probability. Even when total-energy scaling is observed in the measurements of trapping probabilities, it is, nevertheless, almost certain that parallel momentum is dissipated less rapidly than perpendicular momentum. In order to evaluate the potential physical significance of particles that are executing this type of free translational motion in two dimensions, we need to be able to estimate the rate of parallel momentum dissipation, *i.e.* the lifetime of the free particle.

Tully has performed trajectory calculations within the context of a generalized Langevin formalism for Xe scattering from the Pt(111) surface that help us address this interesting and important issue of the lifetime of this freely translating 'precursor' particle.*,[99] Two different groups have measured experimentally the trapping probability of Xe on the Pt(111) surface. One group[79] found that at a surface temperature of 85 K, the trapping probability scaled with $E_i \cos^{1.6} \theta_i$; whereas the other group[100] found that at a surface temperature of 95 K, the trapping probability scaled with $E_i \cos \theta_i$. Although this discrepancy clearly needs to be resolved (a difference of 10 K in the surface temperature should not have this much of a difference on the scaling behaviour), the important point for our discussion here is that normal-energy scaling (with its slow parallel momentum dissipation) was not

---

* We refer to the particle in this state as a 'precursor' because it is a transient intermediate between the reactant (the gas-phase particle) and the product of the reaction (the thermally accommodated ground-state particle on the surface), *i.e.* it is a precursor to the ground-state particle that executes site-to-site adatom hopping on the surface.

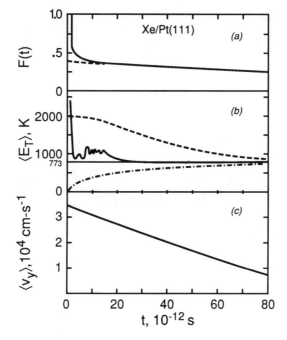

Figure 14 *Calculated time dependence of the thermalization of Xe atoms on the Pt(111) surface, as determined theoretically by Tully.[99] The incident energy is 3.9 kcal mol^{-1}; the incident angle is 45° with respect to the surface normal; and the surface temperature is 773 K. (a) Fraction F(t) Xe of atoms remaining on the surface as a function of time, t. (b) Mean kinetic energy, $\langle E_T \rangle$ (expressed in K), in the x (—·—·—), y (————) and z (———) direction for those Xe atoms remaining on the surface. The z-direction is the surface normal (see Figure 15). (c) Mean of the y-component of velocity, $\langle v_y \rangle$, for those Xe atoms remaining on the surface*
(Reproduced with permission from *Faraday Disc. Chem. Soc.*, 1985, **80**, 291)

observed. Indeed, if the data for Ar scattering from Pt(111) that were discussed earlier are indicative, then we might expect Xe scattering from the same surface to obey total-energy scaling at a surface temperature of 773 K, which is the surface temperature employed by Tully in his calculations. The major conclusion is that we do not expect the rate of parallel momentum dissipation calculated by Tully to be unusually slow. Indeed, quite the contrary.

Pertinent results of the calculations of Tully for Xe trapping on Pt(111) are shown in Figure 14. The incident translational energy of the Xe is *ca.* 0.17 eV with an angle of incidence of 45° onto the Pt(111) surface of which the temperature is 773 K. The fraction of Xe atoms that remains on the surface is shown in Figure 14a. The initial rapid decrease is due to atoms whose perpendicular component of momentum is not dissipated, and, hence, which are not trapped. Notice that this initial decrease has been damped after *ca.*

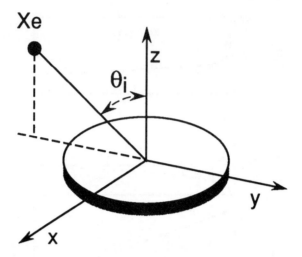

Figure 15 *Co-ordinate system used for the discussion of data presented in Figure 14*

10 ps, and there is a much slower decrease in the fraction of Xe atoms on the surface at longer times. The latter is due to the *desorption* of Xe atoms that have been trapped on the surface. The dashed extrapolation in Figure 14a to $t = 0$ represents the initial trapping probability, which is *ca.* 0.4 under these conditions. The solid curve in Figure 14b shows the mean kinetic energy, expressed as a temperature, of the Xe atoms in a direction perpendicular to the surface (the z-direction). The very rapid approach to the thermal energy of the solid quantifies the rate of exchange between the perpendicular component of the momentum of the Xe and the Pt substrate, and it tracks closely the rapid decrease in the fraction of Xe atoms remaining on the surface, *cf.* Figure 14a. These two curves quantify the time scale of trapping and of accommodation perpendicular to the surface. Notice that the rate of dissipation of parallel momentum is much slower than that of perpendicular momentum, *cf.* the dashed and dot-dashed curves of Figure 14b. The parallel component of the translational energy in the y-direction is initially equal to 1960 K. This is shown in the dashed curve in Figure 14b, and the co-ordinate system is shown in Figure 15. The parallel component of the translational energy in the x-direction [*cf.* Figures 14b and 15] is initially zero. There is a slow net rate of exchange of parallel momentum with the substrate compared to the rate of exchange of perpendicular momentum.

The important conclusion to be drawn from this discussion is that there is a significant period of time in which the atom is trapped in a direction perpendicular to the surface but yet is executing essentially free translational motion parallel to the surface. Notice that the y-component of the velocity of the trapped Xe atom is shown in Figure 14c, and the time scale for its being trapped into localized sites as dictated by the two-dimensional corrugation of the interaction potential parallel to the surface, is of the order of 100 ps. Insofar as motion parallel to the surface is concerned, there are two different

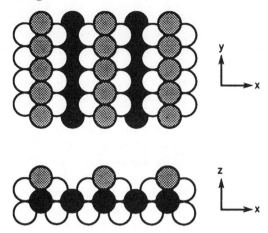

Figure 16 *Schematic diagram demonstrating the missing-row structure of the Ir(110)–(1 × 2) reconstructed surface*

time scales. On a short time scale, there is free translational motion; whereas on a longer time scale, adatom hopping from site to site occurs. The rates of surface diffusion and bimolecular reactions (*e.g.* the Langmuir–Hinshelwood mechanism of Section 3) are invariably discussed in terms of adatom (or admolecule) hopping on the surface. We are forced to conclude from the discussion here, however, that (at least in principle) a second mechanism involving freely translating particles in the two dimensions parallel to the surface must also be considered. The issue is not whether such states exist. Clearly, they do. The issue is rather whether their lifetime is sufficiently long to permit any physical manifestations of their existence to be observed. We note that their presence has been inferred in the case of atomic hydrogen adsorption and atom–atom recombination and desorption on the Si(100)–(2 × 1) surface.[101,102] Their existence has also been inferred in measurements of surface diffusion coefficients,[101–106] and an appropriate analysis involving Monte Carlo simulations which includes these precursors has been presented.[107,108] Finally, we note that the existence of these precursors was anticipated years ago in connection with the formulation of a two-dimensional atomic band structure of hydrogen adatoms on the Ni(111) surface.[1,109,110]

5 Example of a Trapping-Mediated Surface Reaction: Dissociative Chemisorption of Ethane on Ir(110)–(1 × 2)

We shall illustrate many of the important concepts that have been articulated in this chapter with an example, namely, the dissociative chemisorption of ethane on the reconstructed Ir(110)–(1 × 2) surface. This surface of Ir, shown schematically in Figure 16, is highly corrugated; indeed it may be viewed as a series of (111) microfacets inclined 109.5° with respect to one another. The structure of the surface, as determined by low-energy electron diffraction, contains a 'missing' close-packed row in the surface plane in the [011] crystallographic direction.[1,111–115] The 'openness' of this surface might be expected to render it more reactive for certain classes of chemical reactions. Indeed, this has been found to be the case for alkane activation, *i.e.* the dissociative chemisorption of saturated hydrocarbon molecules.

The remarkable reactivity of this surface was first noticed almost ten years ago: following molecular adsorption at *ca.* 100 K, C—H bond cleavage was observed at a threshold temperature of *ca.* 130 K for every alkane with the exception of methane, which desorbed well below 130 K.[72] By contrast, on the smooth Ir(111) surface, no small alkane could be similarly activated.[116] Indeed, even cyclopropane, which adsorbs dissociatively on Ir(110)–(1 × 2) below 100 K,[117] only adsorbs and desorbs molecularly and reversibly on the Ir(111) surface.[116] These observations give us two important pieces of information. First, there is a significant dependence on the surface geometrical structure for the activation of alkanes.* In addition, it appears that alkane activation (except for methane) on the Ir(110)–(1 × 2) surface is unactivated with respect to the vacuum zero of energy (the molecular alkane infinitely far from the surface and at rest). If there is an activation energy of reaction with respect to this zero of energy, then we would expect to observe no measurable reaction following molecular adsorption at low-surface temperature with subsequent heating. Only molecular desorption would be expected, since it would be favoured both for energetic reasons (the Maxwell–Boltzmann factor) and for entropic reasons (the magnitudes of the expected pre-exponential factors of the two competing rate coefficients). This point is discussed more fully later in this section.

Consequently, we conclude tentatively (and we shall verify later) that the potential energy of the reaction

$$C_2H_6(g) \underset{k_d}{\overset{\xi F}{\rightleftharpoons}} C_2H_6(p) \xrightarrow{k_r} C_2H_5(c) + H(c) \qquad (60)$$

along the reaction co-ordinate on Ir(110)–(1 × 2) is qualitatively consistent with the solid curve of Figure 17. If the potential curves for the physical adsorption of ethane and for the chemisorption of an ethyl radical and a hydrogen atom do indeed cross below the zero of energy, as indicated in Figure 17, this has an important qualitative consequence on the probability of dissociation of an impinging ethane molecule, namely, if the surface temperature is increased, then the reaction probability should decrease. This statement can be rendered quantitative by considering the *initial* probability of dissociative chemisorption of ethane, *i.e.* the reaction probability on a clean surface. If the surface temperature is sufficiently high that the con-

---

* This statement can be rendered quantitative for the case of ethane activation on the similarly reconstructed Pt(110)–(1 × 2) surface *vs.* the close-packed Pt(111) surface. On Pt(110)–(1 × 2) the activation energy of C—H bond cleavage of ethane is 2.8 kcal mol^{-1} (with respect to the vacuum zero of energy),[118] whereas on Pt(111) it is *ca.* 8.9 kcal mol^{-1}.[119]

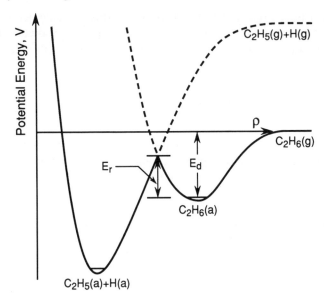

Figure 17 *One-dimensional potential energy diagram (along the reaction co-ordinate, ρ) proposed for the dissociative chemisorption of C_2H_6 on Ir(110)–(1 × 2)*

centration of physically adsorbed ethane is negligibly small, then a pseudo-steady-state approximation implies that:

$$\frac{d\theta_p}{dt} = \xi F - (k_d + k_r)\theta_p \cong 0 \tag{61}$$

where θ_p is the fractional surface coverage of the physically adsorbed ethane, ξ is the trapping probability of molecular ethane into the well for physical adsorption, F is the impingement flux of ethane (on a per site basis), and k_d and k_r are the rate coefficients for molecular desorption and reaction (dissociative chemisorption), respectively.

From equation (61) we find that the pseudo-steady-state fractional coverage of physically adsorbed ethane on the surface is given by:

$$\theta_p \cong \frac{\xi F}{k_d + k_r} \tag{62}$$

and the reaction rate is given by:

$$R_r = k_r \theta_p = \xi F \left(\frac{k_r}{k_r + k_d}\right) \tag{63}$$

Note that the form of equation (63) is intuitively predictable, namely, the rate of the reaction is given by the rate of production of reactants (ξF) multiplied by the rate coefficient for reaction, which is normalized by the sum of the rates of reaction and desorption for this two-channel 'reaction'. The probability that an impinging ethane molecule adsorbs dissociatively is simply the rate of reaction divided by the impingement flux, as defined above, *i.e.*:

$$P_r \equiv \frac{R_r}{F} = \xi \left(\frac{k_r}{k_r + k_d} \right) = \frac{\xi}{1 + k_d/k_r} \tag{64}$$

The form of equation (64) supports our earlier assertion that for the potential energy diagram of Figure 17, the reaction probability decreases as the surface temperature increases. Since both rate coefficients are of the Polanyi–Wigner form, namely:

$$k_i = k_i^{(0)} e^{-E_i/k_B T_s} \tag{65}$$

then equation (64) can be written as follows:

$$P_r = \frac{\xi}{1 + \frac{k_d^{(0)}}{k_r^{(0)}} e^{-(E_d - E_r)/k_B T_s}} \tag{66}$$

Since $E_d > E_r$ in this case, the second term in the denominator becomes progressively smaller as the surface temperature decreases, which gives rise to a larger value of the reaction probability. The corresponding maximum and minimum values of the reaction probability for the case when $E_d > E_r$ are given by:

$$\left. \begin{array}{l} P_r^{max} \to \xi \quad \text{in the limit of low } T_s \\ P_r^{min} \to \dfrac{\xi}{1 + k_d^{(0)}/k_r^{(0)}} \quad \text{in the limit of high } T_s \end{array} \right\} \tag{67}$$

It is apparent from equation (64) that the reaction probability is composed of two different ingredients: a 'dynamics' part describing the gas–surface collision (embodied by ξ) and a surface 'kinetics' part describing the competing dissociative chemisorption and desorption reactions (embodied by k_d/k_r). The dissociative chemisorption of ethane on the Ir(110)–(1 × 2) surface is the system that has been studied the most thoroughly with the aim both of verifying the assumptions inherent in the derivation of equation (64) and of determining the parameters of fundamental physical significance contained therein, namely, ξ, $k_d^{(0)}/k_r^{(0)}$, and $E_d - E_r$. These results are summarized next.

Mullins and Weinberg[120] employed a method similar to the reflectivity technique of King and Wells[121] to measure the initial probability of trapping of ethane on the Ir(110)–(1 × 2) surface at 77 K as a function of impact energy and angle of incidence. The experiments were conducted in a supersonic molecular beam scattering machine with various seedings of the ethane with He, Ar, and He–Ar mixtures, and with variable nozzle temperatures in

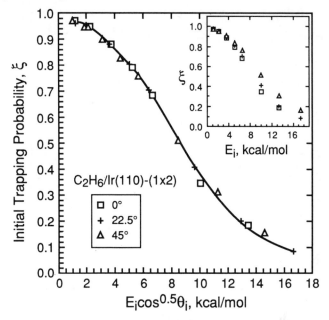

Figure 18 *Initial trapping probability, ξ, of C_2H_6 on $Ir(110)$–(1×2) as a function of $E_i \cos^{0.5} \theta_i$ at a surface temperature of 77 K. The dependence of ξ on impact energy, E_i, is shown in the inset for comparison*
(Reproduced with permission from *J. Chem. Phys.*, 1990, **92**, 3986)

order to vary the impact energy. The results are shown in Figure 18, and it is clear that the initial trapping probability of the ethane scales with $E_i \cos^{0.5} \theta_i$ on this surface. As would be expected, the trapping probability approaches unity at low impact energies, and it decreases monotonically at larger values of $E_i \cos^{0.5} \theta_i$. The fact that the scaling relation is closer to total-energy scaling than to normal-energy scaling indicates that there is rather rapid redistribution of parallel momentum into perpendicular momentum, as would be expected on this microscopically rough surface. Under these conditions ($T_s = 77$ K) where the adsorption is molecular and irreversible[73] (the small fraction of ethane that experiences 'direct' dissociation at the highest value of $E_i \cos^{0.5} \theta_i$ has been subtracted from the data of Figure 18, *vide infra*), those molecules that are trapped are obviously also accommodated to the surface temperature.

It is now possible to test the validity of equation (64) by measuring $P_r(T_s)$ since the trapping probability that appears in that equation has been determined separately, *cf.* Figure 18. We should consider in passing, however, two points related to this issue. First, the probability of dissociative chemisorption will be measured at surface temperatures between 150 and 500 K (not at 77 K). We shall employ the values of the trapping probability that were measured at 77 K, however, since it has been shown that ξ is not a strong function of surface temperature, especially over the limited range of surface temperature that we shall consider here.[50] Furthermore, we shall assume that

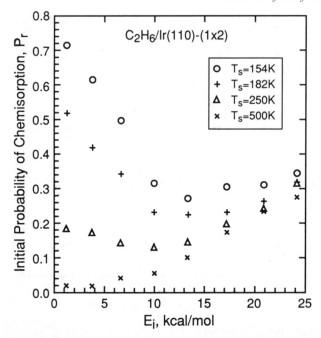

Figure 19 *Initial probability of dissociative chemisorption, P_r, of C_2H_6 on Ir(110)–(1 × 2) as a function of impact energy, E_i, at variable surface temperatures*
(Reproduced with permission from *J. Chem. Phys.*, 1990, **92**, 4508)

those ethane molecules that are trapped at the surface also accommodate to the surface temperature, *i.e.* we shall use the measured surface temperature in our data analysis employing equation (64). Since the rate coefficient of desorption of physically adsorbed ethane has been measured to be:[72]

$$k_d = 10^{13} \, e^{-(7.7 \text{kcal mol}^{-1})/k_B T_s} \quad \text{s}^{-1} \tag{68}$$

we can assess this assumption. At 500 K, the highest surface temperature employed, the residence time of the ethane on the surface is *ca.* 230 ps, from equation (68), and this should be more than sufficient for complete accommodation of both the perpendicular and the parallel components of the momentum, *cf.* Figure 14.

Using the same reflectivity technique and molecular beam scattering machine mentioned above, Mullins and Weinberg[122] measured the initial probability of dissociative chemisorption of ethane on the Ir(110)–(1 × 2) surface as a function of impact energy, parametric in surface temperature.* The results of these measurements for surface temperatures between 154 and 500 K, and impact energies between 1.1 and 24.3 kcal mol^{-1} are

---

* Since the polar angle of incidence is zero degrees in these measurements, the energy-scaling relationship is irrelevant.

shown in Figure 19. The first thing to notice in Figure 19 is that for $T_S \lesssim 250$ K, there is a minimum in $P_r(E_i)$. Below ca. 12–15 kcal mol^{-1}, P_r decreases with increasing E_i; whereas above this impact energy, P_r increases with increasing E_i. At lower values of the impact energy, the dissociative chemisorption is dominated by the trapping-mediated mechanism, and is a strong function of the surface temperature, as may be seen in Figure 19.* At higher impact energies, the probability of dissociative chemisorption increases because the 'direct' mechanism of chemisorption dominates in this regime. At sufficiently high impact energies, where the trapping probability approaches zero (cf. Figure 18), only the direct channel contributes, and the probability of chemisorption is approximately independent of temperature. This regime is being approached at the highest impact energies in Figure 19. Indeed, at a surface temperature of 500 K, there is very little trapping-mediated chemisorption at any impact energy, and, hence, these data allow us to assess the relative importance of the two channels (trapping-mediated and direct) at the other surface temperatures as a function of the impact energy. We should also note that the minimum in $P_r(E_i)$ moves to lower impact energies as the surface temperature increases (cf. Figure 19) because the probability of the direct reaction is not a significant function of surface temperature, whereas the probability of the trapping-mediated reaction decreases rapidly with increasing surface temperature.

Measurements were also made of the probability of dissociative chemisorption at $T_s > 650$ K where there is no detectable contribution of the trapping-mediated channel at any impact energy.** This allows a subtraction of the direct component from all of the data of Figure 19 (for each temperature at each impact energy), which then permits a strong test of the validity of equation (64) insofar as its proposed description of trapping-mediated chemisorption is concerned. For this purpose we rewrite equation (64) as follows:

$$\frac{\xi}{P_r} - 1 = \frac{k_d}{k_r} \qquad (69)$$

and we note that both ξ and P_r have been measured independently. If equation (69) is valid, then a plot of $\ln[(\xi/P_r) - 1]$ as a function of reciprocal surface temperature should be linear with a slope equal to $-(E_d - E_r)/k_B$ and an intercept equal to $\ln[k_d^{(0)}/k_r^{(0)}]$. The data of Figure 19 are plotted in this Arrhenius form in Figure 20, and a straight line is observed over the entire temperature range employed, 150–500 K. This result validates the

---

* Note that the fact that the initial probability of dissociative chemisorption decreases with increasing surface temperature proves our assertion made earlier in this section that $E_d - E_r$, e.g. in equation (66), is positive; and the potential energy diagram in Figure 17 describes this system qualitatively.
** The 'resolution' of the reflectivity technique for measurements of the probability of adsorption in this molecular beam machine is ca. ±0.015.

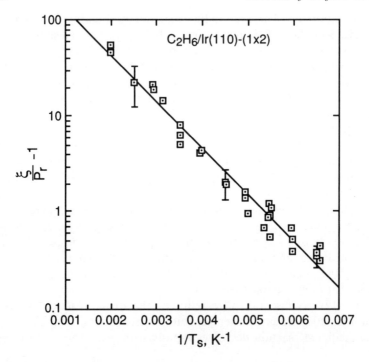

Figure 20 *Verification of the trapping-mediated mechanism proposed in equation (64) for the dissociative chemisorption of C_2H_6 on Ir(110)–(1 × 2)* (Reproduced with permission from *J. Chem. Phys.*, 1990, **92**, 4508)

simple model that was employed to describe the trapping-mediated chemisorption, *cf.* equations (61)–(67).

The slope and intercept of Figure 20 imply that the following rate parameters describe the trapping-mediated dissociative chemisorption of ethane on the Ir(110)–(1 × 2) surface:

$$E_d - E_r = 2.2 \text{ kcal mol}^{-1} \tag{70}$$

and

$$k_d^{(0)}/k_r^{(0)} = 390 \tag{71}$$

where there is an uncertainty (one standard deviation) of ± 0.2 kcal mol^{-1} in the difference in the activation energies, and an uncertainty of ± 150 in the ratio of the pre-exponential factors. Since the rate coefficient of desorption of physically adsorbed ethane was measured independently, *cf* equation (68), the rate coefficient of the elementary surface reaction involving C—H bond cleavage is found to be:

$$k_r \cong 3 \times 10^{10} \, e^{-(5.5 \text{kcal mol}^{-1})/k_B T_s} \text{ s}^{-1} \tag{72}$$

Table 1 *Rate parameters describing both the desorption and the dissociation of physically adsorbed ethane on Ir(110) – (1 × 2), cf., Figure 17*

$k_r^{(0)}$ s^{-1}	E_r/kcal mol^{-1}	$k_d^{(0)}$ s^{-1}	E_d/kcal mol^{-1}
3×10^{10}	5.5	10^{13}	7.7

These rate parameters are summarized in Table 1. Note that the activation energy of the surface reaction, $E_r = 5.5$ kcal mol^{-1}, is measured with respect to the bottom of the physically adsorbed well. It is clear from Figure 17 that the potential curves cross, in this case, 2.2 kcal mol^{-1} below the gas-phase energy zero, and that the apparent activation energy with respect to this latter energy zero is -2.2 kcal mol^{-1}.

The fact that $k_d^{(0)}/k_r^{(0)} > 1$ for this reaction is completely expected.[123] Using the formalism developed in Section 3, the ratio of the pre-exponential factors of desorption and reaction may be written as:

$$\frac{k_d^{(0)}}{k_r^{(0)}} = \frac{\langle \kappa_d \rangle \langle \hat{q}_{C_2H_6}^\ddagger \rangle_d'}{\langle \kappa_r \rangle \langle \hat{q}_{C_2H_6}^\ddagger \rangle_r'} \tag{73}$$

where both the electronic part of the partition functions (the Maxwell–Boltzmann factor) and the zero-point energy contribution of the vibrational partition functions have been extracted from the two transition state partition functions that appear in equation (73). Whereas the constrained and localized partition function for reaction, $(\hat{q}_{C_2H_6}^\ddagger)_r'$, will contain twenty-three degrees of vibrational freedom, we expect that the partition function for desorption, $(\hat{q}_{C_2H_6}^\ddagger)_d'$, will contain two or three rotational degrees of freedom (with the remainder of the twenty-three total degrees of freedom being vibrational). On this basis, for a non-spherically symmetric reactant (such as ethane) we would expect $k_d^{(0)}/k_r^{(0)}$ to lie between ca. 100 and 1000 (in the absence of any quantitative information concerning the relative magnitudes of $\langle \kappa_d \rangle$ and $\langle \kappa_r \rangle$). The measured value of $k_d^{(0)}/k_r^{(0)}$ of 390 is completely consistent with this expectation.

Finally, we note that Verhoef, *et al.*[124] have measured the initial probability of direct dissociative chemisorption of ethane on Ir(110)–(1 × 2) at a sufficiently high surface temperature ($T_s > 600$ K) that the trapping-mediated channel makes no detectable contribution. These results are shown in Figure 21, where the initial probability of chemisorption is plotted as a function of normal-impact energy, *i.e.* $E_i \cos^2 \theta_i$, for both C$_2$H$_6$ and C$_2$D$_6$.* The observed isotope effect is a consequence of zero-point energy differences with possible contributions also from tunnelling assistance (considering the time scale of the direct interaction which is sub-picosecond). Notice that the

---

* As may be seen in Figure 21, the appropriate scaling relation for dissociative chemisorption of ethane on Ir(110)–(1 × 2) is normal-energy scaling, whereas the scaling relation for molecular trapping is $E_i \cos^{0.5} \theta_i$, *cf.* Figure 18.

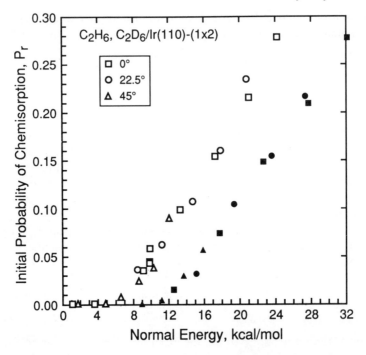

Figure 21 *Initial probability of dissociative chemisorption, P_r, of both C_2H_6 (open symbols) and C_2D_6 (closed symbols) on Ir(110)–(1 × 2) as a function of normal-impact energy $(E_i \cos^2 \theta_i)$ at a surface temperature of* 600 K

threshold value of the normal energy required for detectable direct reaction is *ca.* 8 kcal mol^{-1}, which is *ca.* 10 kcal mol^{-1} greater than the minimum energy pathway from reactant to product (*i.e.* the trapping-mediated mechanism). The reason for this is clear from Figure 1 and was discussed in connection with that figure. In the direct mechanism excess total energy is necessary in order to transfer sufficient energy to the reaction co-ordinate in order for the reaction to occur, *i.e.* for there to be a 'funnelling' of reaction products into their final ground state, rather than a reflection of the reactants from the repulsive wall of the potential. It is for this reason that most of the surface chemistry that occurs in the 'real world' is trapping-mediated rather than direct.

As mentioned in Section 3, one of the few exceptions to this general rule is likely to be the dissociative chemisorption of methane. The reason for this is that methane typically has both quite large energy barriers for activation (with respect to the gas-phase energy zero, *i.e.* $E_r - E_d > 0$ for methane) as well as quite small well depths for physical adsorption, of the order of 4 kcal mol^{-1}, or so, on most transition metal surfaces. For example, consider the interaction of methane with a surface at a temperature of 1000 K (gas temperature equal to surface temperature in a 'bulb' rather than a 'beam' experiment). At this temperature and with a well depth for physical adsorption of 4 kcal mol^{-1}, the residence time of the methane on the surface is only about seven vibrational periods, *i.e.* of the order of 1 ps. On a time scale that is this short, the concept of a trapping-mediated reaction becomes ill-defined, and we would expect significant contributions from the direct channel, *i.e.* the direct reaction of methane molecules in the high-energy tail of the Maxwell–Boltzmann distribution. This is an inefficient way for the molecule to react, but in the case of methane it is the only game in town. We must emphasize, however, that this is an exceptional case. Very few chemical reactants have a binding energy on a catalytic surface which is as small as that of methane (or, more precisely, there are very few systems for which $E_r - E_d$ is as large as that of methane). In all fairness we should point out that this opinion is not universally accepted to be true.[125,126] We are eagerly anticipating future research which will decide this scientifically and technologically important issue unambiguously.

6 Synopsis

The concepts of trapping-mediated and direct surface reactions were defined and discussed in Section 2 of this chapter, and we argued intuitively that trapping-mediated reactions should be more commonly observed. These arguments were quantified in Section 3 within the framework of elementary transition state theory. The specific example of CO oxidation on Rh was considered explicitly to illustrate the dominance of the trapping-mediated channel (the Langmuir–Hinshelwood mechanism) over the direct channel (the Eley–Rideal mechanism). In view of this demonstrated importance of trapping-mediated reactions, the concepts of both trapping and accommodation were discussed in detail in Section 4. Finally, in Section 5, all of the concepts were tied together in a discussion of the trapping-mediated dissociative chemisorption of ethane on the Ir(110)–(1 × 2) surface in the limit of low-surface coverages, *i.e.* the initial rate of ethane activation.

It appears that the next step to be taken in order to achieve a more complete understanding of surface reaction kinetics is to combine dynamic Monte Carlo simulations with proper Hamiltonians of the specific systems in order to formulate an improved transition state theory of surface reaction rates: not just initial rates (as in Section 5), but also rates as a function of surface coverage (or, more precisely, rates as a function of local surface configurations). The classes of reactions which will have to be included in the general case are physical adsorption (trapping), molecular chemisorption, desorption, dissociative chemisorption, surface diffusion, and bimolecular surface reactions. The reactants and intermediates which will have to be described properly include gas-phase reactants, 'intrinsic' precursors (weakly adsorbed species that exist above empty surface sites), 'extrinsic' precursors (weakly adsorbed species that exist above occupied surface sites), as well as more strongly chemisorbed reactants. The general problem contains many very different time scales, which must be treated consistently, for example, within the framework of the theory of Poisson processes.[127] Note that these widely different time scales render the formulation and solution of the problem by dynamic Monte Carlo techniques much more practical than the implementation of molecular dynamics techniques. The first steps of a solution, along these lines, to these very challenging problems of surface rate processes have already been taken.[44,49,73,103,107,128-132]

Acknowledgement. We gratefully acknowledge the financial support of the National Science Foundation (grant no. CHE-9003553), the Department of Energy (grant no. DE-FG03-89ER14048), and the Petroleum Research Fund of the American Chemical Society (grant no. 23801-AC5-C). The valuable assistance of Bill Mitchell and Randy Verhoef in the preparation of this chapter is also very much appreciated.

References

1. M. A. Van Hove, W. H. Weinberg, and C.-M. Chan, 'Low Energy Electron Diffraction', Springer-Verlag, Berlin, 1986.
2. J. B. Pendry, 'Low-Energy Electron Diffraction', Academic Press, London, 1974.
3. L. J. Clark, 'Surface Crystallography: An Introduction to Low-Energy Electron Diffraction', Wiley, New York, 1985.
4. R. O. Jones and P. J. Jennings, *Surf. Sci. Rep.*, 1988, **9**, 165.
5. W. H. Weinberg, *Adv. Colloid Interface Sci.*, 1975, **4**, 301.
6. R. Berndt, B. J. Hinch, J. P. Toennies, and Ch. Wöll, *J. Chem. Phys.*, 1990, **92**, 1435.
7. J. P. Toennies, *J. Vac. Sci. Technol.*, 1987, **A5**, 440.
8. J. P. Toennies, *J. Vac. Sci. Technol.*, 1987, **A2**, 1055.
9. M. W. Cole and D. R. Frankl, *Surf. Sci.*, 1978, **70**, 585.
10. J. N. Smith, Jr., *Surf. Sci.*, 1973, **34**, 613.
11. H. Saltsburg, *Ann. Rev. Phys. Chem.*, 1973, **24**, 493.
12. Y. Kuk and P. J. Silverman, *Rev. Sci. Instrum.*, 1989, **60**, 165.
13. Y. Kuk, F. M. Chua, P. J. Silverman, and J. A. Meyer, *Phys. Rev. B*, 1990, **41**, 12393.
14. Ph. Avouris, F. Bozso, and R. J. Hamers, *J. Vac. Sci. Technol.*, 1987, **B5**, 1387.
15. W. H. Weinberg in 'Methods of Experimental Physics', ed., R. L. Park and M. G. Lagally, Academic Press, Orlando, 1985, p. 23.
16. H. Ibach and D. L. Mills, 'Electron Energy Loss Spectroscopy and Surface Vibrations', Academic Press, Inc., New York, 1982.
17. F. M. Hoffmann, *Surf. Sci. Rep.*, 1983, **3**, 107.
18. Y. J. Chabal, *Surf. Sci. Rep.*, 1988, **8**, 211.
19. R. Weissmann and K. Müller, *Surf. Sci. Rep.*, 1981, **1**, 251.
20. K. Wandelt, *Surf. Sci. Rep.*, 1982, **2**, 1.
21. W. F. Egelhoff, Jr., *Surf. Sci. Rep.*, 1987, **6**, 253.
22. C. Argile and G. E. Rhead, *Surf. Sci. Rep.*, 1989, **10**, 277.
23. J. A. Barker and D. J. Auerbach, *Surf. Sci. Rep.*, 1985, **4**, 1.
24. M. P. D'Evelyn and R. J. Madix, *Surf. Sci. Rep.*, 1984, **3**, 413.
25. T. N. Truong, G. Hancock, and D. G. Truhlar, *Surf. Sci.*, 1989, **214**, 523.
26. B. Jackson, *J. Phys. Chem.*, 1989, **93**, 7699.

27. T. H. Upton, W. A. Goddard III, and C. F. Melius, *J. Vac. Sci. Technol.*, 1979, **16**, 531.
28. T. H. Upton and W. A. Goddard III, *Phys. Rev. Lett.*, 1979, **42**, 472.
29. V. I. Avdeev, T. H. Upton, W. H. Weinberg, and W. A. Goddard III, *Surf. Sci.*, 1980, **95**, 391.
30. D. G. Truhlar and B. C. Garrett, *Ann. Rev. Phys. Chem.*, 1984, **35**, 159.
31. D. G. Truhlar, W. L. Hase, and J. T. Hynes, *J. Phys. Chem.*, 1983, **87**, 2664.
32. B. C. Garrett and D. G. Truhlar, *J. Phys. Chem.*, 1979, **83**, 1052.
33. B. C. Garrett and D. G. Truhlar, *J. Phys. Chem.*, 1979, **83**, 1079.
34. M. S. Daw and M. I. Baskes, *Phys. Rev. B*, 1984, **29**, 6443.
35. G. Wahnström, B. Carmelli, and H. Metiu, *J. Chem. Phys.*, 1988, **88**, 2478.
36. B. Jackson and H. Metiu, *J. Chem. Phys.*, 1987, **86**, 1026.
37. B. Jackson and H. Metiu, *J. Chem. Phys.*, 1987, **85**, 4129.
38. J.-H. Lin and B. J. Garrison, *J. Chem. Phys.*, 1984, **80**, 2904.
39. C. T. Reimann, M. El-Maazawi, K. Walzl, B. J. Garrison, N. Winograd, and D. M. Deaven, *J. Chem. Phys.*, 1989, **90**, 2027.
40. 'Monte Carlo Methods in Statistical Physics', ed., K. Binder, Springer-Verlag, Berlin, 1986.
41. 'Applications of the Monte Carlo Method in Statistical Physics', ed., K. Binder, Springer-Verlag, Berlin, 1984.
42. H. C. Kang and W. H. Weinberg, *J. Chem. Phys.*, 1989, **90**, 2824.
43. L. A. Ray and R. C. Baetzold, *J. Chem. Phys.*, 1990, **93**, 2871.
44. H. C. Kang and W. H. Weinberg, *Phys. Rev. B*, 1988, **38**, 11543.
45. Y. Limoge and J. L. Bocquet, *Acta Metall.*, 1988, **36**, 1717.
46. S. L. Lombardo and A. T. Bell, *Surf. Sci. Rep.*, (to be published), and references therein.
47. K. A. Fichthorn and W. H. Weinberg, *Langmuir*, (to be published).
48. A. F. Voter and J. D. Doll, *J. Chem. Phys.*, 1985, **82**, 80.
49. H. C. Kang, T. A. Jachimowski, and W. H. Weinberg, *J. Chem. Phys.*, 1990, **93**, 1418.
50. C. T. Rettner, E. K. Schweizer, and H. Stein, *J. Chem. Phys.*, 1990, **93**, 1442.
51. C. B. Mullins, C. T. Rettner, D. J. Auerbach, and W. H. Weinberg, *Chem. Phys. Lett.*, 1989, **163**, 111.
52. A. V. Hamza, H.-P. Steinrück, and R. J. Madix, *J. Chem. Phys.*, 1987, **86**, 6506.
53. C. B. Mullins, Y. Wang, and W. H. Weinberg, *J. Vac. Sci. Technol.*, 1989, **A7**, 2125.
54. A. C. Luntz, M. D. Williams, and D. S. Bethune, *J. Chem. Phys.*, 1988, **89**, 4381.
55. H. P. Steinrück and R. J. Madix, *Surf. Sci.*, 1987, **185**, 36.
56. S. L. Tang, J. D. Beckerle, M. B. Lee, and S. T. Ceyer, *J. Chem. Phys.*, 1986, **84**, 6488.
57. M. P. D'Evelyn, H. P. Steinrück, and R. J. Madix, *Surf. Sci.*, 1987, **180**, 47.
58. J. Harris and A. C. Luntz, *J. Chem. Phys.*, 1989, **91**, 6421.
59. P. Alnot, A. Cassuto, and D. A. King, *Surf. Sci.*, 1989, **215**, 29.
60. K. D. Rendulic, A. Winkler, and H. Karner, *J. Vac. Sci. Technol.*, 1987, **A5**, 488.
61. C. R. Arumainayagam, M. C. McMaster, G. R. Schoofs, and R. J. Madix, *Surf. Sci.*, 1989, **222**, 213.
62. B. E. Hayden and D. C. Godrey, *Surf. Sci.*, 1990, **232**, 24.
63. M. B. Lee, Q. Y. Yang, S. L. Tang, and S. T. Ceyer, *J. Chem. Phys.*, 1986, **85**, 1693.

64. M. B. Lee, Q. Y. Yang, and S. T. Ceyer, *J. Chem. Phys.*, 1987, **87**, 2724.
65. T. P. Beebe, Jr., D. W. Goodman, B. D. Kay, and J. T. Yates, Jr., *J. Chem. Phys.*, 1987, **87**, 2305.
66. N. V. Richardson and A. M. Bradshaw, *Surf. Sci.*, 1979, **88**, 255.
67. W. H. Weinberg, *Surf. Sci.*, 1983, **128**, L224.
68. P. A. Thiel, E. D. Williams, J. T. Yates, Jr., and W. H. Weinberg, *Surf. Sci.*, 1979, **84**, 54.
69. G. B. Fischer and S. J. Schmieg, *J. Vac. Sci. Technol.*, 1983, **A1**, 1064.
70. C. T. Campbell and J. M. White, *J. Catal.*, 1978, **54**, 289.
71. P. R. Norton and R. L. Tapping, *Chem. Phys. Lett.*, 1976, **38**, 207.
72. T. S. Wittrig, P. D. Szuromi, and W. H. Weinberg, *J. Chem. Phys.*, 1982, **76**, 3305.
73. H. C. Kang, C. B. Mullins, and W. H. Weinberg, *J. Chem. Phys.*, 1990, **92**, 1397.
74. G. Comsa and R. David, *Surf. Sci. Rep.*, 1985, **5**, 145.
75. C. T. Rettner, E. K. Schweizer, H. Stein, and D. J. Auerbach, *Phys. Rev. Lett.*, 1988, **61**, 986.
76. K. C. Janda, J. E. Hurst, Jr., C. A. Becker, J. P. Cowin, L. Wharton, and D. J. Auerbach, *Surf. Sci.*, 1980, **93**, 270.
77. C. T. Campbell, G. Ertl, H. Kuipers, and J. Segner, *Surf. Sci.*, 1981, **107**, 220.
78. C. T. Rettner, E. K. Schweizer, and C. B. Mullins, *J. Chem. Phys.*, 1989, **90**, 3800.
79. C. T. Rettner, D. S. Bethune, and D. J. Auerbach, *J. Chem. Phys.*, 1989, **91**, 1942.
80. J. E. Hurst, Jr., C. A. Becker, J. P. Cowin, K. C. Janda, and L. Wharton, *Phys. Rev. Lett.*, 1979, **43**, 1175.
81. K. C. Janda, J. E. Hurst, Jr., C. A. Becker, J. P. Cowin, D. J. Auerbach, and L. Wharton, *J. Chem. Phys.*, 1980, **72**, 2403.
82. L. K. Verheij, J. Lux, A. B. Anton, B. Poelsema, and G. Comsa, *Surf. Sci.*, 1987, **182**, 390.
83. R. M. Logan and J. C. Keck, *J. Chem. Phys.*, 1968, **49**, 860.
84. R. M. Logan and R. E. Stickney, *J. Chem. Phys.*, 1966, **44**, 195.
85. J. C. Tully, *J. Chem. Phys.*, 1990, **92**, 680.
86. S. Holloway and J. W. Gadzuk, *J. Chem. Phys.*, 1985, **82**, 5203.
87. J. C. Tully, *Surf. Sci.*, 1981, **111**, 461.
88. C. B. Mullins and C. T. Rettner, (unpublished results).
89. L. B. Thomas in 'Fundamentals of Gas-Surface Interactions', ed., H. Saltsburg, J. N. Smith, Jr., and M. Rogers, Academic Press, New York, 1967, p. 346.
90. L. B. Thomas and F. G. Olmer, *J. Am. Chem. Soc.*, 1942, **64**, 2190.
91. L. B. Thomas and F. G. Olmer, *J. Am. Chem. Soc.*, 1943, **65**, 1036.
92. L. B. Thomas and R. E. Brown, *J. Chem. Phys.*, 1950, **18**, 1367.
93. L. B. Thomas and R. C. Golike, *J. Chem. Phys.*, 1954, **22**, 300.
94. L. B. Thomas and E. B. Schofield, *J. Chem. Phys.*, 1955, **23**, 861.
95. F. O. Goodman and H. Y. Wachman, *J. Chem. Phys.*, 1967, **46**, 2376.
96. C. T. Rettner, C. B. Mullins, D. S. Bethune, D. J. Auerbach, E. K. Schweizer, and W. H. Weinberg, *J. Vac. Sci. Technol.*, 1990, **A8**, 2699.
97. J. E. Hurst, Jr., L. Wharton, K. C. Janda, and D. J. Auerbach, *J. Chem. Phys.*, 1983, **78**, 1559.
98. J. E. Hurst, Jr., L. Wharton, K. C. Janda, and D. J. Auerbach, *J. Chem. Phys.*, 1985, **83**, 1376.

99. J. C. Tully, *Faraday Discuss. Chem. Soc.*, 1985, **80**, 291.
100. C. R. Arumainayagam, R. J. Madix, M. C. McMaster, V. M. Suzawa, and J. C. Tully, *Surf. Sci.*, 1990, **226**, 180.
101. K. Sinniah, M. G. Sherman, L. B. Lewis, W. H. Weinberg, J. T. Yates, Jr., and K. C. Janda, *Phys. Rev. Lett.*, 1989, **62**, 567.
102. K. Sinniah, M. G. Sherman, L. B. Lewis, W. H. Weinberg, J. T. Yates, Jr., and K. C. Janda, *J. Chem. Phys.*, 1990, **92**, 5700.
103. H. C. Kang and W. H. Weinberg, *J. Chem. Phys.*, 1989, **90**, 2824.
104. E. G. Seebauer and L. D. Schmidt, *Chem. Phys. Lett.*, 1986, **123**, 129.
105. X.-P. Jiang and H. Metiu, *J. Chem. Phys.*, 1988, **88**, 1891.
106. A. G. Naumovets and Yu. S. Vedula, *Surf. Sci. Rep.*, 1985, **4**, 365.
107. H. C. Kang and W. H. Weinberg, *Phys. Rev. B*, 1990, **41**, 2234.
108. K. A. Fichthorn and W. H. Weinberg, (in preparation).
109. K. Christmann, R. J. Behm, G. Ertl, M. A. Van Hove, and W. H. Weinberg, *J. Chem. Phys.*, 1979, **70**, 4168.
110. J. Behm, K. Christmann, and G. Ertl, *Sol. State Commun.*, 1978, **25**, 763.
111. H. Bu, M. Shi, F. Masson, and J. W. Rabalais, *Surf. Sci.*, 1990, **230**, L140.
112. H. Bu, M. Shi, and J. W. Rabalais, *Surf. Sci.*, 1990, **236**, 135.
113. W. Hetterich and W. Heiland, *Surf. Sci.*, 1989, **210**, 129.
114. C.-M. Chan, M. A. Van Hove, W. H. Weinberg, and E. D. Williams, *Sol. State Commun.*, 1979, **30**, 47.
115. C.-M. Chan, M. A. Van Hove, W. H. Weinberg, and E. D. Williams, *Surf. Sci.*, 1980, **91**, 440.
116. P. D. Szuromi, J. R. Engstrom, and W. H. Weinberg, *J. Chem. Phys.*, 1984, **80**, 508.
117. T. S. Wittrig, P. D. Szuromi, and W. H. Weinberg, *J. Chem. Phys.*, 1982, **76**, 716.
118. Y.-K. Sun and W. H. Weinberg, *J. Vac. Sci. Technol.*, 1990, **A8**, 2445.
119. J. A. Rodriguez and D. W. Goodman, *J. Phys. Chem.*, 1990, **94**, 5342.
120. C. B. Mullins and W. H. Weinberg, *J. Chem. Phys.*, 1990, **92**, 3986.
121. D. A. King and M. G. Wells, *Surf. Sci.*, 1972, **29**, 454.
122. C. B. Mullins and W. H. Weinberg, *J. Chem. Phys.*, 1990, **92**, 4508.
123. C. T. Campbell, Y.-K. Sun, and W. H. Weinberg, *Chem. Phys. Lett.*, 1991, **179**, 53.
124. R. W. Verhoef, C. B. Mullins, and W. H. Weinberg, (to be published).
125. S. T. Ceyer, *Science*, 1990, **249**, 133.
126. S. T. Ceyer, *Ann. Rev. Phys. Chem.*, 1988, **39**, 479.
127. K. A. Fichthorn and W. H. Weinberg, *J. Chem. Phys.*, (submitted).
128. E. S. Hood, B. H. Toby, and W. H. Weinberg, *Phys. Rev. Lett.*, 1985, **55**, 2437.
129. W. H. Weinberg in 'Kinetics of Interface Reactions', ed., H. J. Kreuzer and M. Grunze, Springer-Verlag, Heidelberg, 1987, p. 94.
130. E. S. Hood, B. H. Toby, W. Tsai, and W. H. Weinberg in 'Kinetics of Interface Reactions', ed., H. J. Kreuzer and M. Grunze, Springer-Verlag, Heidelberg, 1987, p. 153.
131. H. C. Kang and W. H. Weinberg, *Phys. Rev. B*, 1989, **40**, 7059.
132. H. C. Kang, M. W. Deem, and W. H. Weinberg, *Phys. Rev. B*, 1991, (in press).

CHAPTER 6
Thermal Desorption Kinetics
H. J. KREUZER and S. H. PAYNE

1 Introduction

Whereas the equilibrium properties of a large system are controlled by the minimum of its free energy, the kinetics involve questions of energy transfer. To establish the relevant time scales for the adsorption and desorption processes, let us follow a gas particle approaching the surface of a solid. If it rids itself of enough energy within the attractive region of the surface potential, it will get trapped. However, even if it descends all the way to the bottom of the surface potential well, it will eventually evaporate again; thus the concept of absolute trapping is meaningless. For times t_0 required for a particle to traverse the attractive potential well, the particle will remain close to the top of the well within an energy of $k_B T$. In this time there is a fair chance that the particle acquires enough energy from the heat bath of the solid to escape again. If this escape, which we can identify with inelastic scattering, has not happened within a few round trips, the particle will begin its descent to the bottom of the potential well, which, in a quantum picture, corresponds to a cascade of transitions between the bound states of the surface potential, each downward transition accompanied by the emission of phonons into the solid and each upward transition with the absorption of phonons. This adsorption process, characterized by a time scale t_a is, of course, more likely at low temperatures. After it has happened, the particle will try again and again to climb back out of the potential well through a sequence of phonon absorption and emission processes. It will eventually succeed in doing so after a desorption time t_d. If t_a is much shorter than t_d, then adsorption and desorption are statistically independent, and the processes of sticking, energy accommodation (*i.e.* thermalization), and desorption can be well separated. This is most likely the case if the thermal energy $k_B T$ is much less than the depth of the surface potential.

The energy necessary to desorb a particle from the adsorbate can either come from the solid substrate or from some external source. As examples of the latter, lasers or other sources of electromagnetic radiation have been used in photodesorption and photon stimulated desorption.[1,2] Likewise, electron and ion beams are employed to cause electron- and ion-stimulated desorption, respectively.[3] Strong electric fields at field emission tips cause field desorption and evaporation, sometimes used in conjunction with lasers or electrons to produce photon- and electron-stimulated field desorption.[4] Some of these techniques are reviewed elsewhere in this volume.

If the solid itself acts as the reservoir from which the desorption energy is taken, we speak of thermal desorption. Lennard-Jones and co-workers[5] argued that the thermal motion of the lattice should act as a time-dependent perturbation on the surface potential, with which the adsorbate is bound to the surface, and can hence supply the desorption energy. It has been shown in recent years that this picture, in modern parlance called phonon-mediated desorption, is by and large correct, although coupling to the electronic excitations in the case of metal substrates may also be important.[6]

Kinetic theories of adsorption, desorption, and surface diffusion can be grouped into three categories:

(i) At the macroscopic level one proceeds to write down kinetic equations for macroscopic variables, in particular rate equations for the coverage or for partial coverages. This can be done in a heuristic manner, much akin to procedures in gas phase kinetics or, in a rigorous approach, using the framework of non-equilibrium thermodynamics.[7-10]

(ii) If it cannot be guaranteed that the adsorbate remains in local equilibrium throughout desorption, then a set of macroscopic variables is not sufficient and an approach based on non-equilibrium statistical mechanics involving time-dependent distribution functions must be invoked. The kinetic lattice gas model is an example of such a theory. It is derived from a Markovian master equation, but is not totally microscopic in that it is based on a phenomenological Hamiltonian and on postulated transition probabilities that are subject to the principle of detailed balance. For the kinetics in chemisorbed systems, this is as far as theory has progressed.

(iii) Lastly, we realize that a proper theory of the time evolution of adsorption and desorption must start from a microscopic Hamiltonian of the coupled gas–solid system. A master equation must then be derived from first principles with the benefit that transition probabilities are calculated explicitly involving microscopic parameters only. So far this programme has only been completed for phonon-mediated physisorption kinetics, as reviewed in a recent monograph by Kreuzer and Gortel.[6]

This review is structured as follows: In the next section we briefly sketch experimental methods and procedures to the extent they are needed in the ensuing discussions. Next, in Section 3 we present the phenomenological approach for adsorbates that remain in quasi-equilibrium throughout the desorption process, in which case a single macroscopic variable, the coverage θ, and its rate equation are needed. This is followed, in Section 4, by a review of the macroscopic theory based on Onsager's approach to non-equilibrium thermodynamics, as adapted in recent years to surface kinetics by the authors.[7-10] In Section 5 we then proceed to the semi-microscopic level by outlining recent advances in the theory of the kinetic lattice gas model. In Section 6 we very briefly mention the few advances made in physisorption kinetics since 1985. The review concludes with a statement of policy, and the identification of some directions for progress.

Being rather limited in space in writing this review, we have concentrated to report mainly on advances since Menzel's reviews in 1975 and 1982.[11,12]

2 Experimental Preliminaries

Experimental studies of thermal desorption kinetics can be grouped into three approaches (i) isothermal desorption, (ii) temperature jump desorption, and (iii) temperature programmed desorption. In an isothermal desorption experiment one removes the gas phase above the surface, either by reducing the initial pressure P_i to a final pressure $P_f \ll P_i$ by rapid pumping, or by chopping a molecular beam impinging onto the surface. Obviously the removal of the gas phase must be much faster than the subsequent desorption process to avoid problems connected with re-adsorption. This being achieved, one measures the time evolution of the desorbed particles with a mass spectrometer or some other particle detector, like a bolometer suitably placed above the surface.

Rapid removal of the gas phase is often not possible so that one, alternately, creates a non-equilibrium condition by rapidly (on the time scale of desorption) heating (flashing) the substrate from an initial temperature T_i to a final temperature T_f leading to the desorption of the excess particles in the adsorbate. In this temperature jump method, re-adsorption, in particular at the later stages of the time evolution, causes complications. This method is well suited to create and study non-equilibrium conditions within the adsorbate if surface diffusion times are not much faster than desorption times, as it obtains in rare gas adsorbates.

The most frequently employed technique to study desorption kinetics is temperature-programmed desorption or TPD, in which the temperature of the solid is raised as a function of time. Very stable linear temperature ramps can be achieved by direct ohmic heating of the substrate or indirectly by bombarding the back of the substrate with a continuous electron or laser beam. Linear heating rates between 10^{-1} and 10^2 Ks^{-1} can be achieved with typical rates being around 5 Ks^{-1}. Heating the system with a high intensity laser beam leads to large (but rather uncontrolled) heating rates of as much as 10^9 Ks^{-1}.

In this review we can not discuss details of experimental methods for reasons of space and lack of expertise of the authors, but we must spend some time on the methods of data evaluation. To this end we write the rate of change in the coverage:

$$d\theta/dt = R_a - R_d \tag{1}$$

as a balance between the adsorption rate, R_a, and the desorption rate, R_d. Here θ is the coverage in the adsorbate, *i.e.* the total number of particles normalized with the maximum number in a monolayer or the number of adsorption sites. The rate of adsorption is given by:

$$R_a = S(\theta) Pa_s/(2\pi m k_B T)^{1/2} \tag{2}$$

i.e. as the flux of particles of mass m hitting the area, a_s, of one adsorption site from the gas phase at pressure P and temperature T. The sticking coefficient, $S(\theta)$, is a phenomenological transport coefficient that accounts for the efficiency of energy transfer between the substrate and the adsorbate. Without further knowledge about the system, little can be said about the rate of desorption, R_d, short of the fact that it describes an activated process. One frequently rewrites the desorption rate:

$$R_d = r_d(\theta)\theta^x \tag{3}$$

as a process with desorption order x. For atomic adsorption in the low coverage regime (where lateral interactions within the adsorbate are negligible) we have $x = 1$; for dissociative adsorption in the low coverage regime we have $x = 2$; but note that the coverage dependence of the rate constant in general precludes such simple interpretations. It is, indeed, the coverage dependence of the rate constant that contains all the interesting physics and is the reason why desorption experiments are done in the first place. Let us nevertheless, against better judgement, assume that the desorption rate constant is not a function of coverage. This would pertain if the particles within the adsorbate would not experience any mutual interaction. Desorption in this idealized situation is then a strictly first-order process. Let us, in addition, implement the fact that desorption is an activated process by writing the desorption rate constant in the Arrhenius parametrization, in surface science more frequently referred to as the Wigner–Polanyi equation, as:

$$r_d = v\exp(-E_d/k_B T) \tag{4}$$

in terms of an attempt frequency, v, to escape the surface potential well, and an activation or desorption energy, E_d. We will see below that both parameters are, in general, temperature- and coverage-dependent. To describe a TPD experiment we assume a linear temperature ramp, $T(t) = T_0 + \beta t$, which, inserted in equation (1) with the adsorption term dropped, gives with equation (3) for $x = 1$:

$$-\frac{d\theta}{dT} = -\frac{d\theta}{dt}\frac{dt}{dT} = \frac{v}{\beta}\theta\exp(-E_d/k_BT) \qquad (5)$$

The temperature at the peak is then given by the Redhead formula:[13]

$$\frac{v}{\beta} = \frac{E_d}{k_BT^2}\exp(E_d/k_BT) \qquad (6)$$

Conversely, determining the peak temperature experimentally, equation (6) provides one equation for two unknowns, *i.e.* v and E_d. To get a second equation, one can run TPD with a second heating rate differing by one or two orders of magnitude, and solve for v and E_d from the respective peak temperatures. We stress again that this analysis is only valid for systems in which v and E_d are coverage independent, which is rarely the case (and uninteresting). Moreover, this analysis is invalid if two or more inequivalent adsorption sites are present on the surface, in which case a single variable θ does not suffice to characterize the adsorbate, and a rate equation like equation (5) must be written down for each partial coverage. In case particles can convert from one kind of adsorption site to another, additional diffusional terms must be added to the rate equations. This will be discussed below. To finish our discussion of phenomenology, we comment on methods of extraction of the kinetic parameters E_d and v. First, we point out that inspection of TPD data can frequently lead to some (albeit semi-quantitative) insight into the structure and kinetics of the absorbate. Let us assume that the TPD data show one peak. We can then extract a value for E_d assuming that the prefactor v has a 'standard' value of the order of 10^{13} s^{-1}, characteristic of the vibrational frequency of an atom in a potential well with a spatial extent of an angstrom or so. The assumption of a 'standard' frequency is rather dubious as v may vary anywhere from 10^{10} to 10^{20} s^{-1} from system to system and even over several orders of magnitude within one system as a function of coverage. Even so, with the assumption of a standard frequency and a typical heating rate of 5 K s^{-1} one gets, approximately, from equation (6) that:

$$E_d \approx 31\, k_BT_p \qquad (7)$$

If the TPD trace shows several (hopefully well-separated) peaks, this simple formula can be used to estimate different adsorption states, although great caution must be exercised and further analysis is required to substantiate such claims. Fortunately, this Redhead-type analysis, and variations on it,[14] has been largely displaced by the 'complete' analysis of a set of TPD curves, yet one still finds attempts at refinements,[15] based on the coverage-independence of the parameters.

A proper analysis of TPD data starts from the fact that in the desorption rate equation (5), the prefactor and the desorption energy are coverage (and particularly at low temperatures, also temperature) dependent quantities.

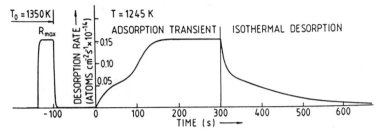

Figure 1 *Typical isothermal adsorption transient (starting at t = 0) followed by an isothermal desorption transient of* Au *on* Mo(110) *at* T = 1245 K *with an incident flux of* 0.15×10^{14} *atoms* $cm^{-2} \cdot s^{-1}$ *and an equilibrium coverage of* 0.86 ML. *The experiment is preceded by a calibration experiment in which the temperature* T_0 = 1350 K *is so high that all arriving atoms are desorbed immediately, which gives* R_{max} = $I(T_0)$. *At* t = −100 s *the shutter is closed and the temperature lowered to* T *as soon as* $I(T_0)$ = 0. *The shutter is opened and closed again at* t = 0 *and* t = 300 s, *respectively, initiating the adsorption and desorption transients*
(Reproduced with permission from *Surf. Sci.*, 1988, **195**, 207)

The method proposed by several groups[16–18] consists of the construction of a series of isosteric (constant coverage) Arrhenius plots of $\ln R_d(\theta)$ [or $\ln(r_d)$ if the reaction order x is assumed] against $1/T$, as interpolated from a family of TPD traces generated by varying initial coverages and/or heating rates. The interpolation, $\theta(T)$, is obtained by integrating under each TPD curve. If the Arrhenius plots are straight lines or consist of sections of straight lines, the

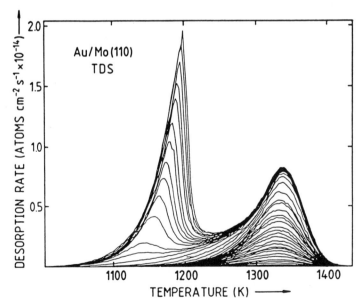

Figure 2 *Thermal desorption spectra of* Au *on* Mo(110) *in the coverage range from* 0 *to* 2 ML. *Heating rate* 5.2 K s^{-1}
(Reproduced with permission from *Surf. Sci.*, 1988, **195**, 207)

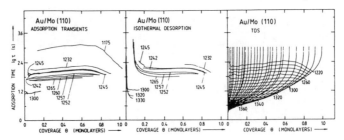

Figure 3 *Adsorption lifetime τ of* Au *on* Mo(110) *in the sub-monolayer range:* (a) *from adsorption transients,* (b) *from isothermal desorption,* (c) *from thermal desorption spectra*
(Reproduced with permission from *Surf. Sci.*, 1988, **195**, 207)

θ-dependence of E_d and ν can be deduced from the variations of slope and intercept, respectively, on these plots. Deviations from straight lines must be expected if (i) more than one adsorption state contributes to a TPD peak, or (ii) if the adsorbate does not remain in quasi-equilibrium throughout desorption, or (iii) if quantum effects produce a temperature dependence in the prefactor, e.g. in physisorbed systems. As for the interpretation of the parameters $E_d(\theta)$ and $\nu(\theta)$, we note that their zero coverage values refer to an isolated particle on the substrate, and their coverage dependence reflects lateral interactions between adsorbed particles. A third and recent method of determining such variations is based on the technique of threshold TPD.[19,20] Here the desorption rate is accurately measured at initial desorption for small changes in coverage and temperature; the initial coverage is then varied and the results fitted to equation (5). Compared to the second method, this extracts the kinetic parameters more directly, and is of advantage for systems away from equilibrium. Any temperature dependence of E_d and ν, however, will not be as easily observed because the desorbing range is so short at each coverage value. We shall return to several points mentioned

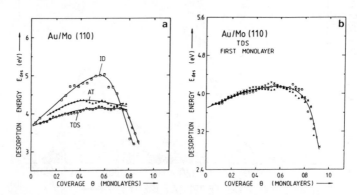

Figure 4 *Desorption energy of* Au *on* Mo(110) *as a function of coverage:* (a) *derived from the data of Figure 3,* (b) *from three different series of TDS experiments*
(Reproduced with permission from *Surf. Sci.*, 1988, **195**, 207)

above in the following sections. In this review we hope to clarify what can be learnt from a proper analysis of TPD data. We reproduce in Figures 1–4 a recent state of the art analysis of TPD data for the system Au/Mo(110).[21]

3 Desorption under Quasi-equilibrium Conditions

We consider desorption from an adsorbate where surface diffusion is so fast (on the time scale of desorption) that the adsorbate is maintained in equilibrium throughout the desorption process. That is to say that at the remaining coverage $\theta(t)$ at temperature $T(t)$ all correlation functions attain their equilibrium values. Thus the adsorbate can be characterized by its chemical potential, $\mu[\theta(t), T(t)]$. For example, if we were to look at desorption of an adsorbate that shows co-existence of a dense and a dilute two-dimensional phase in a certain range of coverage and temperature, then the assumption of quasi-equilibrium implies that at the remaining coverage, $\theta(t)$, the distribution of particles among the two phases is identical to that in equilibrium at coverage θ. In such situations a purely macroscopic description of the desorption process in terms of a single variable, the coverage $\theta(t)$, is sufficient. Its time rate of change has been written down in equation (1) in terms of an adsorption rate, R_a, and a desorption rate, R_d. Assuming now that the adsorbate remains in quasi-equilibrium during desorption, we look at a situation where the gas phase pressure, P, is different from the value, $\bar{P}$, required to maintain an adsorbate at coverage θ. There is then an excess flux to establish equilibrium between gas phase and adsorbate so that we can write:

$$d\theta/dt = S(\theta)(P - \bar{P})a_s/(2\pi mk_B T)^{1/2} \tag{8}$$

Next we express the equilibrium pressure in terms of the gas-phase chemical potential, μ_g:

$$\bar{P} = \left(\frac{k_B T}{\lambda_{th}^3}\right)\exp(\mu_g/k_B T)Z_{int} \tag{9}$$

where

$$\lambda_{th} = h/(2\pi mk_B T)^{1/2} \tag{10}$$

is the thermal wavelength and Z_{int} is the intramolecular partition function accounting for rotations and vibrations. However, in equilibrium, the chemical potential in the gas phase is equal to that in the adsorbate, μ_a, so that we can write the desorption rate in equation (1) as:

$$R_d = S(\theta)a_s k_B T/(h\lambda_{th}^2)Z_{int}\exp(\mu_a/k_B T) \tag{11}$$

We look at a number of examples: first, we consider a non-interacting adsorbate in the sub-monolayer regime. Its chemical potential is given by:

$$\mu_a = -V_0 + k_B T[\ln(\theta/(1-\theta)) - \ln(q_3 q_{int})] \quad (12)$$

where V_0 is the (positive) binding energy of an isolated particle on the surface. Moreover:

$$q_3 = q_z q_{xy} \quad (13)$$

is the vibrational partition function of an adsorbed particle with:

$$q_z = \exp(h\nu_z/2k_B T)/[\exp(h\nu_z/k_B T) - 1] \quad (14)$$

its component for the motion perpendicular to the surface. Likewise, q_{xy} is the partition function for the motion parallel to the surface. We have also made adjustment for the fact that the internal partition function for rotations and vibrations of an adsorbed molecule might be changed from its free gas-phase value Z_{int} to q_{int}, if some of the internal degrees of freedom get frozen out or frustrated. If the adsorbate is not localized, i.e. if the corrugation of the surface potential parallel to the surface is negligible, we can treat it as a two-dimensional ideal gas and have:

$$q_{xy} = a_s/\lambda_{th}^2 \quad (15)$$

Thus equation (11) reads in the high temperature limit, i.e. for $k_B T \gg h\nu_z$, for a non-localized, non-interacting adsorbate:

$$R_d = S(\theta)[\theta/(1-\theta)](Z_{int}/q_{int}) \nu_z \exp(-V_0/k_B T) \quad (16)$$

In particular, if $S(\theta) = S_0(1-\theta)$, then equation (16) is the familiar first-order rate expression, with ν and E_d identified.

Let us next consider a non-interacting, localized adsorbate for which q_{xy} has two factors like equation (14) with frequencies ν_x and ν_y. In this case we find in the high temperature limit:

$$R_d = S(\theta)[\theta/(1-\theta)](Z_{int}/q_{int})(2\pi m a_s/k_B T) \nu_x \nu_y \nu_z \exp(-V_0/k_B T) \quad (17)$$

Since vibrational frequencies ν_x, ν_y, ν_z, are typically of order 10^{12}–10^{13} s^{-1}, the prefactor here can be larger than that in equation (16) by a factor as great as 10^4, depending upon the system under study. Thus (zero coverage) prefactors up to 10^{17} s^{-1} are to be expected even if $Z_{int} = q_{int}$.* Nevertheless, one sees statements in the literature that prefactors as large as 10^{20} s^{-1} are erroneous or unphysical.

* Reference 22 contains a useful list of desorption data; however, the theory and the interpretation of the data are to be taken with a grain of salt.

A word of caution about the high temperature limit: typical vibrational frequencies correspond to vibrational temperatures $T_v \approx 50\text{--}500$ K. Recall that desorption of chemisorbed species, with a heat of adsorption of a few eV, occurs at temperatures well above room temperature so that the high temperature approximation of the partition functions is justified. This is not the case for physisorbed systems with heats of adsorption of less, and frequently much less, than an electronvolt. In such cases the high temperature approximation is not justified. Indeed, for very weakly bound systems, such as rare gases on metals or even N_2 on Ni(110), one might well be in the low temperature regime when desorption occurs. In such cases we write for equation (14):

$$q_z \approx \exp(-h\nu_z/2k_BT) \tag{18}$$

and find, instead of equation (17):

$$R_d = S(\theta)[\theta/(1-\theta)](Z_{int}/q_{int})a_s(k_BT/h\lambda_{th}^2)\exp(-E_d/k_BT) \tag{19}$$

$$E_d = V_0 - h(\nu_x + \nu_y + \nu_z)/2 \tag{20}$$

Thus the effective desorption energy is reduced from its high temperature value by the zero-point energy in the surface potential, a reduction which is significant for some systems. Similarly, the prefactor in equation (19) is reduced (by orders of magnitude) from its value in equation (17) to a quantity independent of substrate vibrations.* While neither the classical nor the quantum limits of the vibrational (and internal) partition functions may apply to a system, they underscore the point that, across systems, there is no fixed relation between the desorption energies and prefactors, as determined by an Arrhenius analysis, and the microscopic binding energies and frequencies of the underlying dynamics. Whereas V_0 and ν_i are parameters in a Hamiltonian, which gives the energy level structure of the substrate binding potential, E_d and ν are strictly phenomenological (but physical) parameters. There is a considerable confusion in the literature on this point.

It is useful at this stage to realize that for systems that remain in quasi-equilibrium throughout the desorption process, the desorption energy is more or less equal to the isosteric heat of adsorption, as derived from the adsorption isotherms:

$$Q_{iso}(\theta, T) = k_BT^2 \left.\frac{\partial \ln P}{\partial T}\right|_\theta \tag{21}$$

---

* Expressions similar to equation (19) have been used frequently in the literature without reference to the fact that they are only valid in the low temperature regime, see e.g. in reference 23.

We first define a differential desorption energy as:

$$E_d(\theta, T) = k_B T^2 \frac{\partial}{\partial T} \ln(R_d)|_\theta \tag{22}$$

corresponding to the instantaneous slope of an isosteric Arrhenius plot. The desorption rate here can be identified from equation (8) in terms of an equivalent pressure that would be maintained if the adsorbate with the instantaneous coverage $\theta(t)$ were in equilibrium at temperature T. We can therefore write equation (22) equivalently as:

$$E_d(\theta, T) = Q_{iso}(\theta, T) - \tfrac{1}{2} k_B T + k_B T^2 \frac{\partial}{\partial T} S(\theta, T)|_\theta \tag{23}$$

Note that the second term is irrelevant in the exponential, and also that the temperature dependence of the sticking coefficient is rather small for most systems, so that the effective desorption energy is given by the heat of adsorption, provided that (i) surface diffusion is fast enough to keep the system in quasi-equilibrium throughout desorption, and (ii) that precursor states are absent. The generalized prefactor, $\nu(\theta, T)$, is now defined via equation (3) and equation (22). In passing we note that a fit to the equilibrium data (isosteric heats, isobars) also provides a sensitive method of deducing V_0 and $\{v_i\}$.

Next we look at the effect of lateral interactions between adsorbed particles which are assumed localized so that, in the simplest model, we can think of the adsorbate as a lattice gas with a nearest neighbour interaction of strength V_2. We illustrate some elementary but important results within the quasi-chemical approximation which gives:

$$\mu_a/k_B T = -V_0/k_B T + (1 - \tfrac{1}{2}c)\ln[\theta/(1-\theta)] - \ln[q_3 q_{int}]$$
$$+ cV_2/2k_B T + \tfrac{1}{2} c \ln[(a - 1 + 2\theta)/(a + 1 - 2\theta)] \tag{24}$$

where c is the site co-ordination number and:

$$a^2 = 1 - 4\theta(1-\theta)[1 - \exp(-V_2/k_B T)] \tag{25}$$

The desorption energy, via equation (11) and equation (22) can be expressed as:

$$E_d(\theta, T) = E_d(0, T) + \tfrac{1}{2} c V_2 \left[\frac{(1 - 2\theta)}{a} - 1 \right] \tag{26}$$

For a large, repulsive ($V_2 > 0$) interaction, for example, $E_d(\theta, T)$ exhibits two distinct and essentially constant values for $\theta \gtrless \tfrac{1}{2}$, implying two peaks in the TPD spectrum for initial coverages $\theta_0 \lesssim 1$. The case of attractive interactions

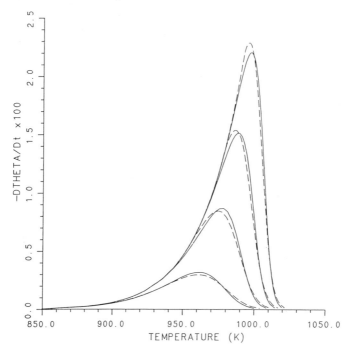

Figure 5 *Temperature programmed desorption calculated for* Ag/Mo(110) *with data from reference 24. Solid lines: quasi-chemical approximation with* $V_0 = 33\,760$ K $= 2.91$ eV, $cV_2 = -4715$ K $= -0.41$ eV, $c = 6$, $T_c = 969$ K, $v_z = 3 \times 10^{12}$ s^{-1}, $v_x = v_y = 3.4 \times 10^{12}$ s^{-1}, $n_s = 0.2$ A^{-2}, $S_1 = S_2 = 1$. *Dashed curves: Bragg–Williams approximation for which* $V_0 = 34\,000$ K, $cV_2 = -4725$ K *and* $T_c = 1074$ K. *Heating rate is* 1.21 K s^{-1}, *initial coverages (top to bottom):* $\theta_0 = 0.75, 0.55, 0.35, 0.15$

is more interesting, especially for temperatures below the critical value, T_c, for the phase transition. Here dilute and condensed phase regions co-exist and equation (24) and equation (26) are replaced, within the co-existence region, by:

$$\mu_a = \tfrac{1}{2}cV_2 - k_B T \ln(q_3 q_{\text{int}}) \qquad (27)$$

$$E_d(\theta, T) = E_d(0, T) - \tfrac{1}{2}cV_2 \qquad (28)$$

Inserting equation (27) into equation (11) we readily get zero-order desorption kinetics within the co-existence region as a result of the coverage-independent chemical potential. As an example we look at the Ag/Mo(110) system for which the isosteric heat of adsorption at zero coverage, identified with E_d, has been measured to be 2.91 eV.[24] A fit to the isothermal desorption data by choosing $v_z = 3.0 \times 12^{12}$ s^{-1}, $v_x = v_y = 3.4 \times 10^{12}$ s^{-1} and $cV_2 = -0.41$ eV produces the TPD traces of Figure 5, their peak heights agreeing, within 10–20%, with those of Figure 4 of reference 24. We note

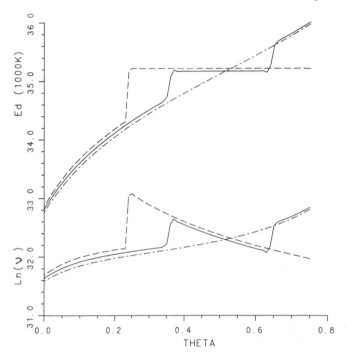

Figure 6 *Upper triple of curves: differential desorption energies, $E_d(\theta, T)$, as a function of coverage, and for three temperatures, derived from the instantaneous slopes of isosteres as calculated from TPD spectra generated in the quasichemical approximation; temperatures are $T_1 = 1000$ K (—·—·—), $T_2 = 950$ K (———), $T_3 = 900$ K (-----). Lower triple of curves: (logarithm of) the corresponding effective prefactors*

that these vibrational frequencies result in an effective prefactor in the Arrhenius parametrization of about 10^{15} s^{-1} as obtained in the evaluation of the experimental data, as discussed in reference 10. The coverage dependence of the kinetic parameters is of particular note. At a fixed temperature, $E_d(\theta, T)$ assumes a constant value, equation (28), if θ lies within the co-existence region, but a quite different value, equation (26), otherwise. Figure 6 shows this dependence for three temperatures spanning T_c ($= 969$ K for V_2 above), but as calculated from the instantaneous slopes of a series of isosteric Arrhenius plots, *i.e.* as if the data were generated experimentally. Note that, because we have imposed first-order kinetics on a zero-order process, the prefactor, $v(\theta, T)$, is not constant in the co-existence region, but decreases like θ^{-1}. This artifact is often contained in the evaluation of experimental data as well. The abrupt changes in E_d and v mark the phase boundary at that temperature, and a complete temperature sequence will map out the phase diagram of the system. As an example we take the work of Kolaczkiewicz and Bauer[25] on the desorption of metals from metals: having increased considerably the sensitivity of their measurements over earlier work,[17] they noted that most Arrhenius plots did not yield one, but two, straight lines. We reproduce their resulting plots of $E_d(\theta)$ and $v(\theta)$ for metals on W(110), in

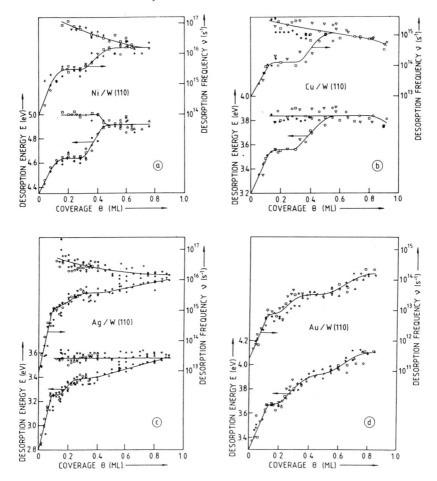

Figure 7 *Desorption energies $E_d(\theta)$ and pre-exponential factors $\nu(\theta)$ of Ni (a), Cu (b), Ag (c), and Au (d) on W(110) as a function of coverage θ. The data are derived from the evaluation of 5, 4, 6 and 7 series of TPD spectra of Ni, Cu, Ag, and Au, respectively, characterized by different symbols. $\theta = 1$ corresponds to 14.12×10^{14} atoms cm^{-2}. The lines through the experimental points are only intended to guide the eye*
(Reproduced with permission from *Surf. Sci.*, 1988, **175**, 508)

Figure 7. For comparison purposes only, we show in Figure 8 the values of E_d and ν obtained by assuming two linear fits to the theoretical isosteres obtained from Figure 5. These curves are the temperature averages over the desorption range of a complete sequence of curves of Figure 6. The double-valuedness of $E_d(\theta)$ and $\nu(\theta)$ is clearly the indicator of desorption through the two-phase region. The results of Figure 7 must be considered as one of the examples of considerable progress in experiment.[26]* Another, we believe, is the results of Klekamp and Umbach[28] who used TPD to obtain similar

---

* There are many examples in the literature where similar features should have been seen, *e.g.* in reference 27.

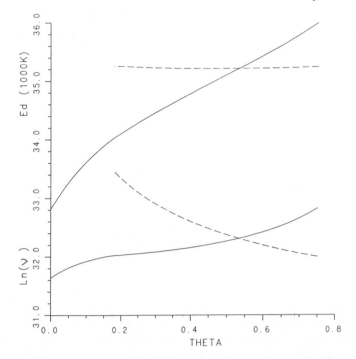

Figure 8 *Coverage dependence of desorption energies (upper pair of curves) and (logarithm of) prefactors in quasichemical approximation, obtained by fitting equation (5) to generated isosteres in the one-phase and two-phase regions separately (solid and dashed curves, respectively)*

plots for the system $SF_6/Ni(111)$ and also have constructed the phase diagram.

Finally, we observe from Figures 7 and 8 the familiar compensation effect, namely an approximate proportionality between the single phase values of E_d and v, as a function of coverage. While this is observed experimentally for systems with attractive interactions (see *e.g.* reference 29 for a further example), and is understood in this case by consideration of entropy change, the argument does not apply for repulsive interactions; nor is the proportionality well founded in lattice gas models. It continues to receive attention, however.[30,31] Any application of a compensation 'law' to the modelling of TPD spectra is suspect.[32]

It is clear from the rate expression, equation (11), that for systems that remain in quasi-equilibrium during desorption, one can extract the chemical potential as a function of coverage and temperature, *i.e.*:

$$\exp(\mu_a/k_B T) = r_d/[S(\theta)a_s k_B T Z_{int}/(h\lambda_{th}^2)] \tag{29}$$

provided the coverage dependence of the sticking coefficient is known. Thus isosteres can be constructed, isosteric heats of adsorption can be calculated, and phase diagrams can be deduced. This procedure of using thermal desorption data to extract equilibrium properties is complementary to direct equilibrium measurements because the latter can usually be performed at low temperatures whereas the former extend to much higher temperatures, e.g. in the case of metals on metals to temperatures above critical. If equilibrium data can be measured independently in addition to thermal desorption, then equation (29) can be used to extract the sticking coefficient. Also note that equation (11) allows one to set up as precise a kinetic theory as the equilibrium properties of the system can be modelled. In this regard the use of transfer matrix methods[33,34] to calculate μ_a, with multiple lateral interactions included, promises to be of value for the study of desorption kinetics.[35]

The quasi-equilibrium theory presented here can easily be extended to systems with more than one adsorbed species and several adsorption sites. If, however, precursors are present as in activated chemisorption, quasi-equilibrium may not obtain during desorption as intermediate steps in the desorption process may be rate limiting. These problems have been discussed very clearly by Cassuto and King[36] at the phenomenological level, and not much has been published since apart from an excellent review by Ehrlich.[37]

We briefly comment on the role of diffusion in desorption kinetics, following a recent analysis of the system $N_2/Ni(430)$ by Payne et al.[38] We assume that there are two distinct adsorption sites on the surface, referred to for simplicity as step and terrace sites, with partial coverages θ_s and θ_t. The phenomenological rate equations in the absence of precursors and lateral interactions are assumed to be:

$$d\theta_s/dt = [S_s a_s P/(2\pi m k_B T)^{1/2}](1 - \theta_s)$$
$$- r_s \theta_s + r_{s \leftarrow t} \theta_t (1 - \theta_s) - r_{t \leftarrow s} \theta_s (1 - \theta_t) \quad (30)$$

$$d\theta_t/dt = [S_t a_t P/(2\pi m k_B T)^{1/2}](1 - \theta_t)$$
$$- r_t \theta_t - r_{s \leftarrow t} \theta_t (1 - \theta_s) + r_{t \leftarrow s} \theta_s (1 - \theta_t) \quad (31)$$

Here S_i for $i = s, t$ are the sticking coefficients at zero coverage with a simple linear decrease assumed as a function of coverage. The rate constants can be obtained from equation (17) as:

$$r_i = (S_i \theta_i Z_{int}/q_{int})(2\pi m a_i/k_B T) v_{xi} v_{yi} v_{zi} \exp(-V_{0i}/k_B T) \quad (32)$$

$$r_{s \leftarrow t} = v_{s \leftarrow t} \exp(-Q_{s \leftarrow t}/k_B T) \quad (33)$$

where $Q_{s \leftarrow t}$ is the activation energy for hopping from the terrace to the step site. Detailed balance demands that:

$$r_{t \leftarrow s} = r_{s \leftarrow t} r_s / r_t$$

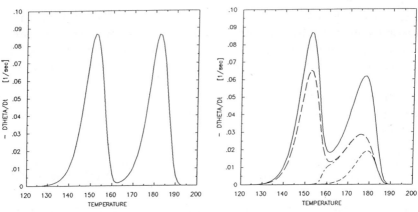

Figure 9 *Temperature programmed desorption spectra for the phenomenological model of equations (30)–(33):* (a) *Parameters from a fitting to equilibrium data of experiment (M. Grunze, private communication); no hopping* ($v_{s \leftarrow t} = 0$), *initial coverage* $\theta_0 = 2/3$, *heating rate* 3 K s^{-1}. (b) *As for* (a) *except* $v_{s \leftarrow t} = 10^8$ s^{-1}, $Q_{s \leftarrow t} = 10$ kJ mol^{-1}. *The solid curve is the sum of the rates from the terrace (long-dashed curve) and step (short-dashed) sites; the net diffusive rate (step → terrace) is the dash-dotted curve. These three components have been scaled down by 0.25*

implying in particular that the activation energy for hopping from the more deeply bound step site to the terrace site is enhanced by the difference in binding energies. To calculate TPD curves we assume that all parameters are such as to produce two distinct desorption peaks in the absence of diffusion, see Figure 9a. Conventional wisdom then has it that, in the presence of fast diffusion, particles trapped in the deeper-bound step sites will diffuse into the terrace sites as the latter desorb. One therefore expects a lowering of the high temperature peak and a shift to lower temperatures, thus filling up the valley between the peaks, as borne out in Figure 9b. We note, however, that the transfer from steps to terraces only sets in at temperatures above the maximum of the lower peak, *i.e.* when the blocking factor $1 - \theta_t$ approaches one, and this for all values of $v_{s \leftarrow t}$, and for all heating rates. Indeed, the effect saturates for hopping prefactors $v_{s \leftarrow t}$ larger than 10^8 s^{-1}. In Figure 9b the solid curve represents total desorption; the short-dashed curve is desorption from the step site, *i.e.* the term $-r_s \theta_s$ in equation (30). Diffusion replenishes the terrace sites maintaining desorption from the terrace sites even at higher temperature, *i.e.* due to the last two terms in equation (30), the sum of which is shown as the dash-dotted curve in Figure 9b. Indeed, this is the fastest route for depleting the population of the step site, as indicated by the terrace site desorption rate, $-r_t \theta_t$ (long-dash curve). The total desorption rate thus shows the high temperature peak reduced and shifted to a slightly lower temperature with considerable filling of the valley. We conclude: if a system's parameters are such that in the absence of diffusion the TPD curves would have two separated peaks, then the two-peak structure remains in the presence of diffusion, no matter how fast it is.

Before closing this section, we want to make contact with another widely used phenomenological approach to adsorption–desorption kinetics based on transition state theory. In this approach one views the desorption process as a chemical reaction where a particle A bound to a surface site Σ, forming a 'molecule' $(A + \Sigma)$, is removed into the gas phase via a transition state or an activated complex $(A + \Sigma)\ddagger$ leaving an empty site behind. One then assumes that the transition state is in equilibrium with the reactants and calculates the corresponding equilibrium constant. Arguing next that the transition state vibrates along the reaction path, one can calculate its partition function, of which one takes the high temperature limit to arrive, eventually, at a formula for the desorption rate constant (which, by the way, no longer contains the vibrational partition function of the transition state). These remarks are not meant to discredit transition state theory as such, which no doubt has its role in chemical reaction kinetics, but to point out that for desorption kinetics simpler arguments, as presented *e.g.* from equation (8) to equation (11), lead to answers in a straightforward way.*

4 Non-equilibrium Thermodynamics of Surface Processes

So far we have assumed that quasi-equilibrium is maintained during the desorption process. This assumption breaks down when the time constant of surface diffusion is not appreciably faster than that for desorption—as happens, for example, in the case of rare gases on metals—in which case non-equilibrium effects within the adsorbate show up in desorption experiments. To deal with such situations it is advantageous to formulate the kinetics of adsorption, desorption, and diffusion using the methods of non-equilibrium thermodynamics. We will demonstrate the approach by considering, in more detail, gas–solid systems in which the adsorbate exhibits co-existence of a dilute, gas-like phase together with islands of a condensed phase, both of course, being two-dimensional (2-d). The 2-d gas phase consists of two parts, adsorbed, respectively, on the bare surface or on top of the condensed islands. Taking the system out of equilibrium by, for example, changing the temperature of the substrate or the ambient gas pressure, will induce surface processes such as adsorption, desorption, and diffusion. Looking at desorption from such a two-phase adsorbate, it can proceed via several channels, namely (i) out of the 2-d gas phase on the bare surface with a time constant t_d, (ii) out of the 2-d gas phase on top of the condensed islands with a time constant $t_{d'}$, and (iii) directly out of the condensed phase with a time constant t_c, a process particularly important around monolayer coverage. On the other hand, as long as desorption proceeds predominantly via the 2-d gas phase, the depletion of the condensate islands takes place by evaporating into the 2-d gas phase, with a time

---

* A case in point is the several pages of arguments needed in reference 39 to arrive at the expression for zero-order desorption, already contained in equation (11) for constant chemical potential.

constant t_{ev}, from where particles then desorb into the 3-d gas phase. This will result in density gradients leading to surface diffusion. Two rather different situations may obtain: (i) if $t_{ev} \ll t_d$, evaporation is so fast that during the desorption process a quasi-equilibrium is maintained between the adsorbed phases. In such a situation there will be a coverage regime where the desorption kinetics is roughly zero order. (ii) If $t_{ev} \gg t_d$ then evaporation is the slowest process in a chain and thus rate determining. With evaporation proceeding via the rim of the condensed islands one expects roughly half-order kinetics. These ideas were first put forward many years ago in a number of papers.[40-45]

From the above discussion it should be obvious that adsorbates under non-equilibrium conditions can no longer be described by a single macroscopic variable, *i.e.* the coverage, but that, as a minimum, one must consider the partial coverages of the various phases and components. To set up a macroscopic theory of surface processes, it is then expedient to follow the Onsager approach to non-equilibrium thermodynamics.[46] Our first task is to identify the proper set of macroscopic variables to describe the adsorbate. To describe typical experiments involving adsorption, desorption, and surface reactions, it is usually not necessary to introduce local densities at the macroscopic level. If the adsorbate forms a single phase, it will remain homogeneous throughout so that diffusion plays no role. If, on the other hand, the adsorbate shows co-existence of two phases, mass transport is limited to particle exchange between the two phases which can be described as 2-d condensation and evaporation. It is then rather dubious to introduce density gradients in the dilute phase, because, for example, at half coverage, the number of molecules in a typical patch of dilute gas is at most a hundred, so that number fluctuations amount to at least ten percent, and the number of particles per patch of dilute gas, and even more so the local particle density, can hardly be taken as macroscopic variables in the two-phase region.

To set up the balance equations controlling energy and mass exchange in a two-phase adsorbate, we consider the total system, consisting of gas phase, adsorbate, and substrate, as closed in a volume V_t and isolated with total energy U_t. As extensive variables we consider the particle numbers $X_1 = N_1$ in the condensed 2-d phase, $X_2 = N_2$ in the dilute 2-d phase on the bare surface, $X_3 = N_{2'}$ in the dilute 2-d phase on top of the condensed islands, $X_4 = N_3$ in the 3-d ambient gas, $X_5 = N_s$ in the substrate, and the respective energy variables $X_6 = U_1, \ldots X_{10} = U_s$. Following Onsager we can write the macroscopic balance equations as:

$$\frac{dX_i}{dt} = \sum_{j=1}^{10} L_{ij} \frac{\partial S}{\partial X_j}\bigg|_{U_t, V_t, N, N_s} \tag{35}$$

Such a description is valid (i) in the linear regime, and (ii) as long as local equilibrium pertains. The latter condition in particular implies that the entropy of the system is given by:

$$S(U_t, V_t, N, N_s) = S_1(U_1, A_1, N_1) + S_2(U_2, A_2, N_2) + S_{2'}(U_{2'}, A_{2'}, N_{2'})$$
$$+ S_3(U_3, V, N_3) + S_s(U_s, V_s, N_s) \qquad (36)$$

Here A_i are the areas occupied by the respective phases. Thus we get in equation (35):

$$\left.\frac{\partial S}{\partial N_j}\right|_{U_j, A_j(V)} = -\frac{\mu_j}{T_j} \qquad (37)$$

$$\left.\frac{\partial S}{\partial U_j}\right|_{N_j, A_j(V)} = \frac{1}{T_j} \qquad (38)$$

introducing the chemical potentials, μ_j, and the temperatures, T_j, in the various phases and components.

Noting mass conservation in the system $dN_s/dt = 0$, $(d/dt)(N_1 + N_2 + N_{2'} + N_3) = 0$, and energy conservation $dU_t/dt = 0$, one gets thirty conditions on the 100 phenomenological coefficients L_{ij}, because the coefficients in front of each thermodynamic force have to vanish independently. Onsager's reciprocity relations $L_{ij} = L_{ji}$ eliminate another forty-three coefficients.

Under isothermal situations the equations simplify considerably and can be written as:

$$\frac{dN_1}{dt} = -\frac{L_{12}}{T}(\mu_2 - \mu_1) - \frac{L_{12'}}{T}(\mu_{2'} - \mu_1) - \frac{L_{13}}{T}(\mu_3 - \mu_1) \qquad (39)$$

$$\frac{dN_2}{dt} = -\frac{L_{12}}{T}(\mu_1 - \mu_2) - \frac{L_{23}}{T}(\mu_3 - \mu_2) - \frac{L_{22'}}{T}(\mu_{2'} - \mu_2) \qquad (40)$$

$$\frac{dN_{2'}}{dt} = -\frac{L_{12'}}{T}(\mu_1 - \mu_{2'}) - \frac{L_{22'}}{T}(\mu_2 - \mu_{2'}) - \frac{L_{2'3}}{T}(\mu_3 - \mu_{2'}) \qquad (41)$$

$$\frac{dN_3}{dt} = -\frac{d}{dt}(N_1 + N_2 + N_{2'}) \qquad (42)$$

Note that equilibrium conditions, i.e. $\mu_i = \mu_j$, imply vanishing fluxes.

The next task is to relate the six Onsager coefficients in equations (39)–(41) to experimentally accessible quantities like sticking coefficients and activation energies. This has been done for the individual processes, such as adsorption, desorption, and 2-d evaporation and condensation.[7,9] The resulting equations are:

$$\frac{dN_1}{dt} = R_{13} + R_{12} + R_{12'} \qquad (43)$$

$$\frac{dN_2}{dt} = R_{23} - R_{12} + R_{22'} \tag{44}$$

$$\frac{dN_{2'}}{dt} = R_{2'3} - R_{12'} - R_{22'} \tag{45}$$

Here

$$R_{i3} = S_i A_i \left[\frac{P\lambda_{th}}{h} - \frac{k_B T}{h\lambda_{th}^2} \exp(\mu_i/k_B T) \right] \tag{46}$$

are the net rates (*i.e.* adsorption minus desorption), of mass exchange between the 3-d gas phase above the surface at temperature T and pressure P and the adsorbed components, $i = 1, 2, 2'$. The chemical potentials, μ_i, are those appropriate for the instantaneous values of $N_i(t)$, and S_i is the sticking probability to i^{th} phase. There is some arbitrariness in deciding whether particles impinging from the 3-d phase onto the condensed islands become instantly part of the condensed phase (sticking coefficient S_1) or equilibrate first with the 2'-gas phase on top of the islands (sticking coefficient $S_{2'}$). This seems irrelevant because adding equations (43) and (45) implies adding equation (46) for $i = 1$ and $2'$, which, with $\mu_{2'} = \mu_1$, results in an overall sticking coefficient $S_1 + S_{2'}$, since $A_1 = A_{2'}$.

The rate of mass exchange between the condensed phase and the 2-d gas phase (2-d condensation minus 2-d evaporation) summarizing diffusive processes, is given, within the co-existence region, by:

$$R_{12} = S_c f A_1^\zeta \frac{\bar{v}}{\pi} \left(\frac{1}{A - A_1} \right) [N_2 - \bar{N}_2(N_a, T)] \theta(N_1) \theta(N_2) \tag{47}$$

i.e. as an excess flux of particles, moving with an average speed $\bar{v}$ and sticking, with probability S_c, to the rim of the condensed islands, the boundary length of which we approximate by fA_1^ζ. Here $\bar{N}_2(N_a, T)$ is the number of particles that would be in the 2-d phase if the adsorbate were at equilibrium with a total number of particles, $N_a = N_1 + N_2 + N_{2'}$, and $\theta(N) = 1$ for $N > 1$ and zero otherwise. Although the phenomenological geometric quantities of f, and ζ, must change as the co-existence region is traversed, approximations are possible, for example if the islands are n in number and circular then $f = 2\sqrt{(n\pi)}$, $\zeta = 1/2$; for irregular perimeters $f = f_0 n^{1-\zeta}$, $0.5 < \zeta < 1$, $f_0 \geq 2\sqrt{(\pi)}$. Likewise, particles in the 2'-phase can be incorporated into the condensed phase at the island rim according to:

$$R_{12'} = S_c' f' A_1^\zeta \frac{\bar{v}}{\pi} \left(\frac{1}{A_1} \right) [N_{2'} - \bar{N}_{2'}(N_a, T)] \theta(N_1) \theta(N_{2'}) \tag{48}$$

Lastly, we get for the rate of equilibrium between the 2'- and 2-phase:

$$R_{22'} = (1 - S_c - R_c) f A_1^\zeta \frac{\bar{v}}{\pi} \left(\frac{1}{A - A_1}\right) [N_2 - \bar{N}_2(N_a, T)] \theta(N_1) \theta(N_{2'}) \quad (49)$$

Here it is assumed that particles skipping along the bare surface will either be reflected back with a probability R_c, or stick to the rim of the condensed islands with a condensation coefficient S_c, or hop on top of them with a probability $1 - S_c - R_c$.

As is evident from the rate equations (39)–(42), the thermodynamic forces responsible for mass transport are differences in chemical potentials. To specify the latter for a two-phase adsorbate one can resort to simple models, for example, by treating a mobile adsorbate as a 2-d van der Waals gas, and a localized adsorbate as a lattice gas with nearest neighbour interactions within the Bragg–Williams and the quasichemical approximations.[7–9] This being specified, the macroscopic theory is complete. It has been used successfully to describe adsorption–desorption kinetics in a number of gas–solid systems, such as metals on metals and rare gases on metals.[7–10] We comment briefly on a few of the results. One example of the new insights gained from the rate equations (43)–(45) that was not contained in equation (11), is the desorption from the co-existence region for a system in which the sticking probability on the bare surface, S_2, is different from that on top of the adsorbate, $S_1 + S_{2'} = S_{1t}$, both being assumed constant, for simplicity. This appears to be the case, e.g., for Xe on Ni and Ru where $S_1 = 0.6$ and $S_2 = 1.0$.[47] In quasi-equilibrium the chemical potentials of the various adsorbate phases are equal, i.e. $\mu_1 = \mu_2 = \mu_a$ in equation (46), so that after summing equations (43)–(45) we get:

$$dN_a/dt = (S_{1t} A_1 + S_2 A_2)[(P \lambda_{th}/h) - (k_B T/h \lambda_{th}^2) \exp(\mu_a/k_B T)] \quad (50)$$

Within the co-existence region one can use the lever rule and re-express equation (50) under desorption conditions, i.e. for $P = 0$, as:

$$\frac{d\theta}{dt} = -[(S_{1t} - S_2)\theta + S_2 \theta_{1c} - S_{1t} \theta_{2c}] \frac{1}{\theta_{1c} - \theta_{2c}} a_s \frac{k_B T}{h \lambda_{th}^2} \exp(\mu_a/k_B T) \quad (51)$$

The new variables here are the lower and upper boundaries to the two-phase region, $\theta_{2c}(T) < 1/2 < \theta_{1c}(T)$. We note that, within the co-existence region and at constant temperature, there is no coverage dependence outside the square brackets. This implies zero-order kinetics, provided that the sticking coefficients are the same for the dilute and the dense phases. However, if this is not the case, quasi-equilibrium in the co-existence region does not imply zero-order kinetics. As an illustration we present in Figure 10, an approximate fit to data on Xe desorbing from Ni(111),[47] calculated in the Bragg–Williams approximation.

Non-equilibrium effects are introduced by increasing the size of condensed-phase islands and decreasing the hopping rate away from the perimeters of these islands, such that $R_{12} \lesssim R_{23}, R_{13}$. The effect on (two of) the

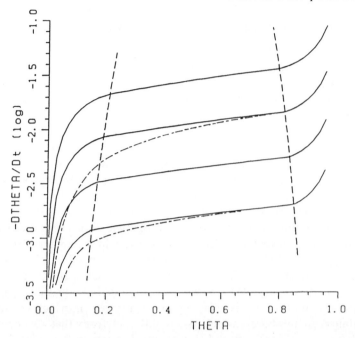

Figure 10 *Isothermal desorption of* Xe/Ni(111) *for temperatures (top to bottom)* T = 82, 80, 78, 76 K *and a two-component adsorbate in quasi-equilibrium (solid lines); and* T = 80, 76 K *for a non-equilibrium situation, c.f. reference 9 (dash-dotted lines). Parameters for the model (Bragg–Williams)* $V_0 = 3200$ K, $cV_2 = -375$ K, $v_z = 5 \times 10^{11}$ s^{-1}, $v_x = v_y = 10^{13}$ s^{-1}, $n_s = 0.2$ A^{-2}, $S_{1t} = 1.0$, $S_2 = 0.6$. *Dashed lines show co-existence boundaries in quasi-equilibrium*

curves of Figure 10 is indicated by the dash-dotted lines, corresponding to imposing the extreme condition $R_{12} \ll R_{23}, R_{13}$. In this case co-existence continues down to $\theta = 0$, *i.e.* it is desorption from the islands that contributes predominantly to the total rate. Such isothermal modelling also shows that the concept of a fixed desorption order [x in equation (3)] is found wanting: in the quasi-equilibrium case, x may not be fixed at zero during co-existence, *c.f.* equation (51) and cannot be unity outside co-existence due to adparticle interactions; in non-equilibrium situations x varies markedly with temperature, and it does not necessarily approach the value of ζ. Even the utility of the Arrhenius parametrization [equation (4)] is questionable when desorption is not a quasi-equilibrium process—isosteric Arrhenius plots are no longer one or more straight lines, but smooth curves. Figure 6 of reference 47 contains an excellent example of this—its significance was overlooked by all. This sounds a caution against the 'blind' fitting of a single line through scattered isosteric data points, so common in the literature. Now the quantity $E_d(\theta, T)$, defined by equation (22) is a necessity if one insists on a parametric analysis of the data. Again, though, the temperature dependence of $E_d(\theta, T)$ can be so great as to render that analysis almost

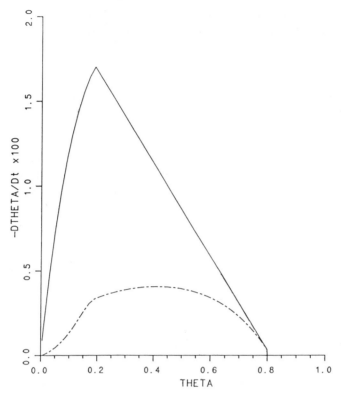

Figure 11 *Desorption at T = 80 K of a two-component adsorbate with desorption from condensed phase suppressed ($S_1 = 0$); quasi- and non-equilibrium situations otherwise as Figure 10. Initial coverage $\theta(t = 0) = 0.8$*

useless, unless a detailed comparison is also made with specific models of the adsorbate. As a final example, we show in Figure 11, the result of setting $S_1 = 0$, $N_{2'} = 0$ in the rate equations (43)–(45). For quasi-equilibrium, c.f. equation (51), there is a linear growth of the rate as θ decreases, as far as $\theta_{2c}(T)$, then a non-linear decrease as the dilute phase desorbs. The extreme non-equilibrium condition produces a variable-order process, however (dash-dotted line).

We conclude this section by listing some other representative developments in both experiment and modelling of thermal desorption. Since the first phenomenological kinetic models describing zero-order desorption and the effect of islanding of the adsorbate were advanced,[40–45] surprisingly little improvement of those models has occurred despite the large number of experiments exhibiting these effects. Of the few groups that have modelled desorption trajectories through a phase co-existence region, Nagai and co-workers[48] have been restrictive in their use of transition state theory (quasi-equilibrium), with a constant prefactor, and no distinction between the co-existing phases, e.g. differing sticking coefficients. That thermal desorption experiments can detect unequal sticking coefficients, as described by references 7–10, has been shown recently by Meyer et al.[49] in their modelling of H/Mo(100). The characteristics of the sticking coefficients at the two-phase interface can also be probed.[50] Experiments clearly illustrating non-equilibrium effects are more scarce. Arthur and Cho[40] observed half-order kinetics long ago for Cu and Au desorbing from graphite. Their Arrhenius plots are not linear but, as has occurred with most experimental analyses since, they derive a desorption energy from the straightest portion of the isosteres. Such fitting procedures, coupled with experimental uncertainties, must be partially responsible for this scarcity. Another, and obvious, reason is that many systems are in the quasi-equilibrium regime at the approach to peak desorption temperatures. Vollmer and Trager[51] have performed a desorption order analysis for Na/LiF(100); unfortunately, they presumed a temperature- and coverage-independent prefactor. Fractional-order desorption has also been advanced for the desorption of particles from fractal surfaces.[52]

Within the framework of thermodynamic models, the inclusion of lateral interactions for sub-monolayer desorption and comparison with experiment has, until recently,[26] progressed to about the level of the quasichemical approximation,[25] or generalizations of it.[53,54] The situation is no better for multi-layered adsorbates. References 7 and 9 show that even a small population of second-layer particles can influence the overall rate when desorption from the islands, on which the particles reside, is restricted. A bilayer model for zero-order desorption has also been advanced by Asada and Masuda.[55] Non-trivial lattice-gas models for multilayers[56,57] may allow a reasonable comparison with experiment, but so far their desorption characteristics have not been tested. Simplistic treatments[58] are of specific value only. The applicability of multilayer models to particular systems will also be guided by a knowledge of the growth mechanisms of the adlayers. In this context we note the work of Tikhov and Bauer,[59] and Knall et al..[60]

There are several recent developments in the understanding of adsorbates which are outside the scope of the presentation here. Phase reconstruction of the adsorbate during desorption,[61] adsorbate-induced reconstruction of the substrate[62-64] and alloying,[65] are examples of surface ordering that strongly influence the kinetic parameters. TPD, in conjunction with LEED, should help to decide on microscopic models for the long-range ordering. Analysis of short-range ordering of specific surface structures, using the desorption of noble gases as a probe, was advanced some time ago.[66] The application of thermal desorption to surface site heterogeneity,[67] multiple adsorption sites,[68,69] and mixed adlayers of species,[70] have all received recent attention. In addition to references 67 and 70, other Monte Carlo studies include those of Lombardo and Bell,[71] and Sales et al..[72] Unfortunately, with the exception of reference 72, there is no, or improper, comparison with well known results of (approximate) analytic models. The growing trend towards 'analysis' by Monte Carlo methods is of little value unless attempts are made to fit experimental data.

5 Kinetic Lattice Gas Model

To go beyond a macroscopic description of time-dependent phenomena, we explore a mesoscopic approach entailing a generalization of the kinetic lattice gas model. It was originally set up in close analogy to the kinetic Ising model for magnetic systems.[73] One assumes that the surface of a solid can be divided into two-dimensional cells, labelled i, for which one introduces microscopic variables $n_i = 1$ or 0, depending on whether cell i is occupied by an adsorbed gas particle or not. The connection with magnetic systems is made by a transformation to spin variables $\sigma_i = 2n_i - 1$. In its simplest form, a lattice gas model is restricted to the sub-monolayer regime and to gas–solid systems in which the surface structure and the adsorption sites do not change as a function of coverage. To introduce the dynamics of the system one writes down a model Hamiltonian:

$$\mathcal{H} = E_s \sum_i n_i + \tfrac{1}{2} V_2 \sum_{\langle ij \rangle} n_i n_j + \ldots \qquad (52)$$

Here E_s is a single particle energy and V_2 is the two-particle interaction between nearest neighbours $\langle ij \rangle$. Interactions between next-nearest neighbours *etc.* and many-particle interactions can be easily added to equation (52). As long as the number of particles in the adsorbate does not change (which is the case for systems in equilibrium, or for diffusion studies) the first term in equation (52) is constant and can be dropped from further consideration. However, if we want to study adsorption–desorption kinetics, the number of particles in the adsorbate changes as a function of time and a proper identification of E_s is mandatory. Arguing that the lattice gas Hamiltonian should give the same Helmholtz free energy as a microscopic Hamiltonian (for non-interacting particles) one can show that the proper identification is given by:

$$E_s = -V_0 - k_B T \ln(q_3 q_{\text{int}}) - k_B T [\ln(\lambda_{\text{th}}^3 P/k_B T) - \ln(Z_{\text{int}})] \qquad (53)$$

Again V_0 is the depth of the surface potential, and q_3 is given by equation (13). To set up the kinetic lattice gas model we restrict ourselves to gas–solid systems in which all relevant processes, like diffusion, adsorption, desorption *etc.*, are Markovian. We introduce a function $P(\boldsymbol{n}; t)$ which gives the probability that a given microscopic configuration $\boldsymbol{n} = (n_1, n_2, \ldots, n_I)$ is realized at time t, where $I = N_s$ is the total number of adsorption sites on the surface. It satisfies a master equation:

$$dP(\boldsymbol{n}; t)/dt = \sum_{\boldsymbol{n}'} [W(\boldsymbol{n}; \boldsymbol{n}') P(\boldsymbol{n}'; t) - W(\boldsymbol{n}'; \boldsymbol{n}) P(\boldsymbol{n}; t)] \qquad (54)$$

where $W(\boldsymbol{n}'; \boldsymbol{n})$ is the transition probability that the microstate $\boldsymbol{n}$ changes into $\boldsymbol{n}'$ per unit time. It satisfies detailed balance:

$$W(n';n)P_0(n) = W(n;n')P_0(n') \tag{55}$$

where

$$P_0(n) = Z^{-1} \exp(-\mathcal{H}(n)/k_B T) \tag{56}$$

is the equilibrium probability. In principle, $W(n';n)$ must be calculated from a Hamiltonian that includes, in addition to equation (52), coupling terms to the gas phase and the solid that mediate mass and energy exchange; this has been done for phonon-mediated adsorption–desorption kinetics.[6] Rather, we follow the procedure initiated by Glauber[74] and guess an appropriate form of $W(n';n)$. We briefly indicate possible forms below, and their consequences for the solution of equation (54).

Assuming that the residence time of an adsorbed particle in an adsorption site is much longer than the transition time into or out of that site, we can write $W(n';n)$ as a sum of terms accounting for individual processes of adsorption, desorption, and diffusion, i.e.:

$$W(n';n') = W_{a-d}(n';n) + W_{dif}(n';n) \tag{57}$$

The simplest choice is what we have termed Langmuir kinetics[75] for which we assume that adsorption into a site i takes place provided that site is empty, irrespective of whether neighbouring sites are occupied. We then get:

$$\begin{aligned}&W_{a-d}(n';n) \\ &= W_0 \sum_i \left[1 - n_i + C_0 n_i (1 + C_1 \sum_a n_{i+a} + C_2 \sum_{a,a'} n_{i+a} n_{i+a'} + \ldots)\right] \\ &\times \delta(n'_{i'}, 1 - n_i) \prod_{l \neq i} \delta(n'_l, n_l)\end{aligned} \tag{58}$$

where a and a' enumerate the neighbours of site i. Similarly we write for diffusion:

$$\begin{aligned}W_{dif}(n';n) = J_0 \sum_{i,a} n_i(1 - n_{i+a})(1 + C_1 \sum_{b \neq a} n_{i+b} + \ldots) \\ \times \delta(n'_{i'}, 1 - n_i)\delta(n'_{i+a}, n'_i\, 1 - n_{i+a}) \prod_{l \neq i, i+a} \delta(n'_l, n_l)\end{aligned} \tag{59}$$

Inserting equation (57) and equation (55) we get:

$$C_0 = \exp(E_s/k_B T) \tag{60}$$

$$C_r = [\exp(V_2/k_B T) - 1]^r \tag{61}$$

Langmuir kinetics, in particular, results in a sticking coefficient:

$$S(\theta) = S_0(1 - \theta) \tag{62}$$

Other choices, and their physical or unphysical implications, have been discussed in reference 75. As long as the master equation (54) is used in Monte Carlo simulations solely to determine the equilibrium properties of the system, it does not matter what choices are made for the transition probabilities $W(n';n)$ as long as they satisfy detailed balance [equation (55)]. However, if we are interested in the adsorption, desorption, and diffusion kinetics of a particular physical system, the choice of transition probabilities $W(n';n)$ is rather narrowed.[75]

To solve the master equation (54), Monte Carlo techniques have been invoked.[76–78] The renormalization group has also been used successfully.[79,80] Likewise, one can derive a hierarchy of coupled equations of motion for average occupations and correlation functions.[74,75,81–85] To this end we define:

$$\langle n_i \rangle = \sum_n n_i P(\mathbf{n}; t) \tag{63}$$

$$\langle n_i n_j \rangle = \sum_n n_i n_j P(\mathbf{n}; t) \tag{64}$$

and obtain from equation (54):

$$d\langle n_i \rangle/dt = W_0(1 - \langle n_i \rangle) - W_0 C_0(\langle n_i \rangle + C_1 \sum_a \langle n_i n_{i+a} \rangle + \ldots) \tag{65}$$

and similar equations for the higher-order correlation functions. Comparing equations (65) and (8) for the case of a homogeneous adsorbate with $\langle n_i \rangle = \theta$, allows one to identify W_0 as:

$$W_0 = S_0 P a_s / (2\pi m k_B T)^{1/2} \tag{66}$$

implying, with equations (60) and (53), that the desorption rate constant is given by:

$$r_d(\theta = 0) = W_0 C_0$$
$$= S_0 a_s k_B T / (h \lambda_{th}^2 q_3)(Z_{int}/q_{int}) \exp(-V_0/k_B T) \tag{67}$$

The hierarchy equation (65) *etc.*, is obviously exact. To solve it, approximations must be made; in particular, the hierarchy must be truncated, for example by invoking the Kirkwood closure approximation, which states that:

$$\langle n_i n_j \ldots n_m \rangle = \langle n_i n_j \rangle \langle n_j n_k \rangle \ldots \langle n_l n_m \rangle / \langle n_i \rangle \langle n_j \rangle \langle n_k \rangle \ldots \langle n_l \rangle \langle n_m \rangle \tag{68}$$

In the simplest (Kirkwood) approximation, and for a homogeneous adsorbate, one is left with two coupled equations for θ and for the average correlation function $\psi = \langle n_i n_{i+a}\rangle$, which read, for a square lattice:[75]

$$d\theta/dt = W_0[1 - \theta - C_0\theta(1 + C_1\psi/\theta)^4] \quad (69)$$

$$\begin{aligned}d\psi/dt &= 2W_0[\theta - \psi - C_0(1 + C_1)\psi(1 + C_1\psi/\theta)^3] \\ &\quad + 2J_0(1 + C_1\psi/\theta)^2[3\theta^2 - 3(1 + C_1)\psi \\ &\quad + C_1\theta\psi + 5C_1\psi^2/\theta - C_1\psi^3(1 + 2\psi/\theta)/\theta^2]\end{aligned} \quad (70)$$

We note that the equilibrium solution of these equations yields, at least in the absence of the diffusional terms, the quasichemical approximation to the isotherms. Other treatments of the closure, in 2-d, correspond to quasichemical, Kirkwood, or maximum entropy approximations.[54,85,86] In 1-d the closure is exact for an immobile gas with nearest neighbour interactions;[87] otherwise it is approximate.[88,89]

In the remainder of this section we briefly survey other applications of the kinetic lattice gas model. It has been used to study non-equilibrium effects on the desorption kinetics. In this context, various choices of transition probabilities have been examined and difficulties with the Kirkwood truncation have been explored.[75] The effects of adatom clustering on desorption rates,[82] and desorption of dimers[83] have also been considered. A specific application, in the one-dimensional case, has been to N_2/Ni(430),[38] including multi-particle interactions and diffusive terms completely. Recently the kinetic lattice gas model has been used to find a time-dependent generalization of the grand canonical ensemble to describe inhomogeneous adsorbates within the co-existence region.[90] If the adsorption–desorption kinetics are precursor-mediated, one can extend the kinetic lattice gas model by introducing three sets of occupation numbers per lattice cell, namely $n_i = 0$ or 1, $m_i = 0$ or 1, and $l_i = 0$ or 1, depending on whether the chemisorbed state, the intrinsic precursor, and extrinsic precursor are empty or occupied, respectively. The Hamiltonian of the system is then:

$$\mathcal{H} = E_s\sum_i n_i + e_s\sum_i m_i + \epsilon_s\sum_i l_i + \tfrac{1}{2}V_2\sum_{\langle ij\rangle} n_i n_j \quad (71)$$

For the transition probability in equation (54) we then write:

$$W(n',m',l';n,m,l) = W_I + W_E + W_C + W_{CI} + W_{CE} + W_{EE} + W_{II} + W_{EI} \quad (72)$$

The significance of the individual terms is illustrated in Figure 12. As an example, we can specify adsorption into, and desorption from, the intrinsic precursor state with:

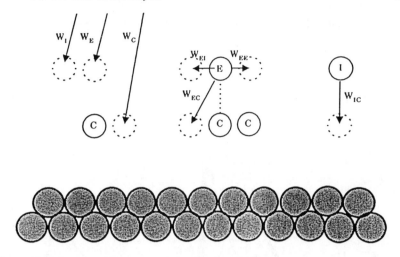

Figure 12 *Schematic illustration of the various processes between extrinsic precursor (E), intrinsic precursor (I), and chemisorbed (C) states included in the kinetic lattice gas model*

$$W_I(n,m',l;n,m,l) = w_I \sum_i (1 - n_i)(1 - m_i + D_I m_i)\delta(m'_i, 1 - m_i) \prod_{k \neq i} \delta(m'_k, m_k)$$
(73)

where the first factor ensures that the intrinsic precursor only exists over an empty site. The full theory of precursor-mediated adsorption–desorption kinetics is published elsewhere.[91]

6 Physisorption Kinetics

A monograph on physisorption kinetics was published by Kreuzer and Gortel which covers the field up to 1985. The book starts with a review of the gas–solid interaction of physisorption. Next the master equation is derived and the transition probabilities are calculated for phonon-mediated adsorption and desorption processes. On this basis desorption times are calculated as well as time-of-flight spectra of the desorption flux. This is followed by a discussion of sticking and energy accommodation. In the final chapter the Kramers equation is derived, from which the macroscopic laws are obtained, together with microscopic expressions of the friction coefficient. Thus the complete programme of non-equilibrium statistical mechanics, as envisaged by Boltzmann, has been completed for physisorption, namely, starting from a microscopic Hamiltonian, one derives kinetic equations and, finally, calculates macroscopic transport coefficients. The only other example where this programme has been completed before is the theory of dilute gases. This is not to say that there does not remain a large amount of work still to be done. In particular, the above theory is largely restricted to the low coverage

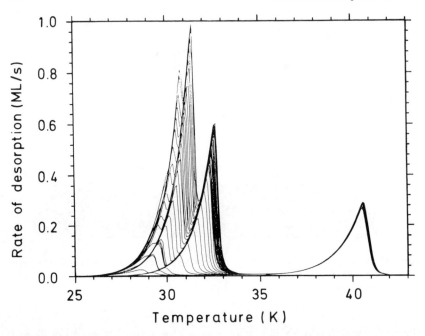

Figure 13 *TPD of* Ar *multilayers on* Ru(001) *for initial coverages between one and close to five monolayers. Heating rate* 0.5 K s^{-1}. *Even the fifth monolayer can be resolved*

regime, except for helium where the kinetics has been worked out for multilayer adsorbates. Although non-equilibrium thermodynamics and the kinetic lattice gas model are being applied to study the kinetics of physisorbed multilayers, little progress has been made on its microscopic foundations. Brenig[92] has reviewed the kinetics and dynamics of the gas–surface interaction in the zero coverage limit stressing the importance of the principles of detailed balance and unitarity. Jack *et al.*[93] have derived the Fokker–Planck equation for physisorption kinetics, again in the zero coverage limit, and presented a general formulation for its solution under isothermal desorption conditions.

The good news in physisorption kinetics comes from Menzel's group where an exhaustive experimental study of rare gases on ruthenium has just been completed. We are fortunate to be able to present some samples of what has been or will be published in the near future to draw attention to this major progress in the field.[94–99]

In their high resolution thermal desorption spectrometry of mono and multilayers of rare gases on ruthenium, Menzel and co-workers can measure a coverage and desorption rate range of five orders of magnitude with a reproducibility of a few percent. Temperatures can be measured in the range of 5–1600 K with a resolution of 25–70 mK and an absolute accuracy of 0.1–0.5 K. Their linear temperature ramps for TPD go from 10^{-2} to 20 K s^{-1} and temperature jumps for isothermal desorption are very well controlled.[94–96]

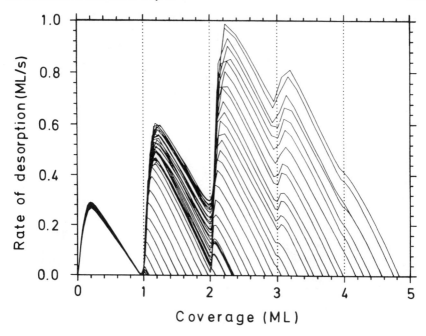

Figure 14 *Replot of the data of Figure 13 as a 'layer plot', i.e. desorption rate vs. coverage, the latter obtained by (partial) integration over time of the rate. Note that the temperature increases from right to left. The breaks in the rates occur at the complete monolayers*

As for the adsorption kinetics, they measured sticking coefficients as a function of coverage, surface temperature, and gas temperature. For all rare gases (and also for H_2 and D_2) they find an increase with coverage which is attributed to enhanced sticking for collisions with the adsorbate rather than with the bare metal.[94,97-98] For neon at 300 K the sticking coefficient increases by three orders of magnitude from $\theta = 0$ to $\theta = 0.5$. At vanishing coverage there is a strong dependence of the sticking coefficient on the gas temperature but very little on the surface temperature, which has been explained as a quantum effect due to zero phonon scattering.[97] Collision induced desorption and replacement, and exchange during desorption is also seen.

The high resolution attained in desorption kinetics allows the discrimination of at least five successive layers, see Figures 13–16. The coverage dependence is stressed in a 'layer plot' in which the rate is plotted against its integral, see Figure 14. Such plots for the rare gases show that the number of particles in the first and higher layers are identical within the accuracy of the measurement of a few percent. Interestingly, this is not the case for physisorbed molecules such as hydrogen for which the first layer contains about 12% more than the higher layers.[99] The wide range accessible is best utilized by plotting the logarithm of the rate *vs.* inverse temperature. We refer to these as high resolution temperature programmed desorption (HRTPD) plots;

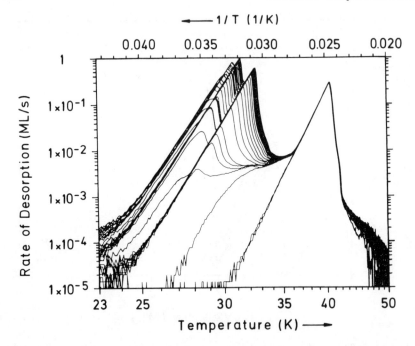

Figure 15 *HRTPD-plots of* Ar *on* Ru(001) *(logarithm of desorption rate vs.* T^{-1}*) for starting coverages between* 0.03 *and* 1.30 ML. *Heating rate* 0.5 K s^{-1}

illustrative examples are shown in Figures 15 and 16. Since the coverage is virtually unchanged over the first two to three decades of the rate, an apparent zeroth order behaviour results, which can be evaluated directly to obtain Arrhenius parameters. Other evaluations, *i.e.* via isotherms, isosteres, and peak shape analysis, have been investigated. Strong non equilibrium effects, *i.e.* deviations from quasi-equilibrium distributions within the adsorbate, due to limited diffusion have been found depending on preparation and annealing, *i.e.* initial coverage and temperature. There is clear evidence for phase transitions and for different two- and three-dimensional growth modes which have also been investigated with LEED. Very strong influences of small amounts of surface contamination and of temperature gradients across the sample have been found which may well be the reason for most, if not all, of the 'roughening transitions' reported in the literature for rare gas layers.

7 Concluding Remarks

In this review we have tried to give a concise overview of the physical processes underlying thermal desorption, the formulation of the phenomenological approach, and methods of evaluation of experimental data; and have outlined the current status of kinetic lattice gas models. We have only made passing reference to experimental aspects of thermal desorption,

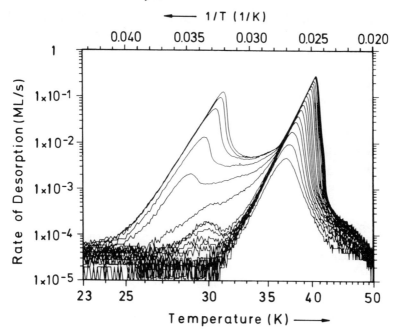

Figure 16 *HRTPD-plots of the data of Figure 13*

although we have chosen to highlight some recent work as exemplifying a new standard. Noting the margin by which such good experimental technique leads the many over-simplified theoretical models and analyses of data elsewhere in the literature, we are not concerned by our bias. To the contrary, it is time the simple but fundamental phenomenological concepts and results outlined in Sections 2 and 3 were understood by all interpreters of TPD data. Specific examples bear re-emphasis: the assumption of a constant prefactor is no longer tolerable, nor must high temperature limits be presumed; nor should a peak temperature analysis of experimental TPD data be considered sufficient—a complete analysis is mandatory if sense is to be made of the phenomenological parameters, E_d and v, including their coverage and temperature dependence. The complete theory at the macroscopic level, including non-equilibrium effects, has now been formulated, at least for the sub-monolayer regime: apart from restrictions on one's knowledge of the chemical potential of the adsorbate, there is no need for further oversimplified modelling of TPD spectra.

The macroscopic theory of multilayer desorption is incomplete. Progress here is required as adsorption into the second layer can be important long before a monolayer is completed. A penalty associated with this formulation will be the introduction of more *ad hoc* parameters. As one should, ultimately, make connections with microscopic concepts, the kinetic lattice gas model provides an alternate, and perhaps more satisfactory description of multilayer effects. One hopes for further analytic work on this model in the monolayer regime, in any case, and not just a swift conversion to Monte Carlo simulation methods as a means of understanding thermal desorption spectra. Progress in the application of the spectroscopy, and in conjunction with other techniques, is also expected in other areas, as mentioned at the end of Section 4, *e.g.* surface/adsorbate reconstruction, determination of sticking coefficients, adlayer growth mechanisms, *etc.*. The advent of high resolution TPD will be an important factor in this progress.

Acknowledgement. We thank Professor D. Menzel for furnishing us with results prior to publication. This work was supported in part by a grant from CEMAID, the Centres of Excellence in Molecular and Interfacial Dynamics.

References

1. T. J. Chuang, *Surf. Sci. Rep.*, 1983, **3**, 1.
2. P. Piercy, Z. W. Gortel, and H. J. Kreuzer, 'Advances in Multiphoton Processes and Spectroscopy', 1987, **3**, 105.
3. 'Desorption Induced by Electronic Transitions, DIET III', ed. R. H. Stulen and M. L. Knotek, Springer series in Surface Sciences, Vol. 13, Springer-Verlag, Berlin, 1988.
4. J. H. Block, in 'Chemistry and Physics of Solid Surfaces IV', ed. R. Vanselow and R. Howe. Springer–Verlag, Berlin, 1982, p. 407.
5. J. E. Lennard-Jones and A. F. Devonshire, *Proc. R. Soc. London, A*, 1936, **156**, 6; 29.
6. H. J. Kreuzer and Z. W. Gortel, 'Physisorption Kinetics', Springer–Verlag, Berlin, 1986.
7. H. J. Kreuzer and S. H. Payne, *Surf. Sci.*, 1988, **198**, 235.
8. H. J. Kreuzer and S. H. Payne, *Surf. Sci.*, 1988, **200**, L433.
9. S. H. Payne and H. J. Kreuzer, *Surf. Sci.*, 1988, **205**, 153.
10. S. H. Payne and H. J. Kreuzer, *Surf. Sci.*, 1989, **222**, 404.
11. D. Menzel, in 'Interactions on Metal Surfaces', ed. R. Gomer, Springer–Verlag, Berlin, 1975, p. 102.
12. D. Menzel, in 'Chemistry and Physics of Solid Surfaces IV', ed. R. Vanselow and R. Howe, Springer–Verlag, Berlin, 1982, p. 389.
13. P. A. Redhead, *Vacuum*, 1962, **12**, 203.
14. A. M. de Jong and J. W. Niemantsverdriet, *Surf. Sci.*, 1990, **233**, 355.
15. E. Tronconi and L. Lietti, *Surf. Sci.*, 1988, **199**, 43.
16. D. A. King, T. E. Madey, and J. T. Yates, Jr., *J. Chem. Phys.*, 1971, **55**, 3236.
17. E. Bauer, H. Poppa, G. Todd, and F. Bonczek, *J. Appl. Phys.*, 1974, **45**, 5164.
18. J. L. Taylor and W. H. Weinberg, *Surf. Sci.*, 1978, **78**, 259.
19. E. Habenschaden and J. Küppers, *Surf. Sci.*, 1984, **138**, L147.
20. J. C. Miller, H. R. Siddiqui, S. M. Gates, J. N. Russell, Jr., J. T. Yates, Jr., J. C. Tully, and M. J. Cardillo, *J. Chem. Phys.*, 1987, **87**, 6725.
21. A. Pavlovska, H. Steffen, and E. Bauer, *Surf. Sci.*, 1988, **195**, 207.
22. E. G. Seebauer, A. C. F. Kong, and L. D. Schmidt, *Surf. Sci.*, 1988, **193**, 417.
23. H. Ibach, W. Erley, and H. Wagner, *Surf. Sci.*, 1980, **92**, 29.
24. M. Paunov and E. Bauer, *Surf. Sci.*, 1987, **188**, 123.
25. J. Kolaczkiewicz and E. Bauer, *Surf. Sci.*, 1986, **175**, 487; 508.

26. E. Bauer, *Appl. Phys. A*, 1990, **51**, 71.
27. J. W. Niemantsverdriet, P. Dolle, K. Markert, and K. Wandelt, *J. Vac. Sci. Technol.*, 1987, **A5**, 875.
28. A. Klekamp and E. Umbach, *Surf. Sci.*, (in press).
29. X. Guo and J. T. Yates, Jr., *J. Chem Phys.*, 1989, **90**, 6761.
30. P. J. Estrup, E. F. Greene, M. J. Cardillo, and J. C. Tully, *J. Phys. Chem.*, 1986, **90**, 4099.
31. H. C. Kang, T. A. Jachimowski, and W. H. Weinberg, *J. Chem. Phys.*, 1990, **93**, 1418.
32. J. W. Niemantsverdriet and K. Wandelt, *J. Vac. Sci. Technol.*, 1988, **A6**, 757.
33. P. A. Rikvold, J. B. Collins, G. D. Hansen, and J. D. Gunton, *Surf. Sci.*, 1988, **203**, 500.
34. L. D. Roelofs and R. J. Bellon, *Surf. Sci.*, 1989, **223**, 585.
35. A. V. Myshlyavtsev, J. L. Sales, G. Zgrablich, and V. P. Zhdanov, *J. Stat. Phys.*, 1990, **58**, 1029.
36. A. Cassuto and D. A. King, *Surf. Sci.*, 1981, **102**, 388.
37. G. Ehrlich, 'Physics and Chemistry of Solid Surfaces VII', ed. R. Vanselow and R. Howe, Springer–Verlag, Berlin, 1988, p. 1.
38. S. H. Payne, A. Killinger, and H. J. Kreuzer, *Surf. Sci.*, (submitted).
39. K. Nagai, *Surf. Sci.*, 1986, **176**, 193.
40. J. R. Arthur and A. Y. Cho, *Surf. Sci.*, 1973, **36**, 641.
41. J. R. Arthur, *Surf. Sci.*, 1973, **38**, 394.
42. M. Bienfait and J. A. Venables, *Surf. Sci.*, 1977, **64**, 425.
43. G. Le Lay, M. Manneville, and R. Kern, *Surf. Sci.*, 1977, **65**, 261.
44. M. Bertucci, G. Le Lay, M. Manneville, and R. Kern, *Surf. Sci.*, 1979, **85**, 471.
45. R. Opila and R. Gomer, *Surf. Sci.*, 1981, **112**, 1.
46. H. J. Kreuzer, 'Non-equilibrium Thermodynamics and its Statistical Foundations', Oxford University Press, Oxford, 1981.
47. D. Menzel, in 'Kinetics of Interface Reactions', ed. M. Grunze and H. J. Kreuzer, Springer–Verlag, Berlin, 1987, p. 2.
48. K. Nagai and A. Hirashima, *Surf. Sci.*, 1987, **187**, L616, and references therein.
49. J. A. Meyer, I. D. Baikie, G. P. Lopinski, J. A. Prybyla, and P. J. Estrup, *J. Vac. Sci. Technol.*, 1990, **A8**, 2468.
50. D. F. Padowitz and S. J. Sibener, *Surf. Sci.*, 1989, **217**, 233.
51. M. Vollmer and F. Träger, *Surf. Sci.*, 1987, **187**, 445.
52. H. O. Martin and E. V. Albano, *Surf. Sci.*, 1989, **211/212**, 1025.
53. U. Leuthäuser, *Z. Phys.*, 1980, **B37**, 65.
54. H. Pak and J. W. Evans, *Surf. Sci.*, 1987, **186**, 550.
55. H. Asada and M. Masuda, *Surf. Sci.*, 1989, **207**, 517.
56. H. Asada, *Surf. Sci.*, 1990, **230**, 323.
57. K. Pilorz and S. Sokolowski, *Z. Phys. Chem.*, 1984, **265**, 929.
58. L. A. Laxhuber and H. Möhwald, *Surf. Sci.*, 1987, **186**, 1.
59. M. Tikhov and E. Bauer, *Surf. Sci.*, 1990, **232**, 73.
60. J. Knall, S. A. Barnett, and J. E. Sundgren, *Surf. Sci.*, 1989, **209**, 314.
61. S.-L. Chang, P. A. Thiel, and J. W. Evans, *Surf. Sci.*, 1988, **205**, 117.
62. V. P. Zhdanov, *J. Phys. Chem.*, 1989, **93**, 5582.
63. R. Schmiedl, W. Nichtl–Pecher, K. Heinz, K. Müller, and K. Christmann, *Surf. Sci.*, 1990, **235**, 186.
64. J. Prybyla, P. J. Estrup, and Y. J. Chabal, *J. Vac. Sci. Technol.*, 1987, **A5**, 791.
65. G. A. Attard and D. A. King, *Surf. Sci.*, 1987, **188**, 589.

66. J. W. Bartha, U. Barjenbruch, and M. Henzler, *J. Vac. Sci. Technol.*, 1985, **A3**, 1588.
67. J. L. Sales and G. Zgrablich, *Surf. Sci.*, 1987, **187**, 1.
68. H. Kasai, S. Enomoto, and A. Okiji, *J. Phys. Soc. Jpn.*, 1988, **57**, 2249.
69. L. J. Whitman and W. Ho, *J. Chem. Phys.*, 1989, **90**, 6018.
70. D. Gupta and C. S. Hirtzel, *Surf. Sci.*, 1989, **210**, 322.
71. S. J. Lombardo and A. T. Bell, *Surf. Sci.*, 1988, **206**, 101.
72. J. L. Sales, G. Zgrablich, and V. P. Zhdanov, *Surf. Sci.*, 1989, **209**, 208.
73. For a review of the kinetic Ising model, see K. Kawasaki, in 'Phase Transitions and Critical Phenomena', Vol. 2, ed. C. Domb and M. S. Green, Academic Press, New York, 1972, p. 443.
74. R. J. Glauber, *J. Math. Phys.*, 1963, **4**, 294.
75. H. J. Kreuzer and Zhang Jun, *Appl. Phys. A*, 1990, **51**, 183.
76. 'Monte Carlo Methods', ed., K. Binder, Springer–Verlag, Berlin, 1986.
77. J. D. Gunton, M. San Miguel, and P. S. Sahni, in 'Phase Transitions and Critical Phenomena', Vol. 8, ed. C. Domb and J. L. Leibovitz, Academic Press, New York, 1983, p. 267.
78. J. D. Gunton, in 'Kinetics of Interface Reactions', ed. M. Grunze and H. J. Kreuzer, Springer–Verlag, Berlin, 1987, p. 238.
79. E. Oguz, O. T. Valls, G. F. Mazenko, and J. H. Luscombe, *Surf. Sci.*, 1982, **118**, 572.
80. J. Luscombe, *Phys. Rev. B*, 1984, **29**, 5128.
81. J. W. Evans, *J. Chem. Phys.*, 1987, **87**, 3038.
82. J. W. Evans and H. Pak, *Surf. Sci.*, 1988, **199**, 28.
83. J. J. Luque and A. Cordoba, *Surf. Sci.*, 1987, **187**, L611.
84. A. Surda and I. Karasova, *Surf. Sci.*, 1981, **109**, 605.
85. A. Surda, *Surf. Sci.*, 1989, **220**, 295.
86. S. Sundaresan and K. R. Kaza, *Surf. Sci.*, 1985, **160**, 103.
87. J. W. Evans, D. K. Hoffman, and H. Pak, *Surf. Sci.*, 1987, **192**, 475.
88. A. Cordoba and J. J. Luque, *Phys. Rev. B*, 1982, **26**, 4028.
89. D. J. W. Geldart, H. J. Kreuzer, and F. S. Rys, *Surf. Sci.*, 1986, **176**, 284.
90. H. J. Kreuzer, *Phys. Rev. B*, (to be published).
91. H. J. Kreuzer, *Surf. Sci.*, 1990, **238**, 305.
92. W. Brenig, in 'Kinetics of Interface Reactions', ed. M. Grunze and H. J. Kreuzer, Springer-Verlag, Berlin, 1987, p. 19.
93. D. B. Jack, Z. W. Gortel, and H. J. Kreuzer, *Phys. Rev. B*, 1987, **35**, 468.
94. H. Schlichting, Ph.D. Thesis, Techn. Univ. München, 1990.
95. H. Schlichting and D. Menzel, to be published.
96. H. Schlichting and D. Menzel, submitted to *Phys. Rev. Lett.*
97. H. Schlichting, D. Menzel, T. Brunner, W. Brenig, and J. C. Tully, *Phys. Rev. Lett.*, 1988, **60**, 2515.
98. M. Head-Gordon, J. C. Tully, H. Schlichting, and D. Menzel, to be published.
99. H. Schlichting and D. Menzel, 'Proc.3S'91 Workshop', Obertraun (in press).

CHAPTER 7

Electron Stimulated Desorption and its Application to Chemical Systems

R. D. RAMSIER and J. T. YATES, JR.

1 Introduction

The study of the behaviour of atoms and molecules adsorbed on surfaces constitutes a major scientific area of inquiry at the present time. Information about the chemical and physical properties of adsorbed layers is needed for the development of fundamental concepts in the field. Such information is of inherent scientific interest, but also has immense practical importance in the technologies. Among the technological areas strongly influenced by surface phenomena are semiconductor and electronic materials science, catalytic science, corrosion prevention, tribology and many areas of metallurgy and materials science.

Much research in surface science deals with the thermal activation of elementary surface chemical processes such as chemisorption, surface migration, surface reaction, desorption, and diffusion to and from the bulk. This is a natural area for chemists who are often concerned with thermally activated processes.

Another means for activation of surface processes involves electronic excitation of surface species by the use of photon, electron, or ion bombardment. Here one can have the opportunity to initiate high energy processes which are thermally inaccessible. Energetic species may be ejected from the surface as a result of electronic excitation. The importance of electron stimulated desorption (ESD) processes is supported by the fact that since 1960, approximately 1000 research publications have appeared in this field.[1]

Portions of this paper are taken from a more comprehensive review of the field of electron stimulated desorption.[1]

One of the most interesting developments in the area of electron stimulated desorption involves the ESDIAD phenomenon, where ESDIAD is an acronym for 'electron stimulated desorption ion angular distribution'. In 1974[2] it was found that adsorbed species on single crystal substrates could eject narrow beams of positive ions during electron bombardment. These ion beams exhibited rather sharp angular distributions which were often in registry with certain crystallographic azimuths of the substrate. The repulsive potential surface responsible for this phenomenon is determined primarily by the directionality of the chemical bonds being broken, and observation of ion angular distributions therefore gives information concerning the orientation of surface bonds of the parent adsorbed molecules. More than seventy studies involving the ESDIAD technique have been published in sixteen years.

Recently, the ESDIAD method has been used to study the dynamical behaviour of surface species.[3-7] Adsorbed molecules thermally oscillate to-and-fro and/or rotate at their adsorption sites, and ESDIAD can measure the range of angular disorder induced by these motions, thus giving one the ability to rather directly observe the dynamical properties of surface species.

Experimental studies via ESDIAD underwent a marked improvement with the introduction of digital data acquisition methods.[8-10] The digital method permits accurate angular distribution measurements to within about 1°. In addition, it has recently been found that both electronically excited neutral[11,12] and anionic[13,14] species can be produced via ESD, opening even more vistas for study of the character of surface bound species. For the emission of electronically excited neutral species, the term ESDIAD will also be employed, since the basic principles of ion ejection also apply to the production of uncharged desorbing species.

2 Models of Electron Stimulated Desorption (ESD) Processes

When considering collisions between an incident low energy particle ($E_i \sim 500$ eV) of mass m and a free particle of mass M, one can estimate the order of magnitude of the maximum energy transferred (ΔE) during the process with classical kinematics. For hard-sphere scattering the elementary result is:[15]

$$\Delta E/E_i = (2mM)(1 - \cos\theta)/(m + M)^2 \qquad (1)$$

where θ is the scattering angle in the centre-of-mass frame of reference. One can see that for $m \ll M$, corresponding to electrons impinging on atoms or molecules, equation (1) reduces to:

$$\Delta E/E_i \approx 2m/M \qquad (2)$$

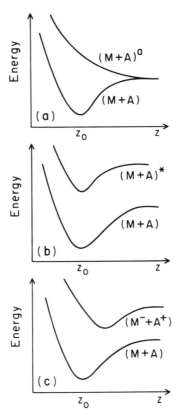

Figure 1 *Graphical representation of possible potential energy curves within the MGR framework. (M + A) represents the ground state configuration, (a) $(M + A)^a$ antibonding, (b) $(M + A)^*$ metastable and (c) $(M^- + A^+)$ ionic states (Reproduced with permission from Surf. Sci. Rep., 1991, **12**, 243)*

where the expression is averaged over the interval $0 \leq \theta \leq \pi$. So the fractional energy transfer is of the order of $2/1840 \approx .001$ for electron–atomic hydrogen collisions, implying that less than .5 eV would be partitioned to the H atom in this case. However, ESD experiments often observe desorbing molecules, ions, and molecular fragments with mass greater than one a.m.u. and with kinetic energies in the range of 2–10 eV. This indicates that direct momentum transfer is not usually dominant in electron–adsorbate collisions, and that electronic energy transfer must be considered. In addition, strong chemisorption bond energies of 1–8 eV are common in many systems of interest, so that normally one treats desorption induced by slow electrons in terms of electronic excitation mechanisms.

The Menzel–Gomer–Redhead (MGR) Model

One of the earliest models proposed to explain ESD from surfaces was advanced independently by Menzel and Gomer[16,17] and Redhead[18] in 1964. The acronym MGR is applied to this early model, although similar ideas were advanced by Ishikawa[19-21] much earlier. This general model incorporates adiabatic approximations and a semi-classical description of possible excitations of an adsorbate(A)–substrate(M) system, and is schematized in Figure 1.

The MGR model assumes that the system is initially in a ground state configuration (M + A) with respect to some reaction co-ordinate (z), often interpreted as distance above the surface. Interaction with an incident electron initiates a Franck–Condon (FC)[22] transition from the (M + A) state to some excited state, possibly; (a) antibonding $(M + A)^a$, (b) metastable atomic $(M + A)^*$, or (c) ionic $(M^- + A^+)$ in nature. These excited states normally involve only valence level excitations in the MGR model. Invoking the FC principle is equivalent to saying that the transition is vertical with respect to z, or in other words that the excitation is purely electronic and occurs so swiftly that nuclear motion is negligible during the excitation time ($\sim 10^{-15}$ s). This view implies that electronic and nuclear terms within the system Hamiltonian decouple, allowing for separation of the state wavefunction into a product of two wavefunctions, one describing the electronic and the other the nuclear state of the system.

The curves in Figure 1 should be considered illustrative, as each merely represents one of a large family of curves describing many possible excited states within the adsorbate–substrate system. Starting from the ground state configuration centred about some equilibrium position (z_0), an FC transition promotes the system to one of these possible higher energy states. After excitation, nuclear motion may occur over a time scale of $\sim 10^{-13}$ s, converting potential energy of the newly sampled potential surface into translational kinetic energy.

Figure 2 superimposes curves of Figure 1 to illustrate the possibility of curve crossings during subsequent nuclear motion. Curve crossings represent Auger processes and resonant tunnelling with the Fermi sea of the substrate, which may result in reneutralization of ionic species and/or de-excitation pathways. Although detailed knowledge of the curves depicted in Figure 2 is not generally available, this descriptive picture of a desorption sequence gives powerful insight into the possible physical mechanisms involved.

The generality of the MGR picture has allowed for its widespread application over the years. Naturally, multilevel excitation and de-excitation processes, electronic differences in adsorption sites, non-isotropic recapture mechanisms, and a host of other variables will often play a role in causing deviations from the model.

The Antoniewicz Model

In 1980,[23] Antoniewicz proposed a modification of the original MGR model which has thus far been successful in describing ESD from physisorbed

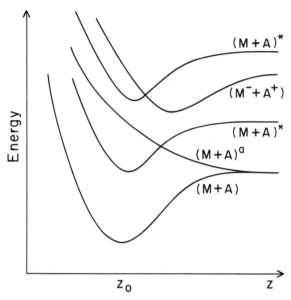

Figure 2 *Superposition of curves from Figure 1 to illustrate possible curve crossings during nuclear motion of desorbing species in the MGR picture. Crossings represent tunnelling and Auger de-excitation and reneutralization processes* (Reproduced with permission from *Surf. Sci. Rep.*, 1991, **12**, 243)

layers.[24-26] Antoniewicz considered an adsorbate which, by interaction with an incident electron, becomes instantaneously positively ionized (with respect to time scales of nuclear motion). The ion experiences a screened image charge potential, which attracts it toward the surface. Pauli repulsion is also diminished for the ion (as compared to the neutral) since it has a smaller atomic radius. These two effects allow the ion to move very close to the surface, which in turn dramatically increases the probability of resonant tunnelling or Auger neutralization. Upon reneutralization, the image potential vanishes, leaving only the Pauli repulsion. The reneutralized species, still moving toward the surface, is repelled, effectively bouncing off the substrate to escape as a desorbing neutral particle. Figure 3 illustrates this proposed sequence leading to neutral particle desorption.

Antoniewicz also included a more complex two-electron process to explain the desorption of positive ions. This is illustrated in Figure 4, where the ground state configuration (M + A) is promoted to an excited ionic curve (M + A$^+$)*. Again the excited ion moves toward the surface, where tunnelling neutralizes the ion, placing the system high on the ground state (M + A) curve. Pauli repulsion, which suddenly dominates at very small separations, will deflect the hot neutral from the substrate, causing it to escape. Since the probability of electron hopping processes at short distances is finite, the neutral species may be reionized by resonant electron tunnelling into the substrate during its escape, yielding ionic desorbing species, as shown by a curve crossing with the (M + A$^+$) curve.

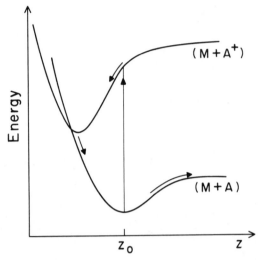

Figure 3 *Illustrative representation of the Antoniewicz picture of neutral particle desorption. The ground state (M + A) is excited to an ionic (M + A$^+$) state by a vertical FC transition. Nuclear motion toward the substrate ensues until electron tunnelling reneutralizes the ion, which then escapes the surface along a neutral particle curve*
(Reproduced with permission from *Surf. Sci. Rep.*, 1991, **12**, 243)

Walkup and co-workers[27] have recently provided further analysis of the Antoniewicz motion of noble gas ions near a metal surface. Using local density functional calculations, the authors predict that the attractive force on a positive ion is substantially less than one would calculate using standard image potential methods. The actual electron charge density distribution at the surface is polarized by the ion and moves (with respect to the plane of the surface) as the ion moves. As the ion approaches the substrate, this image-like surface charge density is repelled by the electron cloud of the ion (Pauli repulsion), pushing the substrate charge density and thus the effective image plane in a negative z direction. Therefore, the 'actual' ion-image distance is larger than one would intuitively assume, leading to a reduction in the attractive forces as compared to those calculated using a static image plane model. Figure 5 illustrates this point in simple terms. This phenomenon, when incorporated into calculations of desorption cross-sections, seems to improve the agreement between theory and experiment.

The Knotek–Feibelman (KF) Model

Another mechanism which has received widespread application in describing ESD phenomena was proposed in 1978 by Knotek and Feibelman.[28] Originally the model was formulated to explain experimental observations concerning ESD from highly ionic maximal valency oxide surfaces, such as TiO_2, V_2O_5, and WO_3, in which all metal valence electrons are formally

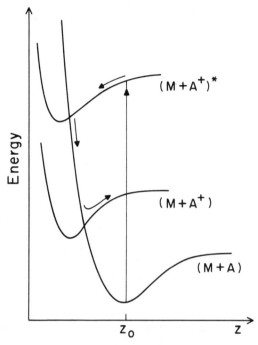

Figure 4 *Schematic diagram of potential energy curves in the two-electron Antoniewicz picture. Initial excitation to an excited ionic state $(M + A^+)^*$ is followed by nuclear motion toward the substrate. Tunnelling places the ion high on the ground state $(M + A)$ curve, and a second tunnelling process results in ionic desorption along the $(M + A^+)$ curve*
(Reproduced with permission from *Surf. Sci. Rep.*, 1991, **12**, 243)

donated to the lattice. Very small total desorption yields were observed for incident electron energies below about 20 eV, where the MGR picture would predict bonding–antibonding transitions leading to significant neutral desorption. However, O^+ desorption yields increased dramatically when the bombarding electron energy reached 25–35 eV, corresponding to ionization energies of metal core levels of the metal oxide substrate. Another feature of ESD from these systems was that since oxygen resides in a charge configuration O^{x-} ($1 < x < 2$), the observation of desorbing O^+ implied that large charge transfer (>2 electrons) was taking place during the desorption sequence.

Figure 6 illustrates the KF mechanism for TiO_2 proposed to explain the above observations, where for ease of illustration $x = 2$. It is envisioned that an incident electron ionizes the highest lying metal atom core level [Ti(3p)], which under normal circumstances would be swiftly filled by an intra-atomic Auger process from a higher lying Ti level. However, the maximal valency configuration of TiO_2 means that the Ti atoms are fully stripped (Ti^{4+}) of valence electrons, thwarting the possibility of this quenching process. Thus an interatomic Auger mechanism was proposed, whereby the metallic core

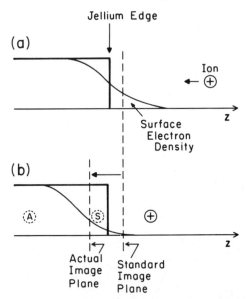

Figure 5 *Schematic illustration of the shift of the effective image plane of a Jellium metal as a positive ion approaches from the vacuum. In (a), the ion polarizes the surface electron density, creating an image-like attractive potential. In (b) the ion is sufficiently close that its electron density presses against that of the surface (Pauli repulsion), pushing the standard image plane position to its actual position. The image charge would appear to be located at S if this phenomenon did not occur, but actually 'resides' at A*
(Reproduced with permission from *Surf. Sci. Rep.*, 1991, **12**, 243)

hole is filled via an Auger process with a neighbouring O^{2-} ion. The oxygen moiety suddenly finds itself in a state O^+ where it experiences a repulsive (instead of attractive) Madelung potential in addition to Pauli repulsions. Thus the O^+ ion is violently ejected from the surface by a mechanism which resembles Coulomb explosion processes observed in gas-phase molecular dissociation.[29]

Thus the KF model explains the large charge transfer needed to produce O^+ ions from O^{2-} moieties via an interatomic Auger process. The KF model also indicates why the oxygen ion desorption threshold corresponds to the Ti(3p) ionization energy (34 eV). Energetic arguments are able to rationalize why only a small O^+ yield is developed for an incident beam energy of 22 eV, coincident with the O(2s) level. The lack of metal valence electrons in TiO_2 and other maximal valency oxides originally made the KF mechanism a system specific description, but the basic ideas have been generalized beyond maximal valency oxides.

Quantum Mechanical Models

Several groups have formulated quantum mechanical models of various aspects of ESD, and these will be discussed qualitatively to stress the

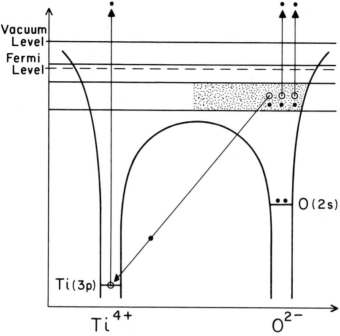

Figure 6 *Energy diagram for the TiO_2 maximal valency configuration and graphical illustration of the KF desorption mechanism. A hole created in the Ti(3p) level is filled via interatomic Auger decay from a neighbouring oxygen site. This leaves the oxygen moiety positively charged, causing it to be Coulombically repelled from its lattice site. Filled circles (●) signify electrons and open circles (○) represent holes*
(Reproduced with permission from *Surf. Sci. Rep.*, 1991, **12**, 243)

physical implications of the results. Brenig's model, which first appeared in 1976,[30] utilized strict Born–Oppenheimer approximations (BOA) concerning the initial excitation of an adsorbate–substrate (M + A) system. These assumptions were equivalent to those made in the MGR, Antoniewicz and KF models already discussed. Thus in Brenig's picture the system undergoes an FC transition upon electron or photon bombardment, with electronic and nuclear terms in the Hamiltonian being separable.

After initial excitation, the time-dependent nuclear motion of the system will be governed by the newly sampled electronic potential energy surface (ionic or neutral). This potential (V') is given the form of an optical potential in our notation as:

$$V' = V + (i\hbar R)/2 \qquad (3)$$

with V and R real functions of nuclear and electronic co-ordinates. Brenig shows that it is the complex part of V' which is responsible for coupling the excitation localized in the (M + A) system to the surface, resulting in de-excitation pathways.

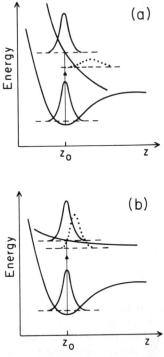

Figure 7 *Schematic illustrations of Brenig's discussion of nuclear motion following an FC transition. In (a) the complex part of the excited state (optical) potential varies significantly over the width of the ground state wavepacket, resulting in a PDD whose centre-of-mass moves much faster than would a classical particle. For case (b) the excited state potential is a slow function of z and the final state PDD is similar to that of the ground state. Case (a) must be treated quantum mechanically whereas (b) may be thought of in semi-classical terms (Reproduced with permission from Surf. Sci. Rep., 1991, **12**, 243)*

As a simple example, assume a normalized Gaussian of 'width' s centred about equilibrium (z_0) for the initial ground state vibrational wavefunction ψ_{M+A}:

$$\psi_{M+A} = (\pi)^{-1/4}(s)^{-1/2} \exp(-\{z - z_0\}^2/2s^2) \quad (4)$$

In standard notation the probability density distribution (PDD) is given by ψ^2.[18] Brenig has shown[31] that the magnitude of the spatial variation of V' within the region spanned by the wavepacket will greatly affect the physical interpretation of the problem. For small (dV'/dz) within the width s, a semi-classical picture of a wavepacket ball rolling atop an excited state potential surface is envisioned. However, if V' varies appreciably in the vicinity of the particle's wavefunction, the time evolution of the system is totally quantum in nature. Even before significant nuclear motion can ensue, the PDD is dramatically altered by excitation to the excited state curve, changing its 'centre-of-mass' and resulting in a skewed kinetic energy distribution (KED). See Figure 7 for an illustration of this discussion.

The width and shape of the KED is dependent on the variation of the excited state potential over the distance s. Brenig's discussion indicates that first-order corrections to the quantum equations of motion for the centre-of-mass of the final state PDD depend on (dR/dz). Thus the slope of $R(z)$ plays the significant role in determining the time evolution of this system. Recall that the PDD peak position represents a point in space about which the wave packet is localized, *i.e.* the position of the particle. A strongly asymmetric KED implies (dR/dz) is large and the PDD has been significantly altered by the transition $(V_0 \rightarrow V')$, so a classical view of nuclear motion breaks down.

Although this model allowed for the possibility of neutral or ionic desorption induced by direct excitation to neutral or ionic states, respectively, strict zeroth order BOAs led to the conclusion that curve crossings were highly improbable events.[30,32] This became a point of heated discussion in the literature,[33–36] since it implied that desorbing neutrals could not have originated from ions reneutralized during escape. Higher order corrections have since shown that curve crossings from ionic to neutral excited states are possible, which may lead to neutral desorption.[31,37,38]

Ueba[39] has introduced a perturbative quantum mechanical approach to investigate the probability of non-adiabatic curve crossings following initial FC excitation within an adsorbate–adsorbent system. The continuum of possible substrate excitations are considered as an electronic energy reservoir which can efficiently conduct energy away from the adsorbate. This continuously distributed spectrum of excited states of the substrate are modelled as elementary bosonic excitations.

Qualitatively speaking, calculations predict that adsorbate coupling to this electronic 'heat bath' increases the total desorption probability, since the curve crossing probability (leading to recapture) is smaller for stronger coupling to the solid. The physical argument presented to explain this seemingly counter-intuitive result is that increased interaction with the substrate provides the adsorbate with the freedom to curve cross between excited states along a continuum of intersections. Thus the system may move from the initial excited state ($|1>$) to some other state ($|2>$) at one point of intersection (z_{12}), but then cross back to $|1>$ at another location (z_{21}), leading to desorption. See Figure 8 for a graphic representation of this discussion.

Clinton and Julita[40] have performed calculations to investigate effects of the final state potential on the observed KED. Following a generalized BOA, they express the matrix elements of the transition operator $\mathcal{T}$ (which takes the system from its ground to some excited state) as:

$$\mathcal{T}_{fi} = \int dz \psi_f^\dagger(z) \psi_i(z) \left\{ \int d\xi \Phi_f^\dagger(\xi, z) \mathcal{T}(\xi) \Phi_i(\xi, z) \right\} \tag{5}$$

where a dagger (†) signifies complex conjugation and subscripts i and f denote initial and final states, respectively. Here the wavefunctions of both

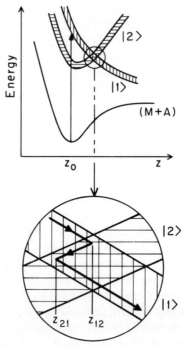

Figure 8 *Ueba's ideas graphically illustrated, where the hatched regions enclose a continuum family of curves which widen with increased coupling to the infinite number of substrate excitations. Multiple curve crossings are possible within the cross-hatched area which, contrary to intuition, actually increase the probability of adparticle desorption*
(Reproduced with permission from *Surf. Sci. Rep.*, 1991, **12**, 243)

states are assumed to be products of the vibrational state vector $\psi(z)$ and the electronic wavevector $\Phi(\xi,z)$, which depends on both electronic (ξ) and nuclear (z) co-ordinates. The assumptions are made that $\mathcal{T}$ operates only on the electronic part of the product wavefunction and that the bracketed integral varies so slowly with z that it can be evaluated at the equilibrium position z_0. Denoting the quantity in brackets as $\mathcal{T}'(z)$ we then obtain:

$$\mathcal{T}_{fi} = \int dz \psi_f^\dagger(z) \psi_i(z) \{\mathcal{T}'(z)\} \tag{6}$$

and using the latter assumption leads to:

$$\mathcal{T}_{fi} \approx \{\mathcal{T}'(z_0)\} \int dz \psi_f^\dagger(z) \psi_i(z) \tag{7}$$

Thus the matrix elements of the transition operator factor into two quantities. $\mathcal{T}'(z_0)$ carries most of the dependence on incident beam energy and electronic configurations of the states. The integral in equation (7), denoted as the Franck–Condon overlap (I_{FC}), largely determines the resulting trajectory and energy of nuclear motion.

To perform explicit calculations of I_{FC}, a good first approximation for the initial state ψ_i takes the form of a Gaussian [as in equation (4)]. The final state wavefunction is usually not known, so a classical argument, termed the reflection approximation (RA), is invoked. This incorporates replacing ψ_f by a Dirac delta function centred at the final state co-ordinate z_f, so that:

$$\mathcal{T}_{fi} \approx \{\mathcal{T}'(z_0)\} I_{FC} \sim \{\mathcal{T}'(z_0)\} \psi_i(z_f) \tag{8}$$

By Fermi's Golden Rule[41] the transition probability (σ_e) under the action of an operator varies as the squared amplitude of the associated matrix elements.
Therefore we see that:

$$\sigma_e \sim |\mathcal{T}_{fi}|^2 \sim |\psi_i(z_f)|^2 \tag{9}$$

Thus the RA results in a reflection of the symmetry of the initial-state vibrational wavefunction through the final-state potential. This means that the KED is mainly determined by the vibrational amplitude of the chemical bond ruptured by ESD. The shape of the resulting KED is, therefore, nearly Gaussian, assuming a slowly varying excited state curve. In most cases investigated by Clinton and Julita,[40] the calculated and experimentally obtained KED had roughly this functional form, with a slight tail to the high energy side. This asymmetry is explained in terms of the variation in the slope of $V(z)$ near the equilibrium position z_0, although effects from anharmonicity in the ground state may also contribute. Note that the RA is equivalent to the case discussed by Brenig[31] when (dV'/dz) is small over the range s, in which the final PDD is not significantly different than the original PDD and nuclear motion can be considered as classical.

Dissociative Attachment (DA) Mechanisms in ESD

In the previously discussed ESD mechanisms, an incident electron merely stimulates an electronic excitation within an (M + A) system. Obviously the energy available for subsequent nuclear motion differs if this electron ends up residing at the vacuum level or in the Fermi sea of the substrate.[16-18] However, in some cases there exists a possibility that the impinging electron is captured for a finite time by the adsorbate, resulting in a dissociative attachment (DA) mechanism of desorption.[42-45]

Consider a diatomic adsorbate (AB) for convenience. A possible DA reaction producing desorption is represented below and in Figure 9:

Attachment

$$M + AB + e \rightarrow M + (AB)^{-*} \tag{10}$$

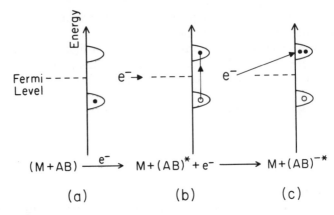

Figure 9 *Illustrative representation of a possible DA surface reaction leading to fragmentation and ESD. Schematized are the AB bonding (lower) and antibonding (upper) density of states, where solid circles (●) represent electrons and open circles (○) represent holes. An incoming electron (e⁻) stimulates a bonding–antibonding transition (a)→(b), then the incident electron is itself captured, producing a transient anion at the surface (c). Strong repulsive forces dissociate the complex to yield charged and electronically excited fragments, each with kinetic energy capable of resulting in the fragment's escape from the surface*
(Reproduced with permission from *Surf. Sci. Rep.*, 1991, **12**, 243)

Dissociation

$$M + (AB)^{-*} \rightarrow M + A^* + B^- \qquad (11)$$

DA processes may involve a molecular valence excitation $(AB)^*$, which has an AB bonding orbital electron promoted to an AB antibonding level via interaction with the electric field of the incident electron. Additionally, the incident electron can be captured in the antibonding orbital, producing an excited anionic state $(AB)^{-*}$. Thus a hole exists in a bonding orbital and two electrons reside in an antibonding orbital, and it is possible that these states may be sufficiently long-lived to allow for desorption from the surface.

Azria, *et al.* have shown[42] that the excess energy (E') following dissociation [equation (11)] can be expressed as:

$$E' = E_i - E_d - E_A + E_B \qquad (12)$$

where E_i, E_d, and E_A are the incident electron, AB dissociation, and A* electronic energies, respectively, and E_B is the electron affinity of the B atom. The ionic (B^-) species carries off a portion (E_{B^-}) of this energy:

$$E_{B^-} = (1 - M^-/M)E' \qquad (13)$$

with M^- and M the ionic (B^-) and molecular (AB) masses, respectively.

Auger Stimulated Desorption (ASD)

In 1979 Franchy and Menzel reported the first experimental findings to substantiate the idea that ESD and photon stimulated desorption (PSD) of ionic species from adsorbate covered metals can be induced and/or enhanced by ionization of deep core levels of the adsorbate.[46] The results clearly show that for the system CO/W(100), C(1s) and O(1s) ionization increases the CO^+ and O^+ yields, respectively. A simplified interpretation of these observations is that following adsorbate core ionization, intra-atomic Auger processes occur, accumulating multiple positive charges on the ionized atom. Localization of multiple holes in either the C—O or M—CO bond will cause violent Coulombic repulsion, and the bond will disintegrate spontaneously, similar to Coulombic explosion mechanisms found in some gas-phase dissociation processes.[29]

The ability of this process to place multiple holes in the original molecular bonds at surfaces has far reaching implications. The rapidity of the desorption sequence implies that the probability of reneutralization events is diminished. Desorbing species should be ejected in directions imposed by the original bonding site geometry. Therefore initial state effects should be most important for ion angular distribution measurements of particles following deep adsorbate core level ionization.[46]

The localization of two holes in a bond for a finite time (long enough to allow desorption) has received significant attention in the literature. Most descriptions are based on configuration interaction (CI) theories developed by Hubbard[47–49] and extended by Cini[50–53] and Sawatsky[54,55] to explain the Auger spectral lineshapes of narrow band metallic solids. We will attempt to summarize briefly the results of the theory, placing emphasis on those parameters having physical significance to ESD phenomena, as discussed by Dunlap, et al.[56]

For simplicity assume an initially full metallic valence band (VB) and an ESD process which instantaneously ejects a core-level electron to the vacuum level. An intra-atomic Auger transition may follow, filling the core hole with a VB electron and ejecting a second electron (for energy conservation) from the VB to the vacuum level. Such a scenario results in two holes appearing in an initially full valence band (or molecular orbital, MO). If the interaction between these two holes is very weak, the total electronic wavefunction is expressible as a product of two single-particle MO wavefunctions. However, when the effective Coulomb repulsion energy (U) between the holes becomes of the same order of magnitude as the average energy available within the VB, their motion becomes correlated.

The bandwidth (W) where the holes reside is a convenient parameter which characterizes the average electronic energy available to the holes, and CI theories usually predict phenomena based on the ratio U/W. When $U \gg W$, a two-hole (2h) state with a well defined (atomic-like) energy forms. The holes are constrained by their mutual repulsion to move together through the crystal, in analogy to the motion of tightly-bound Frenkel excitons.[57] The sharpness of the two-hole state in energy implies that the lifetime of the state is relatively long by application of the Heisenberg uncertainty relation.[41]

Thus we arrive at a physical picture for the existence of long-lived, two-hole states. The only way for the holes to spatially separate from one another would be by losing energy to phonons or electrons, forming two free holes, each of energy $\approx U/2$. But for $U \gg W$, conservation of energy prohibits separation of the holes.[58]

Having provided the physical basis for the existence of two-hole states in a simple metallic VB, we proceed to address the question: what role do such states play in the disruption of chemical bonds at surfaces? The energy difference between correlated two-particle states and two-independent hole states is due to insufficient screening of hole–hole Coulombic repulsions. In the case of metals or ionic solids, weak screening is possible if the holes are placed in an initially full VB as discussed above. The two-hole states thus formed may remain on a single surface site long enough to initiate an irreversible desorption sequence, since the co-ordination number at the surface is less than that of the bulk, reducing the hole hopping probability.[59,60]

Analogously, for the case of such excitations in covalently bonded systems, it has been shown that correlated multi-hole states can exist if the holes reside in an initially full molecular orbital.[60,61] Additional geometrical constraints imposed by bond directionality in covalent systems may isolate holes in a surface bond from the screening effects present in the bulk.[60,61] Without sufficient screening these holes will be correlated, and may remain spatially localized on a single site long enough to allow bond scission to commence.

Feibelman[59] has described the extended lifetime of adsorbate multi-hole localized states in terms of a 'reneutralization bottleneck' for Auger transitions. As seen above, effective Coulombic repulsions (U) force the localized holes to hop together if the inequality:

$$U \gg W \qquad (14)$$

holds, where W can again be thought of as the associated single particle bandwidth. As one- and two-hole hopping rates (R_1 and R_2, respectively) scale as:

$$R_1 \sim W \qquad (15)$$

$$R_2 \sim (W/U) R_1 \qquad (16)$$

it is possible that two-hole states remain localized long enough to increase ESD probabilities.

In his note Feibelman also stresses the idea of reduced intra-site screening in the 2h state. A reduction in the spatial size of the surface species also occurs upon hole creation, decreasing inter-site overlap with wavefunctions centred on adjacent sites. This can decrease R_2 by possibly an order of magnitude, since hole hopping rates depend on the magnitude of this overlap.

Jennison and Emin[62,63] have considered the spatial constraint of multi-hole states on a surface site as analogous to the formation of a small polaron. In the polaron picture, the increase in energy accompanying the trapping of an electron on a single site is in competition with the lowering of energy caused by relaxation of nearby atomic positions. In other words, an electron in a solid produces a local lattice deformation which tends to minimize the total system energy. The electron and this associated 'phonon cloud' move throughout the crystal as a quasi-particle (polaron). A condition sufficient for immediate polaron formation in the bulk is given in our notation as:

$$-(dV/dz)\tau > (2\mu E_\omega)^{1/2} \qquad (17)$$

where the left-hand side is the product of the force exerted on nearby atoms due to a localized electronic excitation and the lifetime of the state, thus giving the momentum transferred to the lattice. The right-hand side represents the vibrational momentum of the lattice atoms, with E_ω the vibrational energy and μ an effective nuclear mass.

In applying these ideas to ESD from surfaces, Jennison and Emin[62,63] indicate that the reduction of the problem from three (bulk) to two (surface) dimensions increases the likelihood of localization of excitations. In general, multi-hole excitations have large lifetimes (τ) and steep energy curves (V), so the inequality of equation (17) is favoured. The authors also point out that it is only essential that the multi-particle state remains locally confined long enough to initiate strain on nearby atoms, causing atomic displacements, which in turn decreases the probability that the holes hop away (positive feedback). If a multi-hole excitation occurs within a surface or adsorbate bond, no restoring forces are present from the vacuum side to oppose strain-induced nuclear motion, and atomic separation (desorption) may occur. The strain-induced lattice distortion can play a critical role in some cases, trapping multi-hole excitations near the surface, thus increasing desorption probabilities.[64]

Using the same approach, Jennison and Emin find that single-hole states are more delocalized than two-hole states. Thus strain initiated atomic motion must be significant in order to spatially trap the one-hole excitation and there may be a time delay involved. These authors have also noted that surface excitations may possibly impart sufficient momentum on nearby atoms to initiate ESD. In addition, defects or local disorder on a surface may act as a nucleus for the trapping of electronic excitations, and thus serve as preferential sites for desorption.

Several other papers have also presented enlightening discussions about the relevance of two-hole localized states in ESD processes.[65–69]

Other Mechanisms of ESD

Avouris, et al.[70,71] have modelled the bonding of F on Al and have produced some interesting results to be highlighted here. Utilizing techniques of density-functional and effective medium theories, the above authors have clearly shown that charge transfer (CT) screening plays a significant role in desorption from ionic systems.

Calculations show that F adsorbs strongly in an anionic $F^-(2s^2 2p^6)$ ground state on Al, bound in a potential well formed by image charge attractions (large z) and Pauli repulsions (small z) as expected. After electronic excitation, the valence excited state $F^0(2s^2 2p^5)$ has no image attraction and is purely repulsive over the FC region. Therefore, charge transfer from the F($2p$) band to the Al valence band or excitation to F^0 by an incident electron should produce neutral fluorine desorption. Additional neutral yield from the core-excited $F^*(2s^1 2p^6)$ state is unlikely, since rapid intra-atomic Auger decay is capable of quenching F^* before it has time to desorb. Thus the neutral fluorine ESD production can be represented by

$$M^+ + F^- + e \rightarrow M + F^0(g) + e \qquad (18)$$

The most interesting conclusions of these studies[70,71] concern the possibility of ejection of F^+ species. Swift CVV Auger decay of the F^* state should produce a surface $F^+(2s^2 2p^4)$ species, which one might intuitively assume to reside in a bound state (resulting exclusively from image attractions and hard-core repulsions) by analogy with the F^- case. Although results show F^+ indeed to be bound to Al, the physical mechanisms for this are quite non-intuitive. Electron-hole attractions in the two-hole F^+ state are sufficiently strong to shift the normally empty F($3s$) and F($3p$) orbitals below the Fermi level (E_F). Thus substrate–adsorbate electron tunnelling places a finite (> 0.5 electron) occupation in these levels, increasing the ionic radius of F^+ beyond that of F^-. Because of this CT process, the net charge of F^+ is reduced, weakening its image attractions. At the same time Pauli repulsions increase, due to swelling of its outer Rydberg orbitals. Thus F^+ is weakly bound on Al in comparison to F^-, and any process (e.g. F^* decay) which provides some nuclear kinetic energy may cause F^+ desorption. The process is schematized below:

Initial Core-Level Excitation
$$M^+ + F^- + e \rightarrow M + F^* + e \qquad (19)$$

Auger Decay and Desorption
$$M + F^* \rightarrow M + F^+(g) + e \qquad (20)$$

Multi-hole states $F^{2+}(2s^2 2p^3)$ accept significant charge from Al, which places about 1.3 electrons in F($3s$) and F($3p$) levels. However, increased Pauli repulsions and image charge screening are insufficient to overcome the substantial image charge attractions in this case, and F^{2+} bonds strongly to the Al surface.

The basic ideas presented for the F/Al system may be applicable to NO adsorption on metals, where large CT occurs, creating very polar [nearly $(NO)^-$] bonding.[71,72] In addition, ionic-like adsorption on small gap semiconductors and high dielectric constant insulators may also involve CT mechanisms similar to those discussed above.[71,72]

Nordlander and Tully[73,74] have investigated theoretically the formation of resonance hybrid states during H adsorption on a metal surface which has direct consequences to ESD. Using a density functional approach and complex scaling techniques, they were able to calculate both energy levels and lifetimes of excited states of adsorbed hydrogen. The results clearly show that in the presence of a clean metal surface, $n = 2$ and $n = 3$ levels undergo degeneracy splitting, forming mixed resonance states near the surface. As the metal–H distance decreases, splitting becomes more pronounced and the states become more weakly bound.

The lifetimes of these adsorbate hybrid states (H*) formed from mixtures of atomic H states vary greatly within the same principal quantum number (e.g. $n = 2$). This can be explained in terms of spatial localization of electron density associated with each hybrid state. Those states which have significant overlap with metallic electron wavefunctions are prone to resonant tunnelling de-excitation via the solid. On the other hand, H* states, which locate most of their electron density far from the surface, will be longer-lived, increasing desorption probabilities.

These theoretical investigations may prove useful in the interpretation of experimental observations described by Johnson and co-workers,[75] where ESD of H^+ and H* species from a hydrogen dosed, alkali-promoted Ni(001) surface are reported. Since ionic and excited state desorption processes occur at different threshold energies, the implication is that the species are produced via different channels, although resonance ionization may still play a role. H* desorption was monitored by spectroscopically observing Lyman α, β, and γ emission lines resulting from radiative de-excitation of desorbing H* species. Narrow observed linewidths indicate that H* radiative transitions occur far from the surface. The relative emission intensities indicate that surface excitation to $n = 2$ states is either more probable than excitation to higher energy states, or that $n > 2$ states have relatively low survival probabilities.

Johnson, et al.[76] have calculated the energy levels and lifetimes of H* states for H co-adsorbed with potassium on a metal substrate, extending previously discussed reports.[73–75] Calculations show that in the presence of K on a metal, H* states undergo degeneracy splitting as in the case of a clean metal, but that the levels shift to higher binding energies. This 'pulling down' of the $n = 2$ and $n = 3$ levels is largely due to the non-uniform electric fields experienced by H atoms in close proximity to the K co-adsorbates. These calculations offer further insights into the observed H^+ and H* desorption characteristics.[75]

Another interesting report to be highlighted in this section is the work of

Hoffman, et al.[77] on the O_2/Pd system. Experimentally, the authors observed that O_2 on Pd(111) at 87 K undergoes both molecular desorption and dissociation to O(a) when exposed to a low flux of electrons. These results could be rationalized under the framework provided by the MGR model. However, for atomic oxygen on Pd(111) an anomalously large enhancement (factor of five) in the total oxygen ESD cross-section was observed for incident beam energies near 10 eV. As no satisfactory explanation for such behaviour was available using standard ESD models, the authors proposed a new mechanism to explain their results, which were later developed in more detail by Gadzuk and Clark.[78]

Consider the chemisorbed atomic oxygen species as existing in a fractionally charged state O^{x-} ($0 < x < 1$) on the Pd surface. Incident electrons ($0 < E_i < 500$ eV) can stimulate desorption of oxygen species (positive ions, negative ions, neutrals, and metastables) via processes described in the previous sections. However, when $E_i = 10 \pm 2$ eV, it is possible to trap the incoming electron, forming a Feshbach or shape resonance anion at the surface. This negative ion in a charge state $O^{(1+x)-}$ feels an increased attraction toward the surface due to image charge forces and moves accordingly, seeking a new equilibrium position nearer the substrate, similar to the Antoniewicz motion of a positive ion.

After some finite lifetime, the resonance decays, releasing the trapped electron and leaving the O^{x-} species with momentum directed towards the surface and high on the repulsive part of the ground state curve. Thus the oxygen species can energetically rebound from the substrate and escape. This temporary resonance state increases the ESD cross-section by providing a new channel for the escape of atomic oxygen from the system, and occurs only over a narrow range of electron energies. Preliminary Hartree–Fock calculations led the authors to suggest that such resonance states may involve a 2p excitation of atomic oxygen.[77,78]

3 Experimental Observation of ESD

It is important to keep in mind that the usefulness of ESD data derives from the ability of the techniques to probe the local properties of an adsorbate–substrate system via direct excitation processes. When indirect processes become significant, elucidating the excitation mechanism from resultant data (thresholds, KED, *etc.*) becomes difficult. However, the ESDIAD phenomenon (see Section 4) may not depend strongly on the excitation mechanism, as long as bond scission occurs. Therefore ion angular distribution data may still provide useful information about adsorbate orientation and dynamics, even for systems in which both direct and indirect excitation processes occur.

Measurement of Ion Yield, Threshold Energy, and Ion Identity

Three or more decades of work involving ESD from surfaces have resulted in the development of some very elegant and sensitive experimental techniques.

There are two general categories under which most ESD related experimental studies can be placed. The first category utilizes methods to measure properties of the desorbing species, whether they be anionic, cationic, or neutral. These could involve measurements of total desorption yields, energy and angular distributions, mass spectra, radiation emitted by electronically excited metastables, and many other possibilities. The second broad category would be those studies which investigate changes in characteristics of the surface due to electron bombardment. These may involve methods for work function and elemental composition determination, observations of differences in low energy electron diffraction (LEED) patterns and temperature programmed desorption (TPD) spectra caused by electron irradiation, vibrational spectroscopic studies of surface bound ESD products, and a variety of other types of investigations. Here we will focus attention on methods for the direct study of ESD using particle detection methods during electron bombardment of surfaces.

Positive ions are the easiest and most widely studied ESD products. Most of the earliest studies involved positive ion detection via magnetic sector mass spectrometry[79–81] or measurements of the total ion flux incident on a collector subtending a large solid angle in front of the sample surface.[18] Quadrupole mass spectroscopy (QMS) of desorbing ions has been exploited,[82] as was an r.f. mass analysis technique[83] and ion energy distribution (IED) measurements of mass analysed positive ions.[84,85] Reviews which include detailed discussions of early experimental endeavours are available.[86]

An illustrative example of the early application of QMS techniques for ESD measurements comes from Ashcroft, et al.[85] Figure 10a is a schematic diagram of the ESD measurement system employed by these workers. Electrons from a commercially available electron gun strike the target (T) at an angle of 45° off normal. The target (W single crystal) can be heated from behind by electron bombardment with a cathode/filament arrangement (F), with the temperature monitored by a thermocouple attached at the crystal edge.

Positive ESD ions desorbing from the crystal are energy analysed by a series of hemispherical grids (G_1–G_3) before striking a hemispherical ion collector (C). Attempts to maintain overall spherical symmetry within the analyser were made, including the use of a guard hemisphere (H) matched in size with, and symmetrically located opposite to, the collector. Electrostatic shielding (hemisphere S) and carefully constructed electrical connections allowed for the measurement of very low ESD ion currents.

Figure 10b illustrates the applied voltage configuration at different locations within the analyser. Most secondary electrons are decelerated substantially between the target T and grid G_1, whereas the region between grids G_1 and G_2 takes care of more energetic back-scattered primary electrons. The majority of detectable particles to pass through grid G_2 and reach the collector C are positive ESD ions, and this region is utilized for kinetic energy analysis by retarding potential techniques. Although the crystal

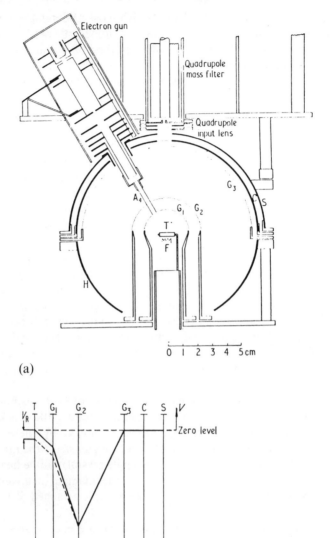

Figure 10 (a) *Schematic diagram of system for kinetic energy and mass analysis of positive ESD ions and* (b) *voltage configurations at various locations within the apparatus*
(Reproduced with permission from *J. Phys. E: J. Sci. Instrum.*, 1972, **5**, 1106)

holder, target, and electron gun distort electric fields by breaking the spherical symmetry, energy resolution of about 1 eV can be achieved.

Mass spectroscopy of energy-analysed ESD ions is then performed by allowing a fraction of these species to pass through a small hole in the collector and into the lensing array of a QMS. Ion detection by a high-sensitivity d.c. channeltron proved to be adequate for ESD measurements. Ashcroft, et al.[85] assessed the performance of this apparatus by comparison of their data with that obtained previously by other workers.

More recently, the widespread availability of hemispherical grid and cylindrical and ellipsoidal mirror analysers have increased the ease with which one may experimentally probe positive ion ESD from solid surfaces.[87-90] Time-of-flight (TOF) techniques can also be applied in ESD studies.[91-94] Negative ion detection in the presence of bombarding and secondary electrons is an intricate problem that has also been overcome.[95,96]

In some cases, desorbing electronically excited metastable species have long enough lifetimes to enable their detection via measurement of the secondary electrons produced when these particles strike the detector surface.[97,98] Neutral desorption products may be detected by post-ESD laser-induced fluorescence[99,100] or ionization.[101-104] Accurate partial pressure measurements have also been used to elucidate the presence of desorbing neutrals.[105] Several articles which include detailed discussions of such experimental techniques are available to the interested reader.[106,107]

Measurement of Angular Distributions of Ions and Electronically Excited Neutrals Produced in ESD

Measuring the angular distribution of desorbing particles (ESDIAD) requires extending the above technologies in one of two ways. First, a detector with a small aperture (*e.g.* QMS or channeltron) can be used to accept only a fraction of the species desorbing from the surface along a known solid angle direction. Then the detector or the adsorbate covered crystal are rotated about appropriate axes to map out the desorption flux or KED of particles with respect to spatial co-ordinates.

A good example of this type of technology is provided by the apparatus of Niehus and Krahl-Urban,[10] represented schematically in Figure 11. For ESDIAD measurements, an electron beam strikes the surface of a single crystal at glancing incidence, and a small fraction of the ions ejected from the surface are collected by a channeltron mounted on a motor-driven, computer-controlled goniometer. The detector can traverse a spherical surface and map out ESD ion intensity over about 67% of the solid angle in front of the sample with an angular accuracy near 1°. Digitally recorded data is displayed in three-dimensional form, and digital background subtraction can be utilized to enhance the appearance of ESDIAD patterns.

Another interesting feature of this design is that the electron gun can provide either a continuous or time-gated (pulsed) beam of electrons. In the pulsed mode, TOF analysis can be performed, greatly enhancing the versatility of this equipment. Furthermore, it was shown that spurious background ESDIAD signals can be significantly reduced in the chopped beam mode.[10] Time-gating allows for detection of ions originating at the sample surface and can discriminate against those arising from other surfaces in the ultrahigh vacuum (UHV) chamber (*e.g.* grids). Soft X-ray production due to

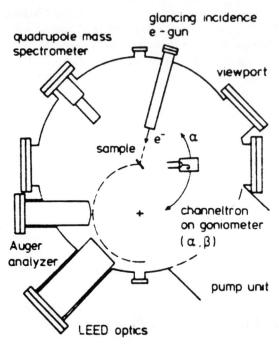

Figure 11 *Illustration of digital ESDIAD chamber with movable ion detector*
(Reproduced with permission from *Rev. Sci. Instrum.*, 1981, **52**, 56)

electron bombardment of the sample may also result in an increased ESDIAD background signal, which can be compensated for digitally.[9,108]

Another approach is to use a spatially resolving detector with a large solid angle of acceptance to collect a significant portion of the species liberated from the surface via ESD.[8,9] A schematic illustration of the digital ESDIAD imaging system developed and currently used by our group is presented in Figure 12.[9] A focused beam of electrons strikes a 1 mm² area of the single crystal sample at an angle of about 54° from the surface normal. The crystal can be resistively heated to > 1000 K and cryogenically cooled to ≈ 90 K, with the temperature monitored by a thermocouple spot welded to the crystal edge.

Particles ejected by ESD and secondary and back-scattered primary electrons encounter a four-grid electrostatic lens array. The first two hemispherical grids are normally held at ground potential, whereas the planar grids are negatively biased to repel electrons during ESDIAD measurements. In this configuration, all positive ion and neutral ESD species pass through the grid system, striking the first of two multi-channel plates (MCP). Electron cascades which follow produce about one million electrons exiting the rear MCP for each positive ion (or excited-state neutral) that hits the first MCP.

This secondary electron pulse is accelerated into a resistive anode,[109] causing an expanding ring of charge, centred at the pulse arrival point, to

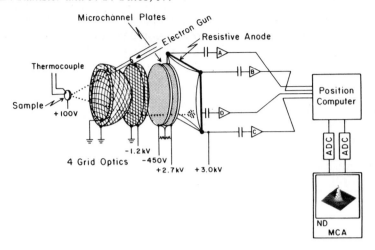

Figure 12 *Digital ESDIAD acquisition system, schematically illustrating the relative positions of the four-grid optics, microchannel plates and resistive anode* (Reproduced with permission from Surf. Sci., 1986, **169**, 91)

propagate across the thin conductive film. The amplitude of the signal arriving at each corner of the anode is used to calculate the initial position (in x and y) where the secondary pulse struck. Analog-to-digital conversion is performed, with the output displayed on the screen of a multi-channel analyser. Used in this pulse counting mode, the position analysis computer is gated to discriminate against signals resulting from multiple-particle impacts. The number of counts in any channel $\{x, y\}$ is then translated into a z co-ordinate for three-dimensional image representation.

Data are digitally stored and may be smoothed and corrected for soft X-ray background effects.[9,108] Further data manipulation is sometimes useful for qualitative interpretation and presentation, and these methods of image processing are available to the interested reader.[8]

The grid-detector system depicted in Figure 12 has a few other features which should be mentioned. First of all, a positive bias placed on the second hemispherical grid can be used to repel positive ions which escape from the crystal surface during ESD. This allows for KED measurements, as well as for the collection of ESDIAD patterns exclusively from metastable neutral ESD-produced species. These patterns can then be digitally subtracted from the ion + metastable patterns (recorded with the first three grids grounded), thus yielding two sets of ESDIAD images; one from metastable neutrals and one from positive ions. Since neutral particle trajectories are not significantly altered by image potential or stray electric field effects, comparison of these two sets of ESDIAD data may prove informative.

Secondly, if poor signal-to-noise ratio is a problem, a positive potential may be applied to the crystal. Positive ESD ions will therefore be accelerated toward the first grid (ground potential), increasing the count rate at the detector. Application of this 'compression' field will distort the polar angle of the ion trajectories but does not significantly affect their azimuthal orientation. The angular resolution of our equipment is normally less than 1°. In addition, reversing potentials on the planar grids allows electrons to be accelerated into the MCP array, enabling one to acquire LEED data without disturbing the crystal position.

4 Electron Stimulated Desorption Ion Angular Distribution (ESDIAD)

The introductory remarks concerning ESDIAD, presented above, are now considered in greater detail.

About sixteen years ago the first report emerged that O^+ desorbing from the O/W(100) system produced sharp patterns in registry with the substrate.[2] A subsequent paper by the same authors expanded on these findings[110] and postulated that the origin of the sharp emission cones of O^+ ions was directly related to the rupture of highly directional substrate–adsorbate bonds by ESD. A qualitative discussion was invoked to rationalize that the angular dependent desorption of ions from a surface reflected the symmetry of the adsorption site from which the ion originated. This interpretation raised many conceptually difficult and challenging questions regarding the mechanisms involved in ESD from surfaces.

Soon thereafter, Gersten, et al.[111] provided a theoretical model and calculated the classical trajectories of oxygen ions liberated from a tungsten surface. The model assumed a rigid substrate which contributes an overall potential for ionic motion formed by a superposition of Hartree terms of the free lattice atoms and an image term dependent only on the substrate–adsorbate distance (z). Oxygen atoms initially bound in a harmonic ground state potential were ionized to O^+ and the ensuing asymptotic desorption directions were calculated and plotted.

The results show that ions can be preferentially ejected along specified directions in space, and in some cases the calculations reproduced experimentally observed patterns.[2,110] This model, however, only stressed a final-state effect, that of the potential experienced by an ion after its production, and neglected initial-state effects such as localization of electron density due to covalent surface-oxygen bonds.

In contrast, Clinton[112] proposed a quantum mechanical explanation of ESDIAD phenomena which allowed both initial- and final-state contributions to the angular dependent ionic desorption. Clinton expressed the total desorption cross-section σ_e as

$$\sigma_e = \sigma_0 P_f |I_{FC}|^2 \tag{21}$$

where σ_0 is the ionization cross-section depending predominantly on electronic co-ordinates, I_{FC} is the overlap integral between initial and final state vibrational wavefunctions (i.e. $|I_{FC}|^2$ is the Franck-Condon factor), and P_f a quantum reneutralization probability.

Utilizing reflection approximation (RA) arguments, Clinton showed that I_{FC} is peaked in the directions of chemical bonds of the initial adsorption configuration; thus σ_e will have a maximum along these directions. Assuming the initial vibrational wavefunction is a three-dimensional Gaussian, the results predict that ion desorption roughly normal to the surface (*i.e.* from atop bonded species) should yield a single central circular ion beam, the width of which is determined by the vibrational amplitude of the chemical bond at the time of rupture. This picture can also be applied to other adsorption geometries (*i.e.* bridge-bonded species) where now the ion beams may form patterns of elliptical cross-section due to different vibrational amplitudes in different directions on the surface.

It is clear that σ_e contains contributions solely attributable to the character of the final-state following excitation. Reneutralization probabilities (*i.e.* P_r) may not be isotropic on an atomic scale, resulting in anisotropic broadening of the resulting ion trajectories. Observed ESDIAD patterns contain information regarding adsorbate geometry and dynamics before excitation convoluted with effects of the environment experienced after excitation.

In 1977 Janow and Tzoar provided a quantitative theoretical kinetic picture of ESDIAD.[113] Conservation of the probability density distribution function was described by the Boltzmann transport equation. Explicit analytical expressions for the total ionic and neutral ESD cross-sections were derived using only a few general assumptions, making the results widely applicable to ESDIAD pattern simulation for many systems.

The authors then proceeded to apply the model to O/W(111), which had been studied experimentally by Madey, *et al*.[114] The calculations assumed a parabolic potential (V_0) for the initial ground state configuration and a final state potential (V) formed by an image term and a superposition of atomic Hartree terms centred on the tungsten atom positions. It was found that the dominant factor determining the trajectory (asymptotic angle of desorption) of high energy ions was the repulsive contribution to V along the nearest-neighbour O—W direction, implying that in this case ESDIAD patterns reflect the original site symmetry of the adsorbate–substrate system. The remaining terms in V (next-nearest-neighbours, *etc.*) contribute significantly only for very low energy ions, tending to defocus the beams, resulting in hazy patterns rather than well-defined, sharp ion beams. The model successfully reproduced many of the O^+ patterns observed by Madey and co-workers,[114] substantiating their interpretation that the ESDIAD patterns reflect the bonding site geometry of oxygen at the surface. However, the model underestimated peak ion energies, thus requiring a modification to include the possibility of substrate excitations which could affect the amount of energy available for desorption.

Janow and Tzoar further applied this model to hydrogen chemisorption on tungsten surfaces.[115] In all cases considered, the model adequately reproduced the width and peak position of IED obtained experimentally by others.[2] Thus, in H/W systems, surface excitations may be swiftly delocalized and not significantly affect the partitioning of kinetic energy to desorbing H^+ ions. These early theoretical descriptions and experimental reports of ESDIAD phenomena made it quite clear that the technique offered a promising means for structural determination of adatom geometries in chemisorption systems.[116]

Following improved quantitative measurements of ion angular distributions (IAD) and IED for O/W,[117-120] Preuss extracted experimental values for parameters to be inserted in his theoretical model of O^+ ESD from tungsten surfaces.[121] Repulsive terms in V were derived from screened Coulomb rather than the Hartree fields used by previous authors.[111,113,115] Calculated IED and IAD results were in very good agreement with those obtained experimentally, with the model incorporating physically reasonable adsorbate bond lengths, vibrational amplitudes, and surface charge densities.

A factor that will influence the trajectory and therefore the ESDIAD pattern produced by ionic desorption is the image potential (V_I). This long range potential ($\sim -1/z$) produces an attractive Coulomb force on the escaping ion which retards the z-component of the ion's momentum long after the short range repulsive forces responsible for initial ion ejection have ceased to contribute significantly. The image force is a final-state effect and is anisotropic in the sense that the x and y components of momentum of the desorbing ion are unaffected. Thus for ideal planar substrates, only the polar angle θ and not the azimuthal angle ϕ of desorption is affected by V_I. The image force and its effect on ESD ion trajectories has been addressed theoretically.[122-126]

5 Chemical Systems Studied by ESD Techniques

An appreciation of the ability of the ESDIAD method to yield information about the behaviour of molecular adsorbates is best obtained by a summary of selected experimental results. Two categories of experiments involving chemisorbed species on single crystal surfaces have been realized by ESDIAD techniques. They are:
(1) Determination of chemical bond directionality in adsorbate/substrate systems.
(2) Observation of the dynamics of surface-bound molecules.

These are illustrated with specific examples below.

CO/Ru(001)

The dynamical behaviour of surface species is an important issue where ESD studies may provide invaluable information. The CO/Ru(001) chemisorption system provides an early example of the application of ESDIAD methods in this regard.[127-129]

CO bonds to Ru(001) through the carbon atom with the molecular axis perpendicular to the substrate for low to near half-saturation coverages,

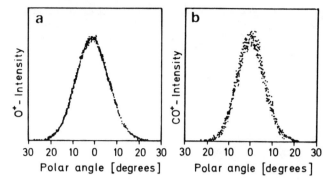

Figure 13 *Polar angle distributions of* (a) O^+ *and* (b) CO^+ *ions liberated from a* $(\sqrt{3} \times \sqrt{3})R30°$ *layer of CO on* Ru(001) *by* 280 eV *electron bombardment at* 100 K
(Reproduced with permission from *Surf. Sci.*, 1985, **163**, 39)

where it forms a $(\sqrt{3} \times \sqrt{3})R30°$ overlayer. ESDIAD studies of this system have observed CO^+ and O^+ ions ejected in narrow cones normal to the surface, as seen in Figure 13. It was found that ion angular distributions broaden with increasing CO coverage at fixed temperature. This was interpreted as indicating an increase of inter-adsorbate repulsions, causing adsorbed CO molecules to tilt an average of 5° from the surface normal at saturation coverage.[128,129]

Madey reported that CO^+ and O^+ ESDIAD images also exhibited a temperature dependent broadening at fixed CO coverage.[127] Patterns remained circular in cross-section and normal to the surface as temperatures increased from 90 to 300 K, with the half-width at half maximum (HWHM) of the ESDIAD patterns rising by about 4°. After correcting for image potential effects Madey argued that most of the observed broadening of ion ejection patterns could be interpreted in terms of surface bending vibrations of CO. The identification of molecular surface dynamics via observation of changes in ion patterns following stimulated desorption provides a good example of the power of such experimental methods.

CO/Pt(111)

A more recent example of the application of ESDIAD to the study of molecular dynamics and lateral inter-adsorbate interactions comes from a systematic study of the CO/Pt(111) system.[12,130–134] The first paper of this series concentrated on characterizing the particles ejected during ESD in this system.[12] It was found that CO^+, O^+ and an electronically excited metastable $(a^3\Pi)$ CO* species were liberated by bombardment with electrons of energy 200–900 eV.

The yield of each species was investigated as a function of CO coverage, temperature, and incident electron energy. Most importantly, it was shown

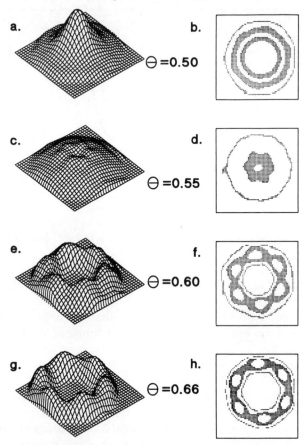

Figure 14 *Three-dimensional and contour plots of* CO^+ *ESDIAD patterns from* CO/Pt(111) *for* CO *coverages of* 0.50–0.66 ML. *Incident electron energy was* 260 eV *and temperature was* 90 K
(Reproduced with permission from *Surf. Sci.*, 1988, **205**, 215)

that no correlation existed between observed CO^+ and CO^* yield data. This implied that CO^* was produced via a direct excitation mechanism and did not result from reneutralization of CO^+ ions during their escape from the surface. This result has far reaching implications, since the trajectory of a neutral species is not significantly altered by electrostatic effects (except by weak dipole image forces), whereas strong image forces act on desorbing ions. Thus CO^* ESDIAD patterns can be assumed to best reflect the actual bonding symmetry of CO at the surface.

At coverages up to 0.50 monolayers (ML), ESD products (CO^+, O^+, CO^*) are ejected normal to the surface, forming circular ESDIAD images.[130] As CO coverage increases to 0.67 ML, this pattern evolves into six beams, with a simultaneous intensity decrease in the normally directed beam, as shown for CO^+ in Figure 14. Because of the image effect, CO^+ beams show larger polar angles than CO^* beams.[130] The six beams are directed along six

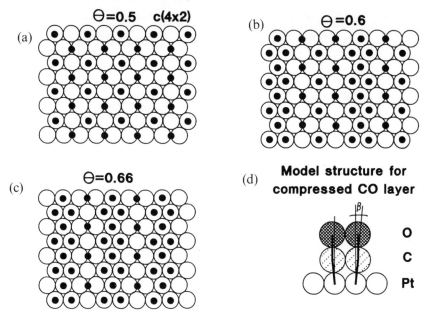

Figure 15 (a)–(c) *Structural model proposed for densely packed CO layers on* Pt(111)
(d) *Schematic of tilting CO molecules responsible for six beam* CO^+ *ESDIAD patterns*
(Reproduced with permission from *Surf. Sci.*, 1988, **205**, 215)

equivalent {110} crystal azimuths of the hexagonal (111) surface. These six-fold symmetric images were proposed to result from tilted CO molecules (about 6° from the surface normal) within densely packed regions in the CO overlayer, due to intermolecular steric repulsions between terminally-bound CO molecules which separate less dense domains. A geometrical model was proposed for the adsorption site symmetry of CO on Pt(111) which followed that of Avery.[135] This is reproduced in Figure 15 for one of three domain orientations.

CO/Pt(111) ESDIAD images for both low (circular) and high (six-beam) CO coverages underwent vibrational broadening between 90–230 K.[130] Referring to Figure 16, low coverage patterns were seen to broaden isotropically in azimuthal directions with increasing temperatures, remaining normal to the surface and circular in cross-section. In sharp contrast, anisotropic broadening of the six-beam pattern was observed as the temperature rose, evolving these into halo-type images at 230 K, as in Figure 17.

This behaviour at high CO coverages was interpreted as evidence for larger vibrational amplitudes perpendicular to the tilt plane of densely packed CO species ({211} equivalent azimuths) than for those oscillations within the tilting plane. Steric squeezing of tilted CO adsorbates between neighbour CO species and neighbour Pt atoms toward which CO tilts was postulated to narrow the potential well along tilt reaction co-ordinates. Thus vibrational energy levels would rise, causing a smaller Boltzmann popu-

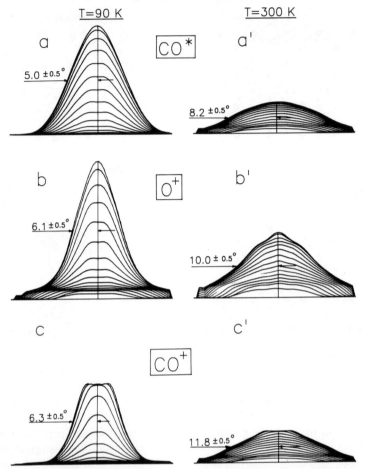

Figure 16 *Polar angular distributions for* (a) CO*, (b) O^+, *and* (c) CO^+ *ESD products from* 0.50 ML CO/Pt(111). *HWHM values increase and yields diminish as the temperature is increased from* 90 *to* 300 K. *Relative intensities for* a:b:c *are* 5:2:1
(Reproduced with permission from *Surf. Sci.*, 1988, **202**, L559)

lation at any given temperature, when compared to levels in the relatively unperturbed potential well along co-ordinates perpendicular to the tilting directions.

CO/Se/Pt(111)

A study of CO co-adsorption with Se on Pt(111) has been carried out.[13] Thermal decomposition of H_2Se left selenium in three-fold hollow sites on the Pt surface, forming a p(2 × 2) structure at 0.25 ML Se coverage. Subsequent CO adsorption at 220 K was seen to modify the Se overlayer, driving the pre-adsorbed selenium atoms into a $(\sqrt{7} \times \sqrt{7})R19.1°$ structure at CO

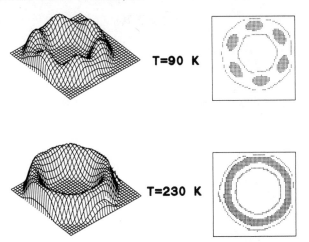

Figure 17 *Temperature induced anisotropic broadening of CO^+ ESDIAD patterns from 0.62 ML CO/Pt(111)*
(Reproduced with permission from *Surf. Sci.*, 1988, **205**, 215)

coverages near 0.15 ML. This phenomenon was observed via two methods; LEED, which verified the structure of the Se overlayer, and ESDIAD, which measured the vibrational amplitude of CO at the surface. Both of these are illustrated in Figure 18.

As is evident in this figure, CO* ESDIAD patterns actually broadened with increasing CO coverage, with temperature and Se coverage held constant, an intuitively unexpected result. This beam broadening was accompanied by a simultaneous structural phase transition of the Se adlayer, as shown by the LEED patterns. It was proposed that a reduction in the CO–Se repulsive forces were responsible for these observed changes. A broadening of ESDIAD patterns implies that CO molecules have more freedom to vibrate at the higher CO coverages, since the Se atoms are further away from each CO site following the overlayer transition. A geometrical model was offered, consistent with the data, for the structure of CO–Se systems on Pt(111) at both low and high CO coverages. In this model, CO–Se interatomic distances increase from 3.2 to 4.2 Å in going from the $p(2 \times 2)$ to the $(\sqrt{7} \times \sqrt{7})R19.1°$ configuration, thus giving CO the opportunity for greater lateral motion during low frequency oscillations. In the more open structure, six Se atoms, each at 4.2 Å distance, surround each CO. In the more closed structure, only three Se atoms are nearest neighbours to each CO, but at a distance of only 3.2 Å.

CO/Pt(112)

Studies of the adsorption and dissociative chemistry of small molcules on stepped metal crystal surfaces have been of interest because of the role of surface defects in controlling surface chemical processes. Indeed, the role of

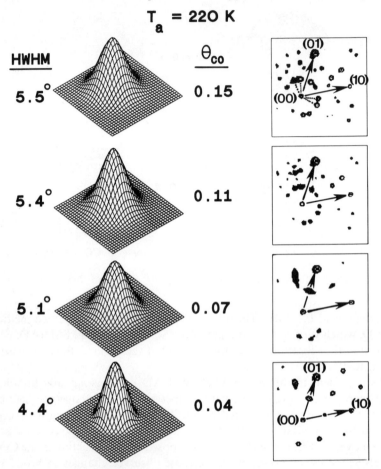

Figure 18 CO* ESDIAD (left) and LEED (right) patterns from various coverages of CO adsorbed on a p(2 × 2)-Se overlayer at 200 K. LEED patterns show that a Se re-ordering transition is induced by addition of CO, which is reflected in a broadening of the ESDIAD images.
(Reproduced with permission from *Phys. Rev. Lett.*, 1988, **61**, 2875)

surface defects in heterogeneous catalysis has been a subject for discussion since the original proposal for the existence of active sites on metals by Taylor in 1925.[136] ESDIAD provides an ideal method to obtain specific information about chemisorption on stepped single crystal surfaces, where the presence of periodic steps provides a physical basis for investigating molecular behaviour at defect sites.

Studies of the chemisorption of CO on Pt(112) using ESDIAD imaging of the CO* species have been carried out.[3-5] This surface is schematized in Figure 19 and consists of three Pt atom wide terraces of (111) orientation, separated by atomic steps of (100) orientation. ESDIAD studies of CO chemisorption on this surface are carried out using two angles of view, View 1 and View 2, as shown in Figure 19.

VIEW 1	VIEW 2
MCP PARALLEL TO THE CRYSTAL FACE AND NORMAL TO [112].	CRYSTAL FACE ROTATED 20° AWAY FROM THE ANALYZER AXIS.

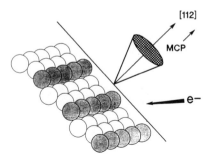 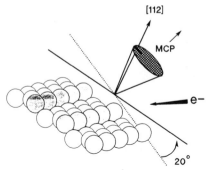

MCP = MICROCHANNEL PLATE DETECTOR

Figure 19 *Schematic presentations of the three-dimensional CO* ESDIAD data from two 'views' which focus on terrace and step CO*
(Reproduced with permission from *J. Chem. Phys.*, 1989, **91**, 7245)

Figure 20 shows the progressive development of the CO* ESDIAD patterns as the coverage is increased. It may be seen in View 2 (which images molecules on the step sites) that initially the only CO* beam is a single beam, oriented in a direction $-20°$ from the normal to the macroscopic plane. This indicates that the adsorbed CO molecules on the step sites are inclined by about 20° in the downstairs direction. As the coverage increases, this beam is attenuated, and beams inclined left and right by $\pm 13°$ appear. At still higher CO coverages, the left and right beams are extinguished and two beams at 0° and $-38°$ develop, indicating tilt directions orthogonal to those observed at intermediate coverages. View 1 is arranged to more readily observe CO molecules originating from the terrace sites. At high coverages, a beam of CO^+ at $+13°$ is characteristic of CO molecules on the terrace sites, probably on Pt atoms directly adjacent to the step atoms.

A model showing the structures proposed is given in Figure 21. At a coverage of 0.19 ML, a $(2 \times n)$ unit cell is observed by LEED, and it is proposed that a single species, inclined in the downstairs direction on the step sites is produced. Here the step sites are one-half filled. As the steps fill to three-quarters coverage, 'triplets of CO' are formed, exhibiting CO* beams in three directions as CO molecules adjacent to the central molecule are repelled in directions parallel to the steps, producing the beams at $\pm 13°$. Finally, when the steps are completely filled, orthogonal tilt directions up and downstairs are established. Development of a CO* beam originating from the terraces can be observed at a coverage of 0.34 ML as shown in View 1. This $+13°$ angle of inclination, from the normal to the Pt(112) macroscopic plane, indicates that CO molecules on the terrace sites are repelled backwards from the step sites containing adsorbed CO species.

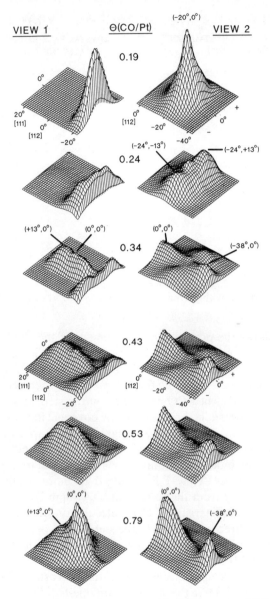

Figure 20 CO* *ESDIAD patterns from stepped* Pt(112) vs. *initial* CO *coverage*
(Reproduced with permission from *J. Chem. Phys.*, 1989, **91**, 7245)

The information obtainable by ESDIAD has been exploited in experiments designed to measure the dynamical behaviour of adsorbates. For this purpose, CO adsorbed at low coverages on step sites was employed. The step sites offer unsymmetrical potential energy surfaces which govern the vibrational amplitudes of chemisorbed CO species. Figure 22 illustrates how a CO* ESDIAD pattern from CO on the step site is analysed in order to

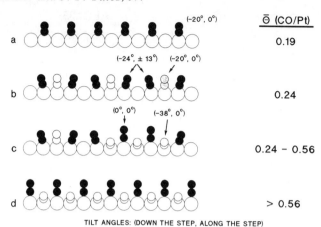

Figure 21 *Proposed geometry of CO on steps of Pt(112). Coverage, $\bar{\theta}$, is with respect to the total number of surface Pt atoms*
(Reproduced with permission from *Chem. Phys. Lett.*, 1990, **168**, 51)

observe anisotropies in the vibrational motion. The pattern is sliced by planes perpendicular and parallel to the step edge direction. As shown in Figure 23, the cross-sections of the ESDIAD patterns increase in width as the temperature is raised, indicating that thermal excitations are being observed. The profiles in the 'up-down' direction are asymmetrical with the larger half-width in the upstairs direction. The profiles in the 'along' direction are symmetrical, as would be expected since the step adsorption site is symmetri-

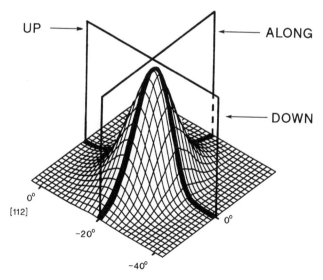

Figure 22 *CO* ESDIAD pattern from CO on Pt(112) at a coverage corresponding to one-half filled step sites. The plane along the step edge direction is orthogonal to that normal to the step edge*
(Reproduced with permission from *J. Chem. Phys.*, 1989, **91**, 7255)

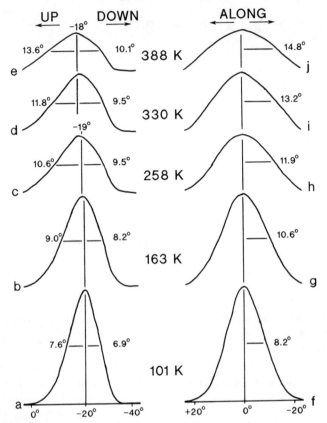

Figure 23 *Cross-sections of* CO* *ESDIAD patterns normal to* (a)–(e) *and parallel to* (f)–(j) *the step edges of* Pt(112) *at various temperatures. Peak half-widths are inset in the figure*
(Reproduced with permission from *J. Chem. Phys.*, 1989, **91**, 7255)

cal in its relationship to neighbouring Pt atoms as one moves along the step. The thermal broadening in the various directions is shown in Figure 24.

These results are interpreted by considering the frustrated translational modes of chemisorbed CO. These modes have a frequency of about 50 cm^{-1}.[137] The largest vibrational amplitudes are observed in the 'along' direction and indicate that at low coverages, the CO molecules have the softest frustrated translational modes in this direction. The smallest vibrational amplitudes are observed in the 'downstairs' direction and may reflect the interaction of the electron cloud of the inclined CO molecule with electron density which has spilled out onto the terrace from the step site.[138,139]

CO/Ni(110)

Interest in the chemisorption of CO on Ni(110) is related to the fact that CO thermally desorbs from this crystal in two binding states which are easily

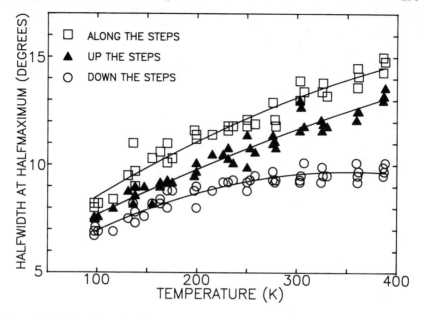

Figure 24 *Half-width at half-maximum measurements from CO* ESDIAD patterns along (☐), up (▲) and down (○) the steps of* Pt(112) vs. *temperature* (Reproduced with permission from *J. Chem. Phys.*, 1989, **91**, 7255)

separated using TPD methods. These two states, shown in Figure 25, are designated a_2 and a_1,[140] and are observed to develop sequentially as the coverage increases. ESDIAD studies of the O^+ angular distribution as a function of CO coverage have revealed that dramatic changes occur in the CO bonding mode as the coverage is increased, which results in an increase of the repulsive forces between the chemisorbed CO molecules.

Figure 26 shows interesting ESDIAD behaviour observed as a function of surface coverage.[141] For electron energies of 300 eV, a normally directed O^+ pattern develops, then decays into a weak and ill-defined pattern at full coverage. A similar experiment with an electron energy of 1000 eV shows clearly that the normal O^+ ESDIAD beam decreases with increasing coverage and is replaced by a two beam O^+ pattern at the highest coverage. This behaviour, best seen at electron energies near 1000 eV, is related to the tilting of the CO molecules at high coverage on the Ni(110) surface. Above a coverage of 0.75 ML, the normally-oriented CO molecules begin to tilt as a $c(4 \times 2)$ overlayer converts into a more complex adsorbate pattern involving a glide plane in the overlayer.[142] The sequence of adsorbate structures corresponding to various coverages of CO is shown in Figure 27. It is seen that above 0.75 ML, a highly uniaxially compressed CO overlayer is produced and that this compression causes the CO molecules to move laterally off their adsorption sites and to tilt.

The magnitude of the tilt angle was determined using the ESDIAD analyser in a zero crystal bias condition, and measuring the angular dis-

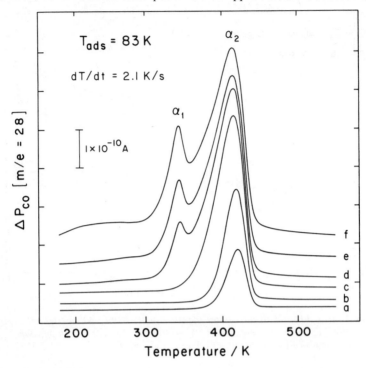

Figure 25 *Thermal desorption spectra* (m/e = 28) *from* CO/Ni(110) *vs. CO coverage. CO coverages were* (a) 0.14; (b) 0.30; (c) 0.64; (d) 0.87; (e) 0.89; (f) 1.00 ML (Reproduced with permission from *Surf. Sci.*, 1986, **165**, 447)

placement of the centre of the two-beam ESDIAD pattern by means of a rotation of the crystal through a known angle, a, as shown in Figure 28. This permitted the distance, ΔR, to be estimated from the known crystal-to-grid distance, R. The tilt angle, $\beta = 19°$ was determined in this measurement. A similar quantitative measurement of $\beta = 19°$ was made by Riedl and Menzel,[143] using a mass spectrometer for detection of the two ion beams. In addition, recent angular resolved X-ray photoelectron spectroscopy (XPS) measurement of the tilt angle yielded a value of 21°.[144] The agreement between these measurements is excellent, showing that positive effects on the value of β, caused by image forces, are either rather small, or that they are cancelled to a large degree by reneutralization effects which would cause β to become somewhat smaller.

CO/Cr(110)

Shinn and Madey have published a series of articles concerning the unique CO/Cr(110) system which will be briefly summarized together.[145-149] CO chemisorbs molecularly on Cr(110) at 90 K in two distinctly different binding states. For CO coverages up to 0.25 ML, a single state, denoted a_1, fills, forming a c(4 × 2) overlayer structure when saturated at 1/4 of a monolayer. Further CO exposure results in the population of a second state,

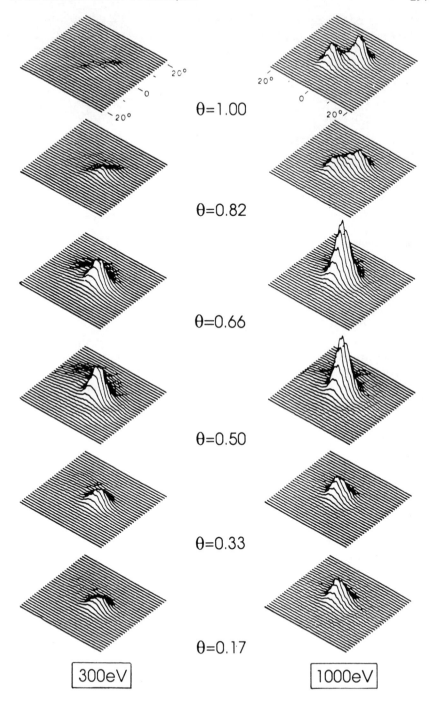

Figure 26 *ESDIAD patterns from* CO/Ni(110) *vs.* CO *coverage for two electron beam energies. Corrections for the effects of 'compression fields' on the angular scale have not been made*
(Reproduced with permission from *J. Vac. Sci. Technol.*, 1986, **A4**, 1446)

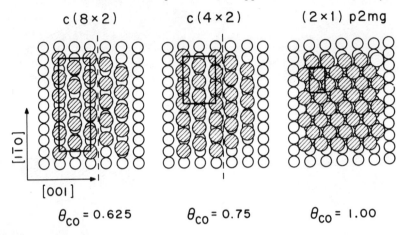

Figure 27 *Proposed geometry of CO on Ni(110) consistent with LEED and ESDIAD measurements*
(Reproduced with permission from *Surf. Sci.*, 1986, **165**, 447)

labelled a_2, until the surface is fully saturated at 0.35 ML CO coverage. In the region, 0.25–0.35 ML, some interconversion ($a_1 \rightarrow a_2$) occurs.

The a_1 state is especially intriguing. The authors report that these CO molecules are thought to bond in a 'lying-down' fashion, with both carbon and oxygen orbitals interacting with the surface. Literature was cited in support of this view,[145–149] based on vibrational and electronic spectroscopy studies. The a_2 state involves terminally bound CO molecules, with Cr—C—O bonds oriented perpendicular to the substrate.

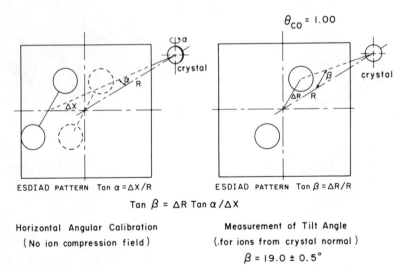

Figure 28 *Measurement of tilt angle of CO on Ni(110) using the indicated functions and measurements*
(Reproduced with permission from *Surf. Sci.*, 1986, **165**, 447)

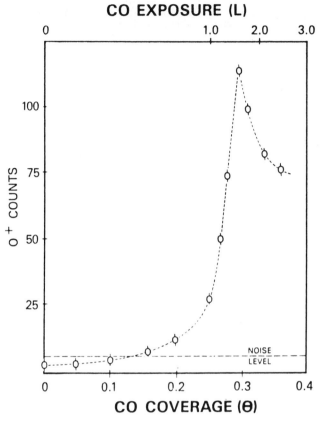

Figure 29 O^+ *ESD yield from* $CO/Cr(110)$ *as a function of* CO *coverage*
(Reproduced with permission from *Surf. Sci.*, 1987, **180**, 615)

Quantitative TOF analysis of ESD ejected particles was employed in this study, in addition to ESDIAD, LEED, electron energy loss spectroscopy (EELS) and Auger electron spectroscopy (AES). Time-of-flight spectra showed large H^+ and F^+ signals, even from the clean Cr surface, which were determined to result from trace impurities with rather high ESD cross-sections. CO residing in the a_1 configuration yielded negligible ESD signals. However, large CO^+ and O^+ yields were recorded from a_2 CO, *i.e.* when the CO coverage exceeded 0.25 ML. This is clearly seen in Figure 29 where the ESD O^+ ion yield is plotted as a function of CO coverage.[145] The O^+ signal decrease at high coverage was presumably due to increased lateral CO–CO interactions as the CO packing density rose.

The cross sections for electron induced positive ion liberation from a_2 CO were measured to be at least fifty times higher than those from a_1 CO. The assignment of a_1 CO as a molecule adsorbed in a lying-down position with its axis parallel to the surface is supported by these ESD results. De-excitation pathways should be greatly enhanced for a molecule so strongly coupled to a metal surface. Ions desorbed from a_2 CO were observed by

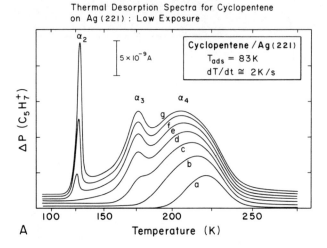

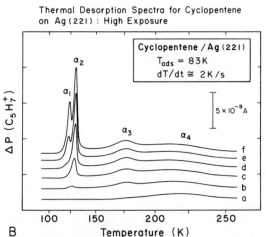

Figure 30 *Thermal desorption spectra for cyclopentene on* Ag(221) *for exposures of* (A) a, 1.2; b, 2.5; c, 3.1; d, 3.7; e, 4.9; f, 6.2; g, 6.3 *and* (B) a, 7.4; b, 8.6; c, 9.3; d, 12.5; e, 15.6; f, 18.8; *in units of* 10^{14} cm^{-2}
(Reproduced with permission from *J. Chem. Phys.*, 1986, **85**, 6093)

ESDIAD to be directed along the substrate normal, in accordance with the proposed terminal bonding configuration of the a_2 state.

Cyclopentene/Ag(221)

Most studies of the effects of step defects on adsorbed molecules have dealt with chemisorbed species where rather strong adsorption bonds control the process. One might wonder whether more weakly bound (physisorbed) species would also sense the presence of step defects by preferential adsorption and orientation at these sites. These ideas were tested using a stepped

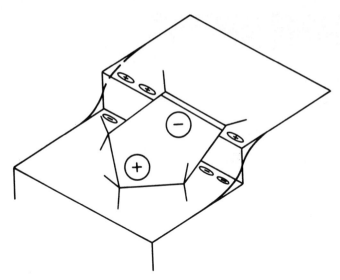

Figure 31 *Schematic diagram of cyclopentene on Ag(221) in the a_4 state. The smoothing of the charge distribution at the step edge is indicated, along with the sign of the molecular dipole*
(Reproduced with permission from *J. Chem. Phys.*, 1986, **85**, 6093)

Ag(221) surface which physically adsorbs an interesting molecule, cyclopentene.[150] Figure 30 shows the TPD spectra taken for cyclopentene on this surface as a function of exposure. For low exposures (top panel) a single binding state, a_4, is first observed, followed by a_3. Both of these states correspond to C_5H_8 species physisorbed on the metal surface. The a_2 and a_1 states are associated with multilayers. No thermal dissociation of this molecule occurs on the Ag(221) surface.

It has been found that the a_4 state is associated with H^+ ESDIAD patterns which are centred 4–5° off normal, pointing in the downstairs direction of this crystal. Control experiments with an unstepped Ag(111) crystal indicated that the a_4 state is associated with H^+ beams which are of maximum intensity within 1° of the surface normal. These results indicate that the cyclopentene molecule is sensing the step structure of the crystal and that the ejection of H^+ ions is an indicator of the presence of a special orientation of the molecule near the step. It is likely that in the case of Ag(111), a random orientation of C_5H_8 molecules leads to the average normal ejection direction for H^+.

The dominant forces which orient the C_5H_8 molecules may involve electric fields at the step site interacting with the C_5H_8 molecule via its dipole moment (0.19 Debye), oriented negative toward the double bond. The step site involves a non-uniform electric field due to local electron density anisotropy near the step, and the interaction between the molecular dipole and this field is probably responsible for the orientation effect. An interaction energy of 0.96 kcal may be estimated assuming an electric field at the step of

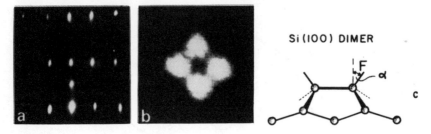

Figure 32 (a) *LEED pattern from* Si(100). (b) F^+ *ESDIAD pattern from* F/Si(100)–(2 × 1). (c) *Schematic diagram of* Si *surface dimer with an adsorbed* F *moiety* (Reproduced with permission from *Surf. Sci.*, 1987, **184**, L332)

magnitude of 0.5 V Å^{-1}.[150] Since thermal energy at the temperature of these experiments is only 0.16 kcal, the orientation of the dipole is expected. A schematic model for the oriented C_5H_8 molecule is shown in Figure 31. H^+ originating from the upward-pointing C—H moieties on the oriented C_5H_8 molecule would be expected to exhibit a downstairs direction of ejection in ESDIAD experiments.

F/Si(100)

Understanding the chemisorption of F on Si(100) is of importance because of the use of fluorine-containing molecules (HF, CF_4, *etc.*) as etchants in silicon-based technologies. The Si(100) surface normally exists as a (2 × 1) structure, caused by Si—Si dimer formation in the first layer. The LEED and ESDIAD patterns exhibited by the Si(100) surface containing chemisorbed F are shown in Figure 32a and b, where the four-fold F^+ ESDIAD symmetry is due to a superposition of Si dimer domains in orthogonal directions on the crystal.[151] Measurements of the angle, a, indicate that it is 35 ± 5° from the normal. This direction corresponds closely to that expected for the Si—F bond angle on this surface, and a schematic diagram of the Si—F bond projecting from the Si dimer sites is shown in Figure 32c.

By using a mass spectrometer as a detector of the F^+ ions produced by ESD, it was possible to measure the threshold electron energy as 27.8 ± 1 eV (corrected for contact potentials) and the measurement of this threshold energy (laboratory voltages scale, uncorrected) is shown in Figure 33. The F^+ ions desorb with a most probable kinetic energy of 2.4 eV, as determined in the ESDIAD analyser using retarding potential techniques.[151] These represent the first ESDIAD measurements made for an atomic species on a semiconductor surface. Similar measurements were also reported for F adsorption on a stepped Si(100) surface in which the Si dimer distribution was preferentially oriented in a single direction.[152] It has been suggested on other grounds that Si—F species are the dominant surface fluorosilyl moieties at low fluorine coverages on Si(100).[153]

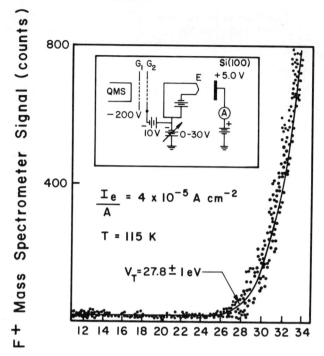

Figure 33 *Threshold measurements for ESD of F^+ from Si(100). The inset illustrates the circuit used for such measurements*
(Reproduced with permission from *Surf. Sci.*, 1987, **184**, L332)

H/Ni(111) (Alkali Metal Co-adsorbates)

An experimental study of the H/Ni(111) system appeared in 1988.[154] The H^+ ESD yield following saturation exposure of a clean Ni(111) surface with hydrogen was negligible. However, if the substrate was precovered with sub-monolayer quantities of alkali adatoms, H^+ production by ESD rose significantly. Figure 34 shows the dramatic increase in H^+ yield from H/Ni(111) under electron bombardment as the initial potassium coverage is increased. Pre-adsorption of Li and Na produced similar effects whereas Rb and Cs did not.

A simple electrostatic model of the interaction between co-adsorbed H and K atoms was presented, describing the system in terms of alkali hydride-like complex formation with H^{x-} adsorbed atop K^{x+} *(x < 0.29 e)*. As seen in Figure 34 the H^+ yield increased linearly with K coverage from 0.04–0.18 ML, indicating that H sensitization to electron impact was due to short range H–K interactions, *i.e.* bonding. It is also evident that for higher initial coverages of K the production of H^+ decreases significantly. This latter observation was proposed to result from several factors, including increased excitation damping and alkali–alkali depolarization effects, both

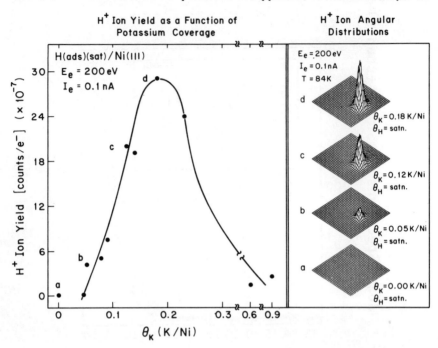

Figure 34 H^+ *ESD yield and corresponding H^+ ion angular distributions from saturation coverages of hydrogen on Ni(111) vs. initial potassium coverage. Incident electron energy and beam current was 200 eV and 0.1 nA, respectively*
(Reproduced with permission from *J. Chem. Phys.*, 1988, **89**, 570)

of which should become more effective as interatomic distances decrease at high K coverages.

$H_2O/Ni(110)$

In 1984, Benndorf and co-workers[155] published a detailed study of the nature of water adsorption on oxygen-predosed Ni(110). Auger electron spectroscopy and TPD indicated that H_2O did not adsorb (dissociatively or molecularly) on the clean surface at 300 K, but the presence of O(a) induced its dissociative adsorption at this temperature, forming surface-bound OH groups. This was verified by isotopic mixing using ^{18}O(a) and $H_2^{16}O$ followed by TPD experiments, which verified the thermal desorption of both $H_2^{16}O$ and $H_2^{18}O$, formed by surface recombination reactions of adsorbed hydroxyl species.

Angular resolved ultraviolet photoelectron spectroscopy (UPS) also prompted the same conclusion; water adsorption on oxygen covered Ni(110) at 300 K leads to the formation of OH(a). In addition, UPS showed that O—H bonds of such moieties were inclined away from the surface normal, and the O—H bonds were azimuthally oriented along {001} equivalent directions.

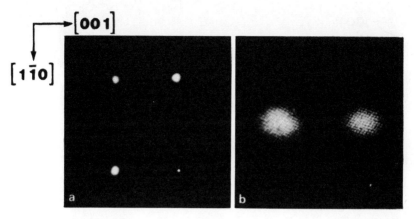

Figure 35 (a) *LEED pattern of clean* Ni(110) (b) H$^+$ *ESDIAD pattern from* OH(a) *moieties, formed by* H$_2$O *dosing of the* 0.2 ML O/Ni(110) *system at* 300 K
(Reproduced with permission from *Surf. Sci.*, 1984, **138**, 292)

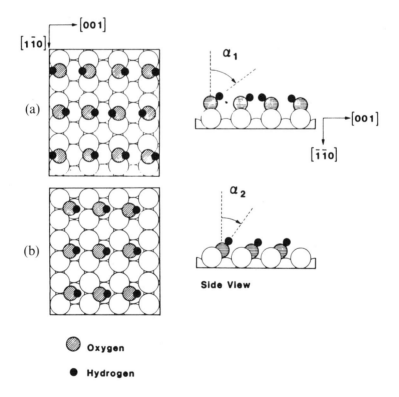

Figure 36 *Proposed models of* OH(a) *bonding to* Ni(110). (a) *Short bridging site involving two* Ni *ridge atoms* (b) *Three-fold site, with* OH *bonded to two ridge and one trough* Ni *atoms*
(Reproduced with permission from *Surf. Sci.*, 1984, **138**, 292)

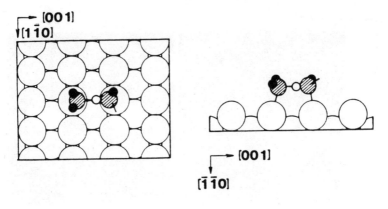

- ● ESD Active Hydrogen
- ○ ESD Inactive Hydrogen
- ◐ H_2O Molecule

Figure 37 *Model for H_2O dimer formation on Ni(110) surfaces*
(Reproduced with permission from *Surf. Sci.*, 1985, **157**, 29)

H^+ ESDIAD patterns from the $H_2O/O/Ni(100)$ system at 300 K verified the above conclusions. Two proton beams, directed along {001} and {00$\bar{1}$} crystal azimuths were observed as shown in Figure 35. This was consistent with OH(a) being bound to Ni through the oxygen atom, with the hydrogen pointing up and away from the substrate. The polar angle at which OH(a) species were tilted with respect to the surface normal was estimated to be 30–35°.

Combining these results with LEED data, the authors proposed the most probable geometrical configuration of hydroxyl groups adsorbed on the Ni(110) surface. This model placed the OH species at three-fold sites, with the oxygen-end co-ordinated to two outer-layer (ridge) and one second-layer (trough) Ni atoms. This has been schematically illustrated in Figure 36b. Another possible adsorption configuration is shown in Figure 36a, which incorporates OH(a) bridge bonded to two Ni ridge atoms.

The same authors published a subsequent study of water adsorption on clean Ni(110) at temperatures less than 150 K.[156] H_2O adsorbs molecularly in this case, and strong evidence for the formation of water dimers via hydrogen bonding was observed for 0.2–0.5 ML coverages. In the dimer configuration, water molecules were proposed to bind to Ni atoms of adjacent {110} ridges through their oxygen lone-pair orbitals. The hydrogen bond (O ··· H—O) forms parallel to the surface across the trough and is azimuthally oriented in the {001} direction (perpendicular to the trough direction). See Figure 37 for an illustration of these ideas. Four H^+ ESDIAD beams were observed from this dimer-covered surface, forming a

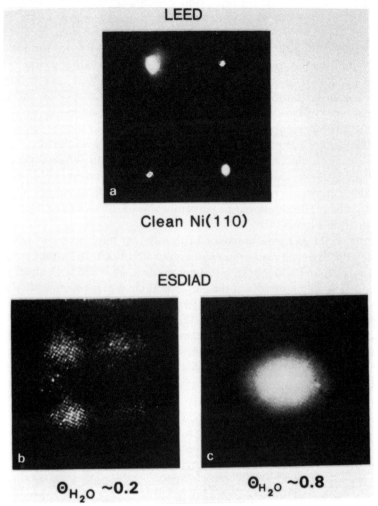

Figure 38 (a) *LEED pattern from clean* Ni(110). (b)–(c) H^+ *ESDIAD pattern from* H_2O/Ni(110) *bound in dimer form* (b) *and hydrogen-bonded layers* (c) (Reproduced with permission from *Surf. Sci.*, 1985, **157**, 29)

rectangular pattern in registry with {001} and {110} crystal azimuths. Angular resolved UPS and LEED data corroborated the ESDIAD results, leading to the proposed bonding scheme illustrated in Figure 37.

As H_2O coverage was increased from 0.5 to 1.0 ML, a c(2 × 2) LEED pattern formed. Proton ESDIAD emission simultaneously transformed from four beams to a single central feature, clearly seen in Figure 38. These observations led the authors to the conclusion that distorted hexagonal ring clusters of water form under high coverage conditions, producing a hydrogen-bonded, bilayer network. The outer layer has water molecules with O—H bonds pointing perpendicular to the surface, resulting in normally-directed H^+ ejection during ESD.

$NO/Pt(112)$

The chemisorption of the NO molecule on the stepped Pt(112) surface offers insight into the use of the ESDIAD method for observation of the hybridization of adsorbates on their adsorption sites. It is well known that NO forms a bridged bond with two Pt atoms at low coverages on various Pt surfaces.[157–160] This type of bonding at a step site would be expected to produce anisotropic frustrated translational motions, in which higher amplitude vibrations occur in the direction perpendicular to the plane of the adsorbed molecule, compared to the in-plane direction. These expectations follow from cluster calculations of Richardson and Bradshaw.[161] For the sp^2 hybridized molecule bridge-bonded to a step site, the direction of highest vibrational amplitude should be perpendicular to the step edge direction, as illustrated in Figure 39.[162]

Digital ESDIAD measurements of this system have been carried out,[162] yielding the expected result shown in Figure 40. In this study, the O^+ species was analysed and a highly elliptical pattern was observed with an ellipticity of about 2.3. The long axis of the ellipse is aligned perpendicular to the step edge. The angular width of the pattern in both directions increases as the temperature is increased, indicating that thermally excited molecular vibrations are being observed. Control experiments, in which O^+ ESDIAD patterns from sp hybridized CO species attached to a single Pt step site were studied, showed that anisotropic image effects on O^+ ions could not produce the elliptical O^+ patterns observed from NO.[163] Thus, ESDIAD was able to elucidate that NO is bridge bonded to step Pt sites.

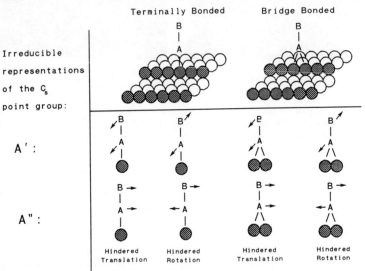

Figure 39 *Schematic illustrating the normal modes of vibration of terminally and bridge-bonded diatomic molecules. Arrows represent the directions of nuclear motion*
(Reproduced with permission from *J. Chem. Phys.*, 1990, **92**, 2208)

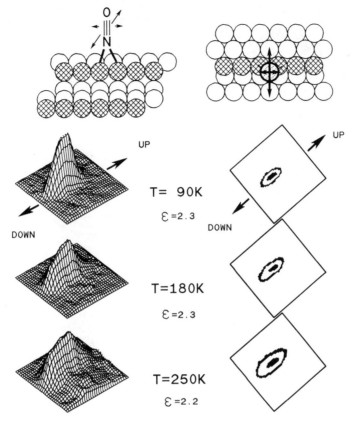

Figure 40 O^+ *ESDIAD pattern from* NO/Pt(112) *vs. temperature for* 0.20 NO *coverage with respect to saturation coverage. Contour lines on the right are at peak maximum and half-maximum values*
(Reproduced with permission from *J. Chem. Phys.*, 1990, **92**, 2208)

$NH_3/Si(100)$

One system of great technological importance is the process of silicon nitride growth via decomposition of ammonia on clean silicon. Avouris and colleagues[164–166] studied the $NH_3/Si(100)$–(2×1) system, reporting that dangling bonds of the Si surface were sufficiently reactive to promote ammonia dissociation at 90 K. H atoms so produced subsequently occupy dangling bond sites, thus passivating surface reactivity. Silicon nitride growth was thus determined to be a self-limiting process; surface passivation following dissociative adsorption of ammonia quenches substrate chemical activity, with further NH_3 exposure resulting in only molecular adsorption. Heating to 700–900 K thermally desorbs H from surface sites, allowing more ammonia decomposition and Si_3N_4 growth.

To circumvent high temperatures, which could be detrimental to microelectronics production, it was found that exposure of the $NH_3/Si(100)$

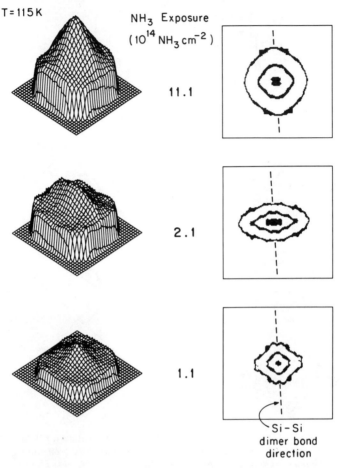

Figure 41 H^+ *ESDIAD pattern at 115 K following various NH_3 exposures on Si(100). Pronounced ellipticity of the patterns is observed from NH_2(a) at intermediate coverages, with the major axis of the elliptical H^+ pattern perpendicular to Si—Si dimer bond directions. Second and successive NH_3 layers adsorb randomly, producing a broad normally directed H^+ beam*
(Reproduced with permission from *Surf. Sci.*, 1989, **218**, 75)

system to an electron beam promoted silicon nitride growth even at 90 K. This was attributed to ESD of H from Si dangling bond sites, which were then reactivated for ammonia decomposition. Thus ESD processes were selectively used to enhance growth rates of a technologically important material without the need for thermal energy. This study also emphasized that the rate limiting step for silicon nitride formation was not thermally activated bulk/surface diffusion, but the desorption of adsorbed hydrogen, which could be accomplished efficiently via ESD without unwanted side effects of high temperatures required to thermally desorb H(a).

Johnson *et al.*[167] published an ESDIAD study involving the adsorption of ammonia on stepped Si(100)–(2 × 1) surfaces. For low NH_3 exposures, they

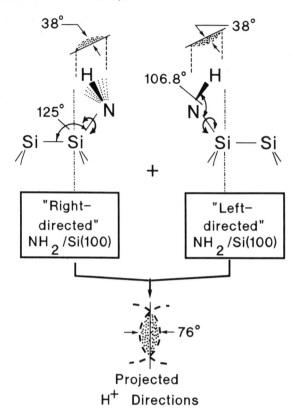

Figure 42 *Torsional motions of* $NH_2(a)$ *about the Si—N bond axis, resulting in an* H^+ *ESDIAD pattern of elliptical cross-section. Two different geometrical configurations are thought to contribute to the observed* H^+ *angular distribution* (Reproduced with permission from *Surf. Sci.*, 1989, **218**, 75)

observed H^+ ESDIAD images in the form of a weak halo, oriented in such a manner that ion emission was predominantly along directions perpendicular to Si—Si dimer bonds. The authors postulated the dissociative chemisorption of ammonia at low coverage, resulting in adsorbed NH_x and H species. These ideas were consistent with those of Avouris and his co-workers.[164-166]

Two subsequent studies[168,169] utilized various techniques, including TPD and digital ESDIAD, to investigate this system in greater detail. Dissociative adsorption of ammonia was observed to occur predominantly at 120 K through the reaction channel $NH_3(g) \rightarrow NH_2(a) + H(a)$. This was verified by an isotopic mixing study involving pre-adsorbed deuterium, which acted to pre-passivate some Si dangling bonds. It was observed that upon heating near 700 K, thermal desorption of ammonia and its deuterated analogues commenced via recombination reactions on the surface. The predominant isotopically labelled species to desorb was NH_2D, implying that NH_2 is the major nitrogen containing fragment at the surface following NH_3 dissociative adsorption.

H^+ ESDIAD patterns obtained following ammonia adsorption at low coverages on clean silicon showed a broad plateau topped by a 'fin-shaped' elliptical feature as seen in Figure 41. The major axis of this elliptical H^+ pattern was oriented perpendicular to Si—Si dimer bonds on the surface. A geometrical model was proposed, consistent with ammonia uptake measurements, to explain these ESDIAD results. It was postulated that upon dissociation, one amino species bonds to a single dimer pair. Estimating bond angles and amplitudes of torsional vibrations, the authors argued that H^+ ejected from NH_2(a) species by ESD should sweep out an elliptical pattern, subtending an angular width of $\sim 76°$ at the minor axis, as illustrated in Figure 42. Thus an intuitively appealing model was developed which qualitatively explained the experimental findings.

Higher NH_3 exposures resulted in multilayer molecular adsorption. H^+ ESDIAD patterns from such overlayers showed a broad central feature oriented normal to the surface as shown in the top portion of Figure 41. It was proposed that disordered overlayers resulting in random ammonia accommodation were responsible for these observations.

$NH_3/Ni(110)$

The adsorption of NH_3 by the Ni(110) surface has been studied by the ESDIAD method[170] with the purpose of observing the chemisorption interaction of a three-fold symmetric molecule with a two-fold symmetric surface. It was found that NH_3 is adsorbed as an undissociated molecule and that the adsorption occurs on atop sites of the Ni(110) surface. These sites permit the adsorbed NH_3 molecule to exhibit rotational symmetry about its three-fold axis, normal to the macroscopic crystal plane. Various site locations have been considered as shown in Figure 43, and the observed H^+ halo pattern is only consistent with occupancy of the atop Ni sites on Ni(110). In particular, two-beam and four-beam patterns from NH_3 molecules associated with (111) hole sites involving both atop and valley Ni atoms are excluded by the observation of the halo H^+ pattern.[170] The results reported above for NH_3 on Ni(110) were obtained by the photographic ESDIAD method; a repetition of this work was carried out with the digital ESDIAD technique,[9] yielding the same basic results, and a digital image of the H^+ halo has been obtained.

It was found that NH_3(a) could be converted to NH_2 species by electron bombardment, and that the NH_2 species produced a two-beam H^+ ESDIAD pattern. The molecular plane of the NH_2 species is accurately perpendicular to the two-fold axis of Ni(110) and as a result, NH_2(a) produces two H^+ beams in {001} equivalent directions.[171] Since NH_2(a) thermally decomposes to produce both H_2(g) and N_2(g), the use of TPD to study the production of these species is informative. Figure 44 shows that two types of deuterium may be detected by TPD following various degrees of ESD damage to an ND_3 layer on Ni(110). Curve 1 shows that almost no ND_3 dissociation occurs without ESD. As the bombardment fluence increases, prominent β-

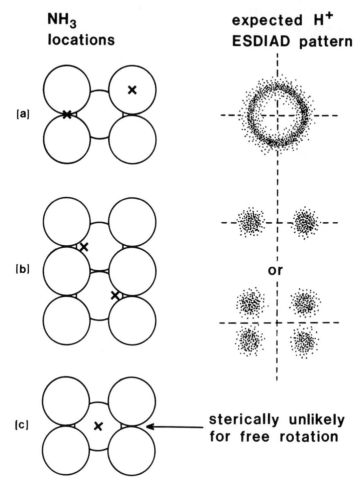

Figure 43 *Possible NH_3 chemisorption sites on Ni(110) and associated H^+ ESDIAD pattern which might be expected*
(Reproduced with permission from *Chem. Phys. Lett.*, 1984, **106**, 477)

and γ-D_2 thermal desorption states are produced. The β-D_2 state originates from D atoms from the ESD decomposition of ND_3; the γ-D_2 state originates from ND_2(a) species (produced from ND_3 upon electron bombardment) which thermally decompose, yielding D_2 at a temperature above the desorption temperature of adsorbed D on Ni(110). The ratio of D in these two states ($\sim 2.3:1$) is indicative that the ND_2(a) species is the major product of ESD at the levels of electron bombardment employed here. The cross-section for this ESD-induced surface decomposition involving adsorbed ammonia is very high, about 1.6×10^{-16} cm^2, for 55 eV electrons.

This study, in which chemisorbed NH_2 species are made from chemisorbed NH_3 species by ESD has been suggested as a method for preparing artificial layers of catalytically important surface species.[171] Studies of high

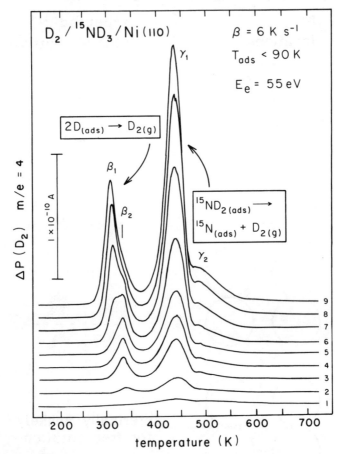

Figure 44 *Thermal desorption spectra of D_2 from Ni(110) after bombardment of $^{15}ND_3$(a) by 55 eV electrons. Curves 1–9 were recorded after exposing the system to 0.00, 0.30, 0.44, 0.95, 1.38, 2.33, 3.68, 5.52, and 7.41 × 10^{19} electrons m^{-2}, respectively*
(Reproduced with permission from *Surf. Sci.*, 1985, **154**, 139)

surface concentrations of these artificially produced species can lead to important insight into their involvement in real catalytic processes, where their surface concentrations are likely to be very low, as in the synthesis of ammonia from N_2 and H_2.

$PF_3/Ni(111)$

ESDIAD proved to be an incisive experimental tool during the study of another system, PF_3/Ni(111), as reported by Alvey et al.[6,7] At low coverages (0.04 ML), six-beam F^+ ESDIAD patterns were observed at 85 K, broadening into a halo at higher temperatures. At coverages near 0.25 ML, six well resolved F^+ beams were also produced by ESD, as seen in Figure 45a. In

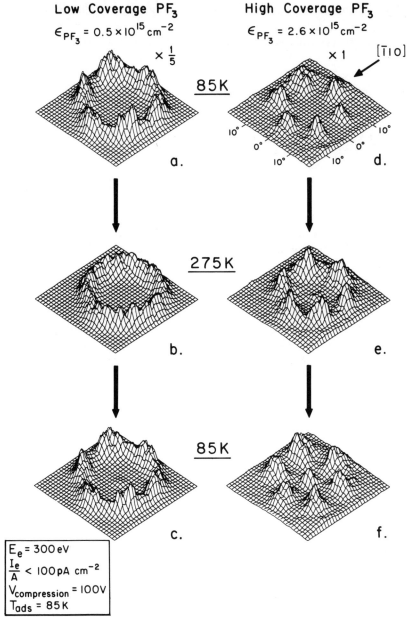

Figure 45 F^+ *ESDIAD pattern from low (left) and high (right)* PF_3 *coverages on Ni(111) and their temperature dependence. Estimated* PF_3 *coverages were 0.04 and 0.25 ML, respectively. A compression voltage (100 V) was used to draw positive ions to the detector, distorting polar but not azimuthal angular distributions*
(Reproduced with permission from *J. Chem. Phys.*, 1987, **87**, 7221)

addition, LEED data collected at this coverage showed that PF_3 formed a $p(2 \times 2)$ overlayer.

These observations led the authors to propose a geometrical model for the bonding of PF_3 to Ni(111) at 85 K. Molecular adsorption at atop sites through the lone electron pair of phosphorus was envisioned, with P—F bonds directed toward nearest-neighbour Ni atoms giving F^+ beams along {110} equivalent azimuths. There are two such orientations possible, explaining the observed six-fold symmetric F^+ ESDIAD patterns.

F^+ ESDIAD images were also collected at a number of temperatures between 85 K and 275 K. At low PF_3 coverages, the six-beam F^+ pattern observed at 85 K broadens into a halo pattern, indicating that PF_3 undergoes thermally accessible rotations about the Ni—P axis. However, F^+ beams from the high coverage system maintained six-fold symmetry even at 275 K, although they were slightly broader than at 85 K. This was taken as evidence for intermolecular interactions at high coverage which increase the energy barrier for rotation. These effects are evident in Figure 45.

The low coverage PF_3/Ni(111) case was modelled as a two-dimensional, quantum-mechanical, hindered rotor. Computer simulations were performed, using a superposition of ninety-nine rotational wavefunctions, yielding a value of 80 ± 20 cm^{-1} for the energy barrier to rotation which best reproduced the temperature dependence of the experimentally observed F^+ ESDIAD patterns. Thus, ESDIAD not only clarified the bonding site symmetry of PF_3 on Ni(111), but provided direct information about surface dynamics and interadsorbate interactions. The proposed bonding of PF_3 to Ni(111) was consistent with TPD and UPS studies by Nitschke et al.[172]

Another very interesting result of this study was the observation of electron induced surface chemistry. Alvey et al.[6] observed that prolonged exposure of $p(2 \times 2)$ PF_3 overlayers to an electron beam significantly altered the F^+ ESDIAD patterns. At 85 K, the original six beams, along {110} equivalent directions, actually rotated 30° about the substrate normal following bombardment by 3×10^{15} electrons cm^{-2}. In addition, an intense central F^+ beam simultaneously formed, as seen in Figure 46.

These observations were interpreted as reflecting the conversion of PF_3, via F^+ ESD, to adsorbed PF_2 and PF species. As shown in Figure 47, PF_2 fragments were assumed to bond at bridging sites on Ni(111), giving two azimuthally directed F^+ ESDIAD beams. There are three equivalent orientations of this adsorption geometry, accounting for the rotated six-beam F^+ pattern. PF species, on the other hand, were assigned to bond at three-fold hollow sites, with the P—F axis perpendicular to the surface. This configuration is responsible for the central ESDIAD beam. These species and site assignments have been verified by extended Huckel calculations of Chan and Hoffmann.[173]

$O_2/Ag(110)$

In 1985, Bange and co-workers reported ESDIAD results from the O_2/Ag(110) system which provided much needed insight into the bonding

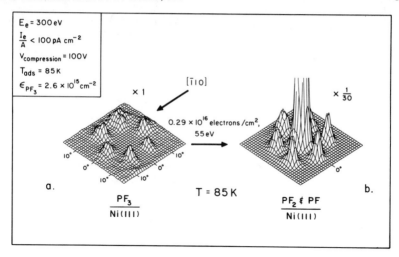

Figure 46 F^+ *ESDIAD pattern from* 0.25 ML PF_3/Ni(111) *vs. electron beam exposure.* PF_3 conversion to PF_x species is readily observable
(Reproduced with permission from *J. Am. Chem. Soc.*, 1988, **110**, 1782)

geometry of O(a) and O_2(a) at the surface.[174] Molecular O_2 adsorption was accomplished with the crystal at 80 K. An important point stressed throughout this paper was that the O^+ signal rose monotonically with the amount of oxygen on the surface at low coverage (< 0.3 ML). This presumably indicated that O^+ ESD originates from the majority of surface bound species and did not result mainly from defect or step sites; however, this interpretation may require revision.[174]

O^+ ESDIAD patterns from 0.33 ML O_2/Ag(110) at 80 K show two distinct beams oriented along equivalent {001} crystal azimuths. This direction is perpendicular to that of Ag atom rows on the (110) surface. The temperature dependence of these beams was investigated by heating the surface and then cooling back to 80 K for data collection.

It was observed that for the range 80–500 K, O^+ ESD beams always emerged along equivalent {001} azimuths; however the polar angle of desorption was not constant with temperature. Figure 48 shows that from 80–150 K, each O^+ beam emerged at 20° off normal. At 155 K the beams converged to make a polar angle of only 6°, finally diverging to stabilize at 12° off normal at 250–500 K.

These observations were rationalized by consulting the literature about O_2 thermal behaviour on Ag. The authors reported that molecular oxygen is stable from 80–150 K on Ag(110). O_2 thermal desorption and dissociation to form O(a) occurs near 175 K, and at 250 K atomic oxygen is mobile enough to form well ordered LEED patterns which are then stable up to 700 K. The observed changes in O^+ ESDIAD polar angles within these temperature regions were thus correlated with previously determined species identity, providing a self-consistent picture of the O_2/Ag(110) system.

A more recent report on O/Ag(110) surface chemistry was provided by

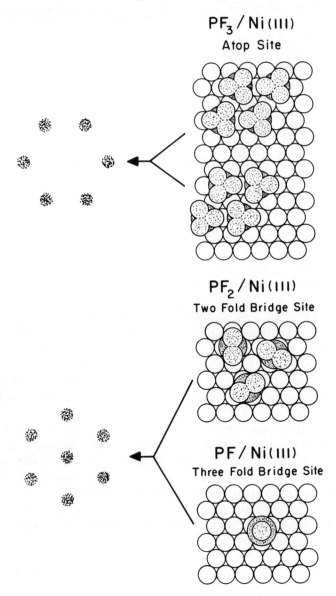

Figure 47 *Summary of PF_3 and PF_x surface structures and resultant F^+ ESDIAD patterns*
(Reproduced with permission from *J. Am. Chem. Soc.*, 1988, **110**, 1782)

Corallo *et al.*[175] Utilizing TOF methods the authors were able to characterize the kinetic energy distributions and identities of positive ions ejected during ESD. Following exposure to O_2 at ~ 300 K, forming O(a) on the clean Ag surface, electron bombardment liberated predominantly OH^+ and H^+, with some O^+ also present. The OH^+ species desorbed with a most

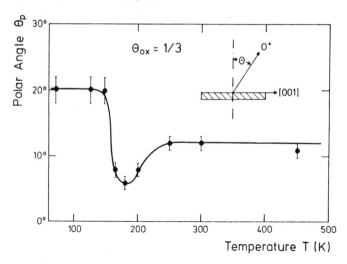

Figure 48 *Polar angle of O^+ ESD along {001} equivalent azimuths as a function of temperature to which the 0.33 ML O_2/Ag(111) system has been heated. Measurements were performed at 80 K under field free conditions* (Reproduced with permission from *Chem. Phys. Lett.*, 1985, **113**, 56)

probable kinetic energy of 4.7 eV, whereas H^+ carried about 3 eV of translational energy, as seen in Figure 49.

The observation of hydrogen (OH^+ and H^+) during ESD was proposed to result from H originally dissolved in the bulk Ag crystal. Repeated annealing of the substrate to drive hydrogen out was effective at lowering, but not extinguishing, subsequent OH^+ and H^+ ESD signals. This observation, combined with good experimental vacuum conditions and the known solubility of H in Ag seemed to support the postulate about bulk-trapped hydrogen.

The $OH^+:O^+$ ESD signal ratio was seen to rise with increased oxygen coverage, indicating that either; (i) OH^+ and O^+ originate from different sites, with those responsible for O^+ desorption filling first upon oxygen exposure, resulting in eventual saturation of O^+ yield, or (ii) an increased oxygen presence drives hydrogen diffusion, drawing more H from the bulk into the near surface region, resulting in an increased OH^+ yield.

Another important observation by these authors was that ion signals were roughly constant for incident electron energies of 100–300 eV and 400–600 eV. However, in the intermediate electron energy region, 300–400 eV, noticeable ion yield fluctuations occurred, indicative of new ESD channels being sampled. Silver (3d) core level excitations occur within this energy range, possibly implying that interatomic Auger processes contribute to ESD in this system.

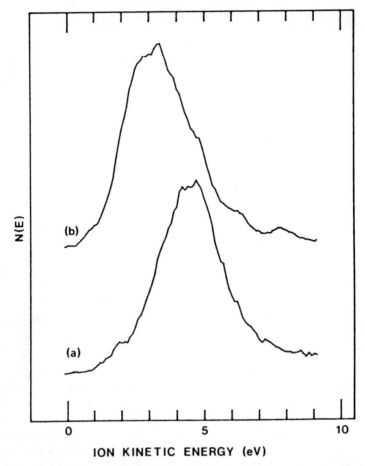

Figure 49 *Time-gated kinetic energy distributions for* (a) OH^+ *and* (b) H^+ *ions liberated from* O/Ag(110) *by* 220 eV *electron impact. Hydrogen was proposed to originate from the bulk of the silver crystal*
(Reproduced with permission from *Surf. Interface Anal.*, 1988, **12**, 185)

Pyridine/Ir(111)

Mack et al.[176] investigated the adsorption of pyridine on the (111) plane of iridium with angular resolved UPS, ESDIAD, LEED, and TPD. Room temperature exposure of Ir(111) to forty Langmuirs (40 μTorr·s) of pyridine vapour produced a $(2\sqrt{3} \times 2\sqrt{3})R30°$ overlayer structure. There are three such domains possible on the (111) surface. Analysis of UPS spectra recorded from these ordered overlayers led the authors to conclude that pyridine bonds to the surface through its nitrogen lone pair orbitals, with the ring plane oriented nearly perpendicular to the surface. This slightly inclined adsorption configuration distorts the molecular C_{2v} symmetry toward C_s character, an effect observable in polarized UPS spectra.

H^+ ESDIAD patterns from the same system strengthen the above conclusions. Figure 50 shows contour plots of the H^+ emission intensity as a

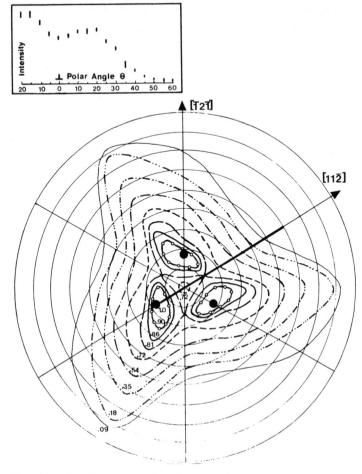

Figure 50 H^+ *ESDIAD contour plots from pyridine on* Ir(111). *Black dots mark locations of maximum signal intensity. Polar angle with respect to the surface normal increases in ten degree increments outward from the centre* (Reproduced with permission from *Surf. Sci.*, 1985, **159**, 265)

function of polar angle. It is clearly evident that proton ejection is concentrated in sharp cones about 20° off normal, corresponding to the tilt plane of the pyridine ring and to the idea that the C—H bond farthest from the N atom in the pyridine ring is being imaged by ESDIAD. The three H^+ lobes result from ESD from each of the three $(2\sqrt{3} \times 2\sqrt{3})R30°$ overlayer domains. Although the absolute pyridine coverage could not be determined, this study clearly elucidated the bonding geometry in these ordered overlayers.

6 Conclusions

In this review we have attempted to illustrate the wide range of theory and experiment which has been focused on electron stimulated desorption during the last thirty years. It has been shown that many models for ESD exist, and that these mechanisms are prime examples of quantum physics at work in the near-surface region.

This chapter also illustrates the wide range of chemical information which can be derived from ESD studies. Primary information about the bonding geometry of surface species can be obtained by observation of ESDIAD patterns from adsorbates on single crystal surfaces. In many cases, the ESDIAD patterns are consistent with known modes of surface bonding, often analogous to bonding in chemical compounds. The observation of unusual surface bonding at defect sites has been demonstrated using stepped single crystal surfaces. Such studies offer the promise of probing defect sites of crystal surfaces where enhanced chemical reactivity is often possible.

Perhaps the most exciting modern development involving ESDIAD has been its application to the study of the thermal dynamics of adsorbed species, where the low frequency vibrational amplitude (and anisotropy) of adsorbates may be monitored. These vibrations may probe the entry channels of surface reactions, and the observation of motions of this type may relate to directional preferences in surface chemical processes such as surface diffusion and bimolecular surface reaction.

In addition to the use of ESDIAD as an investigative tool for observing the behaviour of adsorbed species, ESD may be employed to produce unusual surface species. Thus, the systematic production of high surface coverages of 'adsorbed radical' species such as $NH_2(a)$, $PF_2(a)$, or $PF(a)$ is a route to the observation of synthetically produced catalytic intermediate species at high coverages; these same species are present only at low coverages under actual catalytic synthesis conditions and are generally unobservable.

The future holds great promise for the use of ESD methods to probe surface behaviour. Advances will come from the application of spectroscopic methods to probe the dynamical states of desorbing species. In addition, the use of ESDIAD as a probe of the fine details of chemical reactivity and bonding at surfaces holds great promise for surface chemistry studies on metals and semiconductors. In combination with surface spectroscopies and atomic resolution imaging methods [such as scanning tunnelling microscopy (STM)], the ESDIAD method offers great opportunities at present for understanding surface chemistry at an ultimate level of atomic detail.

It is hoped that this review will stimulate future workers in the field of surface science to consider desorption induced by electronic transitions as a fruitful field of endeavour.

Acknowledgements. The authors express their sincere gratitude to Professor D. E. Ramaker and to Andràs Szabó for critically reading portions of this manuscript. We also appreciate the unwavering secretarial efforts of Ms. M. Augenstein and Ms. M. Merenick. This work has been supported by AFOSR under contract 82-0133.

References

1. R. D. Ramsier, and J. T. Yates, Jr., *Surf. Sci. Rep.*, 1991, **12**, 243.
2. J. J. Czyzewski, T. E. Madey, and J. T. Yates, Jr., *Phys. Rev. Lett.*, 1974, **32**, 777.
3. M. A. Henderson, A. Szabó, and J. T. Yates, Jr., *J. Chem. Phys.*, 1989, **91**, 7255.
4. M. A. Henderson, A. Szabó, and J. T. Yates, Jr., *J. Chem. Phys.*, 1989, **91**, 7245.
5. M. A. Henderson, A. Szabó, and J. T. Yates, Jr., *Chem. Phys. Lett.*, 1990, **168**, 51.
6. M. D. Alvey and J. T. Yates, Jr., *J. Am. Chem. Soc.*, 1988, **110**, 1782.
7. M. D. Alvey, J. T. Yates, Jr., and K. J. Uram, *J. Chem. Phys.*, 1987, **87**, 7221.
8. A. Szabó, M. Kiskinova, and J. T. Yates, Jr., *Surf. Sci.*, 1988, **205**, 207.
9. M. J. Dresser, M. D. Alvey, and J. T. Yates, Jr., *Surf. Sci.*, 1986, **169**, 91.
10. H. Niehus and B. Krahl–Urban, *Rev. Sci. Instrum.*, 1981, **52**, 56.
11. D. R. Sandstrom, M. J. Dresser, and W. D. Dong, *Phys. Rev. B*, 1986, **34**, 5125.
12. M. Kiskinova, A. Szabó, A.-M. Lanzillotto, and J. T. Yates, Jr., *Surf. Sci.*, 1988, **202**, L559.
13. C. E. Young, E. L. Schwietzer, M. J. Pellin, D. M. Gruen, Z. Hurych, P. Soukiassian, M. H. Bakshi, and A. S. Bommannavar, 'Desorption Induced by Electronic Transitions', DIET III, ed. R. H. Stulen and M. L. Knotek, Springer Series in Surface Sciences, Vol. 13, Springer–Verlag, Berlin, 1988, p. 94.
14. S. A. Joyce, A. L. Johnson, and T. E. Madey, *J. Vac. Sci. Technol.*, 1989, **A7**, 2221.
15. H. S. W. Massey, 'Atomic and Molecular Collisions', Halsted Press, New York, 1979.
16. D. Menzel and R. Gomer, *J. Chem. Phys.*, 1964, **40**, 1164.
17. D. Menzel and R. Gomer, *J. Chem. Phys.*, 1964, **41**, 3311.
18. P. A. Redhead, *Can. J. Phys.*, 1964, **42**, 886.
19. Y. Ishikawa, *Rev. Phys. Chem. Jpn.*, 1942, **16**, 83, 117.
20. Y. Ishikawa, *Proc. Imp. Acad. (Tokyo)*, 1943, **19**, 380, 385.
21. Y. Ishikawa, *Proc. Imp. Acad. (Tokyo)*, 1942, **18**, 246, 390.
22. 'Electron-Molecule Collisions', ed. I. Shimamura and K. Takayanagi, Plenum Press, New York, 1984.
23. P. R. Antoniewicz, *Phys. Rev. B*, 1980, **21**, 3811.
24. E. R. Moog, J. Unguris, and M. B. Webb, *Surf. Sci.*, 1983, **134**, 849.
25. Q.-J. Zhang, R. Gomer, and D. R. Bowman, *Surf. Sci.*, 1983, **129**, 535.
26. Q.-J. Zhang and R. Gomer, *Surf. Sci.*, 1981, **109**, 567.
27. R. E. Walkup, Ph. Avouris, N. D. Lang, and R. Kawai, *Phys. Rev. Lett.*, 1989, **63**, 1972.
28. M. L. Knotek and P. J. Feibelman, *Phys. Rev. Lett.*, 1978, **40**, 964.
29. T. A. Carlson, 'Desorption Induced by Electronic Transitions', DIET I, ed. N. H. Tolk, M. M. Traum, J. C. Tully, and T. E. Madey, Springer Series in Chemical Physics, Vol. 24, Springer–Verlag, Berlin, 1983, p. 169.
30. W. Brenig, *Z. Phys.*, 1976, **B23**, 361.
31. W. Brenig, 'Desorption Induced by Electronic Transitions', DIET I, ed. N. H. Tolk, M. M. Traum, J. C. Tully, and T. E. Madey, Springer Series in Chemical Physics, Vol. 24, Springer–Verlag, Berlin, 1983, p. 90.
32. W. Brenig in 'Proc. 7th Int. Vac. Cong. and 3rd Int. Conf. Solid Surf'., ed., R.

Dobrozemsky, F. Rudenauer, F. P. Viehböck, and A. Breth, Vienna 1977, p. 719.
33. B. Bell, M. H. Cohen, R. Gomer, and A. Madhukar, *Surf. Sci.*, 1976, **61**, 656.
34. W. Brenig, *Surf. Sci.*, 1976, **61**, 659.
35. W. L. Clinton, *Surf. Sci.*, 1978, **75**, L796.
36. W. Brenig, *Surf. Sci.*, 1978, **75**, L800.
37. W. Brenig, *J. Phys. Soc. Jpn.*, 1982, **51**, 1914.
38. M. Tsukada, *J. Phys. Soc. Jpn.*, 1982, **51**, 2927.
39. H. Ueba, *Phys. Rev. B*, 1983, **27**, 7393.
40. W. L. Clinton and R. E. Julita, *Phys. Rev. B*, 1985, **31**, 6441.
41. K. Gottfried, 'Quantum Mechanics', W. A. Benjamin, New York, 1966.
42. R. Azria, L. Parenteau, and L. Sanche, *Phys. Rev. Lett.*, 1987, **59**, 638.
43. L. Sanche and L. Parenteau, *Phys. Rev. Lett.*, 1987, **59**, 136.
44. L. Sanche, *Phys. Rev. Lett.*, 1984, **53**, 1638.
45. L. Sanche, L. Parenteau, and P. Cloutier, *J. Chem. Phys.*, 1989, **91**, 2664.
46. R. Franchy and D. Menzel, *Phys. Rev. Lett.*, 1979, **43**, 865.
47. J. Hubbard, *Proc. R. Soc. (London) A*, 1963, **276**, 238.
48. J. Hubbard, *Proc. R. Soc. (London) A*, 1964, **277**, 237.
49. J. Hubbard, *Proc. R. Soc. (London) A*, 1964, **281**, 401.
50. M. Cini, *Solid State Commun.*, 1976, **20**, 605.
51. M. Cini, *Surf. Sci.*, 1979, **87**, 483.
52. M. Cini, *Solid State Commun.*, 1977, **24**, 681.
53. M. Cini, *Phys. Rev. B*, 1978, **17**, 2788.
54. G. A. Sawatzky, *Phys. Rev. Lett.*, 1977, **39**, 504.
55. G. A. Sawatsky and A. Lenselink, *Phys. Rev. B*, 1980, **21**, 1790.
56. B. I. Dunlap, F. L. Hutson, and D. E. Ramaker, *J. Vac. Sci. Technol.*, 1981, **18**, 556.
57. C. Kittel, 'Quantum Theory of Solids', Wiley, New York, 1964.
58. D. E. Ramaker, private communication.
59. P. J. Feibelman, *Surf. Sci.*, 1981, **102**, L51.
60. D. R. Jennison, J. A. Kelber, and R. R. Rye, *Phys. Rev. B*, 1982, **25**, 1384.
61. H. H. Madden, D. R. Jennison, M. M. Traum, G. Margaritondo, and N. G. Stoffel, *Phys. Rev. B*, 1982, **26**, 896.
62. D. R. Jennison and D. Emin, *Phys. Rev. Lett.*, 1983, **51**, 1390.
63. D. R. Jennison and D. Emin, *J. Vac. Sci. Technol.*, 1983, **A1**, 1154.
64. D. E. Ramaker, C. T. White, and J. S. Murday, *J. Vac. Sci. Technol.*, 1981, **18**, 748.
65. D. E. Ramaker, 'Desorption Induced by Electronic Transitions', DIET I, ed. N. H. Tolk, M. M. Traum, J. C. Tully, and T. E. Madey, Springer Series in Chemical Physics, Vol. 24, Springer–Verlag, Berlin, 1983, p. 70.
66. D. E. Ramaker, 'Desorption Induced by Electronic Transitions', DIET II, ed. W. Brenig and D. Menzel, Springer Series in Surface Sciences, Vol. 4, Springer–Verlag, Berlin, 1985, p. 10.
67. D. R. Jennison, E. B. Stechel, and J. A. Kelber, 'Desorption Induced by Electronic Transitions', DIET II, ed. W. Brenig and D. Menzel, Springer Series in Surface Sciences, Vol. 4, Springer–Verlag, Berlin, 1985, p. 24.
68. D. E. Ramaker, *Phys. Rev. B*, 1980, **21**, 4608.
69. D. R. Jennison, J. A. Kelber, and R. R. Rye, *Chem. Phys. Lett.*, 1981, **77**, 604.
70. Ph. Avouris, R. Kawai, N. D. Lang, and D. M. Newns, *Phys. Rev. Lett.*, 1987, **59**, 2215.

71. Ph. Avouris, R. Kawai, N. D. Lang, and D. M. Newns, *J. Chem. Phys.*, 1988, **89**, 2388.
72. Ph. Avouris, R. Walkup, R. Kawai, D. M. Newns, and N. D. Lang, 'Desorption Induced by Electronic Transitions', DIET III, ed. R. H. Stulen and M. L. Knotek, Springer Series in Surface Sciences, Vol. 13, Springer–Verlag, Berlin, 1988, p. 144.
73. P. Nordlander and J. C. Tully, *Phys. Rev. Lett.*, 1988, **61**, 990.
74. P. Nordlander and J. C. Tully, *Surf. Sci.*, 1989, **211/212**, 207.
75. P. D. Johnson, X. Pan, J. Tranquada, S. L. Hulbert, and E. Johnson, 'Desorption Induced by Electronic Transitions', DIET III, ed. R. H. Stulen and M. L. Knotek, Springer Series in Surface Sciences, Vol. 13, Springer–Verlag, Berlin, 1988, p. 73.
76. P. D. Johnson, A. J. Viescas, P. Nordlander, and J. C. Tully, *Phys. Rev. Lett.*, 1990, **64**, 942.
77. A. Hoffman, X. Guo, J. T. Yates, Jr., J. W. Gadzuk, and C. W. Clark, *J. Chem. Phys.*, 1989, **90**, 5793.
78. J. W. Gadzuk and C. W. Clark, *J. Chem. Phys.*, 1989, **91**, 3174.
79. D. Lichtman, R. B. McQuistan, and T. R. Kirst, *Surf. Sci.*, 1966, **5**, 120.
80. M. Nishijima and F. M. Propst, *J. Vac. Sci. Technol.*, 1970, **7**, 410.
81. G. D. Rork and R. E. Consoliver, *Surf. Sci.*, 1968, **10**, 291.
82. D. R. Sandstrom, J. H. Leck, and E. E. Donaldson, *J. Chem. Phys.*, 1968, **48**, 5683.
83. D. A. Degras and J. Lecante, *Nuovo Cimento Suppl.*, 1967, **5**, 598.
84. J. W. Coburn, *Surf. Sci.*, 1968, **11**, 61.
85. K. W. Ashcroft, J. H. Leck, D. R. Sandstrom, B. P. Stimpson, and E. M. Williams, *J. Phys. E: J. Sci. Instrum.*, 1972, **5**, 1106.
86. T. E. Madey and J. T. Yates, Jr., *J. Vac. Sci. Technol.*, 1971, **8**, 525.
87. D. M. Hanson, R. Stockbauer, and T. E. Madey, *Phys. Rev. B*, 1981, **24**, 5513.
88. R. Stockbauer, R. L. Kurtz, and T. E. Madey, 'Desorption Induced by Electronic Transitions', DIET III, ed. R. H. Stulen and M. L. Knotek, Springer Series in Surface Sciences, Vol. 13, Springer–Verlag, Berlin, 1988, p. 126.
89. D. E. Eastman, J. J. Donelon, N. C. Hien, and F. J. Himpsel, *Nucl. Instrum. Meth.*, 1980, **172**, 327.
90. L. A. Larson, F. Soria, and H. Poppa, *J. Vac. Sci. Technol.*, 1980, **17**, 1364.
91. M. M. Traum and D. P. Woodruff, *J. Vac. Sci. Technol.*, 1980, **17**, 1202.
92. K. Ueda and A. Takano, *Technol. Rep. Osaka Univ.*, 1988, **38**, 217.
93. A. M. Hudor, *Rev. Sci. Instrum.*, 1981, **52**, 819.
94. R. E. Gilbert, D. F. Cox, and G. B. Hoflund, *Rev. Sci. Instrum.*, 1982, **53**, 1281.
95. L. Z. Xiang and D. Lichtman, *Surf. Sci.*, 1983, **125**, 490.
96. J. L. Hock, J. H. Craig, Jr., and D. Lichtman, *Surf. Sci.*, 1979, **85**, L218.
97. I. G. Newsham and D. R. Sandstrom, *J. Vac. Sci. Technol.*, 1973, **10**, 39.
98. I. G. Newsham, J. V. Hogue, and D. R. Sandstrom, *J. Vac. Sci. Technol.*, 1972, **9**, 596.
99. G. Betz, E. Wolfrum, P. Wurz, K. Mader, B. Strehl, W. Husinsky, R. F. Haglund, Jr., and N. H. Tolk, 'Desorption Induced by Electronic Transitions', DIET III, ed. R. H. Stulen and M. L. Knotek, Springer Series in Surface Sciences, Vol. 13, Springer–Verlag, Berlin, 1988, p. 278.
100. N. H. Tolk, R. F. Haglund, Jr., M. H. Mendenhall, E. Taglauer, and N. G. Stoffel, 'Desorption Induced by Electronic Transitions', DIET II, ed. W. Brenig

and D. Menzel, Springer Series in Surface Sciences, Vol. 4, Springer–Verlag, Berlin, 1985, p. 152.
101. A. R. Burns, E. B. Stechel, and D. R. Jennison, *Phys. Rev. Lett.*, 1987, **58**, 250.
102. A. R. Burns, E. B. Stechel, and D. R. Jennison, 'Desorption Induced by Electronic Transitions', DIET III, ed. R. H. Stulen and M. L. Knotek, Springer Series in Surface Sciences, Vol. 13, Springer–Verlag, Berlin, 1988, p. 67.
103. A. R. Burns, D. R. Jennison, and E. B. Stechel, *J. Vac. Sci. Technol.*, 1987, **A5**, 671.
104. A. R. Burns, E. B. Stechel and D. R. Jennison, *J. Vac. Sci. Technol.*, 1988, **A6**, 895.
105. L. A. Petermann, *Nuovo Cimento Suppl.*, 1963, **1**, 601.
106. R. Stockbauer, *Nucl. Meth. Instrum. Phys. Res.*, 1984, **222**, 284.
107. T. E. Madey and R. Stockbauer in 'Solid State Physics: Surfaces', Methods of Experimental Physics Vol. 22, ed. R. L. Park and M. G. Lagally, Academic Press, Orlando, 1985, p. 465.
108. R. M. Wallace, P. A. Taylor, M. J. Dresser, W. J. Choyke, and J. T. Yates, Jr., *Rev. Sci. Instrum.*, 1991, **62**, 720.
109. M. Lampton and C. W. Carlson, *Rev. Sci. Instrum.*, 1979, **50**, 1093.
110. T. E. Madey, J. J. Czyzewski, and J. T. Yates, Jr., *Surf. Sci.*, 1975, **49**, 465.
111. J. I. Gersten, R. Janow, and N. Tzoar, *Phys. Rev. Lett.*, 1976, **36**, 610.
112. W. L. Clinton, *Phys. Rev. Lett.*, 1977, **39**, 965.
113. R. Janow and N. Tzoar, *Surf. Sci.*, 1977, **69**, 253.
114. T. E. Madey, J. J. Czyzewski, and J. T. Yates, Jr., *Surf Sci.*, 1976, **57**, 580.
115. R. Janow and N. Tzoar, *Surf. Sci.*, 1978, **75**, L766.
116. E. Bauer, *J. Electron. Spectrosc. Relat. Phenom.*, 1979, **15**, 119.
117. H. Niehus, Proc. 7th Int. Vac. Cong. & 3rd Int. Conf. Solid Surf., ed. R. Dobrozemsky, F. Rudenauer, F. P. Viehböck, and A. Breth, Vienna, 1977, p. 2051.
118. H. Niehus, *Vak. Tech.*, 1978, **27**, 136.
119. H. Niehus, *Surf. Sci.*, 1979, **87**, 561.
120. H. Niehus, *Surf. Sci.*, 1979, **80**, 245.
121. E. Preuss, *Surf. Sci.*, 1980, **94**, 249.
122. W. L. Clinton, *Surf. Sci.*, 1981, **112**, L791.
123. T. E. Madey in 'Inelastic Particle–Surface Collisions' Springer Series in Chem. Phys. Vol. 17, ed. E. Taglauer and W. Heiland, Springer–Verlag, Berlin 1981, p. 80.
124. Z. Misković, J. Vukanić, and T. E. Madey, *Surf. Sci.*, 1984, **141**, 285.
125. Z. Misković, J. Vukanić, and T. E. Madey, *Surf. Sci.*, 1986, **169**, 405.
126. W. L. Clinton, M. A. Esrick, and W. S. Sacks, *Phys. Rev. B*, 1985, **31**, 7550.
127. T. E. Madey, *Surf. Sci.*, 1979, **79**, 575.
128. W. Riedl and D. Menzel, *Surf. Sci.*, 1985, **163**, 39.
129. W. Riedl and D. Menzel, *Surf. Sci.*, 1989, **207**, 494.
130. M. Kiskinova, A. Szabó, and J. T. Yates, Jr., *Surf. Sci.*, 1988, **205**, 215.
131. M. Kiskinova, A. Szabó, and J. T. Yates, Jr., *Phys. Rev. Lett.*, 1988, **61**, 2875.
132. A. Szabó, M. Kiskinova, and J. T. Yates, Jr., *J. Chem. Phys.*, 1989, **90**, 4604.
133. M. Kiskinova, A. Szabó, and J. T. Yates, Jr., *J. Chem. Phys.*, 1988, **89**, 7599.
134. M. Kiskinova, A. Szabó, and J. T. Yates, Jr., *Surf. Sci.*, 1990, **226**, 237.
135. N. Avery, *J. Chem. Phys.*, 1981, **74**, 4202.
136. H. S. Taylor, *Proc. R. Soc. London, A.*, 1925, **108**, 105.

137. A. M. Lahee, J. P. Toennies, and Ch. Wöll, *Surf. Sci.*, 1986, **127**, 371.
138. R. Smoluchowski, *Phys. Rev. B*, 1971, **3**, 1215.
139. J. Tersoff and L. M. Falikov, *Phys. Rev. B*, 1981, **24**, 754.
140. M. D. Alvey, M. J. Dresser, and J. T. Yates, Jr., *Surf. Sci.*, 1986, **165**, 447.
141. M. J. Dresser, M. D. Alvey, and J. T. Yates, Jr., *J. Vac. Sci, Technol.*, 1986, **A4**, 1446.
142. R. J. Behm, G. Ertl, and V. Penka, *Surf. Sci.*, 1985, **160**, 387.
143. W. Riedl and D. Menzel, *Surf. Sci.*, 1985, **163**, 39.
144. D. A. Wesner, F. P. Coenen, and H. P. Bonzel, *Phys. Rev. B*, 1989, **39**, 10770.
145. N. D. Shinn and T. E. Madey, *Surf. Sci.*, 1987, **180**, 615.
146. N. D. Shinn and T. E. Madey, 'Desorption Induced by Electronic Transitions', DIET III, ed. R. H. Stulen and M. L. Knotek, Springer Series in Surface Sciences, Vol. 13, Springer–Verlag, Berlin, 1988, p. 217.
147. N. D. Shinn and T. E. Madey, *J. Vac. Sci. Technol.*, 1985, **A3**, 1673.
148. N. D. Shinn and T. E. Madey, *Phys. Rev. Lett.*, 1984, **53**, 2481.
149. N. D. Shinn and T. E. Madey, *J. Chem. Phys.*, 1985, **83**, 5928.
150. M. D. Alvey, K. W. Kolasinski, J. T. Yates, Jr., and M. Head-Gordon, *J. Chem. Phys.*, 1986, **85**, 6093.
151. M. J. Bozack, M. J. Dresser, W. J. Choyke, P. A. Taylor, and J. T. Yates, Jr., *Surf. Sci.*, 1987, **184**, L332.
152. A. L. Johnson, M. M. Walczak, and T. E. Madey, *Langmuir*, 1988, **4**, 277.
153. N. D. Shinn, J. F. Morar, and F. R. McFeely, *J. Vac. Sci. Technol.*, 1984, **A2**, 1593.
154. A.-M. Lanzillotto, M. J. Dresser, M. D. Alvey, and J. T. Yates, Jr., *J. Chem. Phys.*, 1988, **89**, 570.
155. C. Benndorf, C. Nöbl, and T. E. Madey, *Surf. Sci.*, 1984, **138**, 292.
156. C. Nöbl, C. Benndorf, and T. E. Madey, *Surf. Sci.*, 1985, **157**, 29.
157. B. E. Hayden, *Surf. Sci.*, 1983, **131**, 419.
158. J. L. Gland and B. A. Sexton, *Surf. Sci.*, 1980, **94**, 355.
159. H. Ibach and S. Lehwald, *Surf. Sci.*, 1978, **76**, 1.
160. R. L. Gorte and J. L. Gland, *Surf. Sci.*, 1981, **102**, 348.
161. N. V. Richardson and A. M. Bradshaw, *Surf. Sci.*, 1979, **88**, 255.
162. A. Szabó, M. A. Henderson, and J. T. Yates, Jr., *J. Chem. Phys.*, 1990, **92**, 2208.
163. W. L. Clinton, *Phys. Rev. Lett.*, 1977, **39**, 965.
164. F. Bozso and Ph. Avouris, *Phys. Rev. Lett.*, 1986, **57**, 1185.
165. Ph. Avouris, F. Bozso, and R. J. Hamers, *J. Vac. Sci. Technol*, 1987, **B5**, 1387.
166. F. Bozso and Ph. Avouris, *Phys. Rev. B*, 1988, **38**, 3937.
167. A. L. Johnson, M. M. Walczak, and T. E. Madey, *Langmuir*, 1988, **4**, 277.
168. M. J. Dresser, P. A. Taylor, R. M. Wallace, W. J. Choyke, and J. T. Yates, Jr., *Surf. Sci.*, 1989, **218**, 75.
169. P. A. Taylor, R. M. Wallace, W. J. Choyke, M. J. Dresser, and J. T. Yates, Jr., *Surf. Sci.*, 1989, **215**, L286.
170. C. Klauber, M. D. Alvey, and J. T. Yates, Jr., *Chem. Phys. Lett.*, 1984, **106**, 477.
171. C. Klauber, M. D. Alvey, and J. T. Yates, Jr., *Surf. Sci.*, 1985, **154**, 139.
172. F. Nitschke, G. Ertl, and J. Küppers, *J. Chem. Phys.*, 1981, **74**, 5911.
173. A. W. E. Chan and R. Hoffmann, *J. Chem. Phys.*, 1990, **92**, 699.
174. K. Bange, T. E. Madey, and J. K. Sass, *Chem. Phys. Lett.*, 1985, **113**, 56; T. E. Madey, private communication, 1991.

175. G. R. Corallo, G. B. Hoflund and R. A. Outlaw, *Surf. Interface Anal.*, 1988, **12**, 185.
176. J. U. Mack, E. Bertel, and F. P. Netzer, *Surf. Sci.*, 1985, **159**, 265.

CHAPTER 8
Photochemistry in the Adsorbed State

J. C. POLANYI and H. RIELEY

1 Introduction

It is a welcome sign of glasnost that a volume on gas–surface interactions should be included in a series devoted to advances in gas-phase photochemistry and kinetics. This is as it should be since the interest in molecular dynamics is common to both fields, and there will ultimately be much to learn from comparisons of the dynamics in the two different environments.

In what follows we outline the current state of development of the young field of adsorbate photochemistry. This is considered under four headings: *photodissociation* (PDIS); *photoreaction* (PRXN); *photodesorption* (PDES) and *photoejection* (PEJ). Given the limitations of space, the discussion will stress the first two of these processes as they involve the breaking and making of chemical bonds—the constituent elements of a chemical process. The typical dissociation energies of chemical bonds has necessitated the use of ultraviolet (UV) wavelengths to initiate photochemical processes in the gas-phase species. As we shall note, radiation extending over a wide range of wavelengths induces chemistry in adsorbates.

Historical Perspective

One of the earliest collections of experimental and theoretical work to address photoinduced phenomena at the gas–solid and liquid–solid interfaces can be found in a Faraday Discussion of the Chemical Society held in 1974 entitled, 'Photoeffects in Adsorbed Species'.[1] The section in this Discussion which focused on photochemistry contained mainly work which concentrated on processes at the liquid–solid interface, probed through the shift in UV absorption in organic dyes[2] suggesting charge transfer between the excited adsorbate and the substrate. At this time, although ultra-high vacuum (UHV) conditions were prevalent in studies of UV and X-ray photoelectron spectroscopy (UPS and XPS), they were little in evidence in the photochemical experiments, in which multilayers of adsorbate were, therefore, unavoidable. Theoretical studies were in the early stages of development, focusing on photoemission and photodesorption.[3] These early investigations suggested that chemistry at interfaces could be initiated by light although the involvement of gas-phase chemistry could not, at that date, be excluded.

The widespread availability of laser technology gave an important new impetus to the field, offering the possibility of laser micro-fabrication of solid state devices. Osgood, Ehrlich, and co-workers used ultraviolet radiation generated from either pulsed or continuous wave (CW) lasers to excite dissociative electronic states in volatile organometallic precursors at the gas–solid interface.[4-7] The result was photodeposition in the irradiated regions of the insulator or semi-conductor substrate. Evidence was presented for the occurrence of both photodissociation of gas-phase precursors in the vicinity of the surface and photodissociation at the surface.

Developments in the use of lasers for photodeposition, photoetching, photowriting, and photodoping have been comprehensively reviewed by Osgood.[8] A recent edition of Chemical Reviews: 'Materials for Microelectronics', is dedicated to this field.[9]

With this stimulus lasers began to be used to investigate gas–surface interactions. Initially, investigations centred on the effects of infrared (IR) radiation. Infrared laser-induced desorption of intact parent species was observed. This was consistent with IR absorption by vibrational modes in the adsorbate followed by the flow of energy into the adsorbate–surface bond. A review by Chuang covers research in this developing field.[10]

Other pioneering experiments noted the importance of heat, incident electrons, and light for decomposition on the surface,[11] and revealed surface enhancements of absorption cross-section in SERS (Surface Enhanced Raman Spectroscopy) studied in UHV,[12,13] as well as giving early evidence of the influence of photogenerated charge carriers from the bulk on the surface chemistry.[14,15]

Dynamical studies in UHV soon followed. Initial work on ices[16] and at sub-monolayer coverages on single crystals[17] was able to differentiate photodissociation in the adsorbed state from that in the gaseous state, as well as distinguishing photolytic from pyrolytic events.

Initially, dielectric substrates were used to minimize quenching of the electronically excited adsorbate. It was recognized, however, that the photodissociative event could be many orders of magnitude shorter than a typical electronically excited state lifetime ($\sim 10^{-14}$ s as compared with $\sim 10^{-8}$ s) and that, as a consequence, only the most rapid quenching could compete effectively with adsorbate photodissociation.

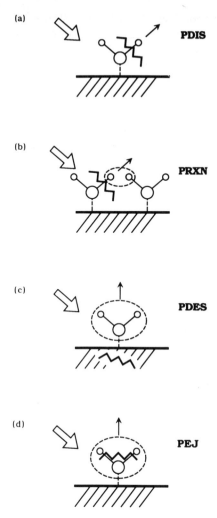

Figure 1 *A pictorial representation of the photoprocesses occurring at the gas–solid interface treated in this review.* (a) PDIS; *photodissociation of an adsorbed molecule,* (b) PRXN; *photoinduced reaction between adjacent adsorbate molecules,* (c) PDES; *substrate mediated photodesorption of intact adsorbate molecules,* (d) PEJ; *adsorbate mediated photoejection of intact adsorbate molecules*

The extension of the range of substrates from insulators to semiconductors and metals occurred over a short space of time, with the majority of studies employing UHV conditions and surface analytical techniques together with a continuous light source. A number of researchers employed CW lasers or arc lamp sources and observed photoreaction with the surface.[18–48] The use of pulsed lasers as the photolytic source made possible an increasing number of dynamical studies of adsorbate photochemistry.[17,49–76]

Knowledge of the dynamics aided substantially in the elucidation of varied photoinduced processes. Thus, observation of energetic photoproducts having non-Boltzmann energy distributions gave clear evidence of non-thermal photochemical pathways. Altered product energy distribution and angular distribution relative to that for the corresponding photodissociation in the gas-phase, together with variation in outcome depending on substrate and coverage, showed that the photoprocesses were occurring in the adsorbed state.[17,49]

Photofragment angular distributions, $P(\theta)$, taken together with translational energy distributions, $P(E_T)$, indicated alignment and orientation in the substrate for CH_3Br,[49,50] H_2S,[51] and HBr[77] on LiF(001). Dependence of yield on the wavelength and polarization of incident radiation has aided in the assignment of the mechanism underlying the photoprocess. The photochemistry of NO_2 on Pd(111)[72] and O_2 on Ag(110)[43] and photodesorption of HBr from LiF(001)[77] were shown in this way to be mediated by absorption of radiation by the substrates.

Surface diagnostics have provided essential complementary information on the products of photodissociation or photoreaction which remain at the surface.

Following this broad survey we now consider, in detail, some of the underlying processes.

2 Photochemical Processes

We begin by defining photodissociation, photoreaction, photodesorption, and photoejection in the adsorbed state. Schematics of these four processes, with acronyms PDIS, PRXN, PDES, and PEJ, are shown in Figure 1.

Photodissociation; PDIS

PDIS refers to photodissociation of an adsorbed molecule; see Figure 1a. This may occur directly or indirectly as noted, for example, in a recent review by White.[78] *Direct* absorption of a photon of sufficient energy results in Franck–Condon transition from the ground to an electronically excited repulsive or predissociative potential energy surface. If the excited state is repulsive, bond fission is prompt ($\sim 10^{-14}$–10^{-13} s), so that dissociation competes on favourable terms with energy transfer to the substrate.

We refer to two types of *indirect* photodissociation of adsorbates, both involving absorption of photons by the substrate. The first of these indirect PDIS events is the analogue of the well known process of sensitized photolysis in gases. In the context of adsorbate PDIS, electronic excitation of defects, plasmons or impurities in the substrate lead, by way of E→E (electronic to electronic) energy transfer, to excitation and dissociation of the adsorbate. The second indirect PDIS pathway, also substrate-mediated, involves the phototransfer of an electron from the substrate to an antibonding orbital of the adsorbate; this charge transfer photodissociation is symbolized CT/PDIS.

Photoreaction; PRXN

PDIS often results in the formation of photofragments whose translational energies are in excess of the energy required to surmount the activation barrier to chemical reaction in the adsorbed state. Under suitable conditions of high, but not necessarily multilayer coverage, inter-adsorbate distances are likely to be reduced and the probability for reactive encounters increased. In the presence of islands, inter-adsorbate distances will be small even at low coverages. Low inter-adsorbate distances and favourable orientations will lead to reaction between a photofragment and a neighbouring (or near by) adsorbate molecule. Reaction of this type, initiated by absorption of radiation in the adsorbate and involving only adsorbed molecules, we term PRXN. The special feature of PRXN is the existence of alignment, orientation, and preferred separation between the reacting species which contributes to defining the reactive collision. It should be noted that the classic Langmuir–Hinshelwood process in which co-adsorbed species A(ad) and B(ad) (which may have been formed photolytically) migrate across the surface ultimately meeting to form product AB(ad) does not constitute PRXN as we have defined it, since all memory of the initial bond alignments, orientations, and molecular separations has been lost.

The PRXN process is shown schematically in Figure 1b. In Figure 1b PRXN involves PDIS to yield an atomic photofragment. When the photofragments are atomic, we designate the reaction PRXN(A). A second category of PRXN has been reported in which photofragments are not involved, but reaction takes place between aligned adjacent molecules; this category of PRXN, being bimolecular, is designated PRXN(B).

Photodesorption; PDES

When absorption of light by the substrate results in the desorption of *intact* adsorbate molecules, and *only* in this case, we refer to the process as PDES [Figure 1c].

It should be noted that elsewhere in the literature the term 'photodesorption' has been used to describe any photoinduced process, substrate-mediated or adsorbate-mediated, that releases any material from the surface, whether it be a photofragment, photoreaction product, or parent molecule (neutral or charged). This has been the usage in the subject area term DIET, for 'Desorption Induced by Electronic Transitions'. The four biannual conferences held so far on the subject of DIET serve to illustrate a wide variety of processes encompassed by surface photochemistry, particularly at short wavelengths, *i.e.* for radiation of high energy, capable of producing ionization.[79–82] In the present review we restrict ourselves to the consideration of lower photon-energies of a few eV, comparable with the energy of a chemical bond. We do this for several reasons. In the first place it would be impractical to cover the full range of incident energies and consequent photoprocesses in one review. Secondly particular interest attaches to the

Photoejection; PEJ

The final process that we seek to distinguish is one in which intact molecules are released (as above) but in which the incident radiation is absorbed by the adsorbate. The adsorbate (rather than undergoing PDIS or PRXN) converts its electronic excitation into nuclear motion [see Figure 1d], in a process termed PEJ. This constitutes a pathway for quenching. By analogy with quenching in the gas-phase the energy conversion is written $E \rightarrow V, R, T$; *i.e.* the electronic excitation is channelled into vibrational, rotational, and translational nuclear motion. The translational energy is commonly substantial; in the electron volt range. The process has been found to become important at coverages in excess of one monolayer. It has been proposed for insulator substrates that it is due to quenching interaction between adsorbate molecules. Quenching of an adsorbate molecule directly below the uppermost layer, it is postulated, could eject the upper layer molecule to give the observed PEJ.

3 Photodissociation; PDIS

Methodologies

Two distinct experimental approaches are used for investigating photodissociation processes at the gas–solid interface. The two are distinguished by the nature of the observable. In one category of study the speed, angular distribution, and (occasionally) internal excitation of the *photofragments* leaving the surface are measured. In the second category of investigation it is the photoproduct left behind at the *surface* that is monitored.

The second category of study makes use of the standard tools of surface science. Until now the favoured light source for the second category of experimental study has been CW, hence, dynamical information has usually been lacking. Even if pulsed light sources had been used, the time-resolution of standard surface diagnostics (*e.g.* Auger or EELS) would not reveal the dynamics, consequently there is no loss of information in using a CW light source. The advantages of such a source are, moreover, substantial; low power, wide tunability, and low cost (if an arc lamp is used, as is customary).

Studies of the first type, that characterize the dynamics of photofragment escape from the surface, are rich in information content. In the case of inert substrates they explore the major pathway for photodissociation. If, however, the substrate is reactive, studies of the escaping species only give a partial description of PDIS. Clearly the same applies in reverse for surface studies in which the fraction of the photoproducts escape the surface and only those that react with the surface are monitored. In addition, as already noted, the use of surface diagnostics gives (at present) relatively little insight into the dynamics.

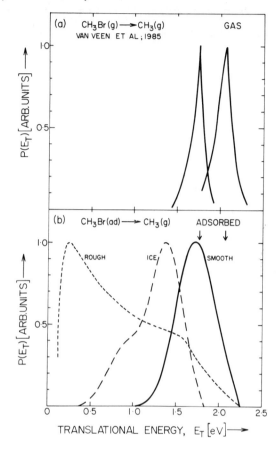

Figure 2 *Translational energy distributions of the methyl (CH_3) photofragment from the 222 nm PDIS of CH_3Br. (a) Gas-phase result derived from G. N. A. Van Veen, T. Baller, and A. E. De Vries, Chem. Phys., 1985, **92**, 59; (b) the solid curve is for CH_3Br on an annealed LiF(001) substrate. The long dashes are for CH_3Br ice and the short dashes are for CH_3Br on an unannealed (i.e. rough) LiF substrate. (The arrows in the lower panel indicate the positions of the gas-phase peaks)*
(Reproduced with permission from *J. Chem. Phys.*, 1988, **89**, 1475.)

Insulators

The crystalline insulator substrate for which there exists the greatest body of data on photoprocesses in the adsorbed state is LiF(001).[17,49–59,65] Early studies on the CH_3Br/LiF(001)[17,49,50] system established that photodissociation in the adsorbed state *did* occur. Bond fission resulted from single-photon absorption of UV radiation by chromophores in the various adsorbate molecules.[17] Photodissociation in the adsorbed state was distinguished by significantly altered translational energy and angular distribution of the photofragments as compared with the gas-phase.[50,51] That both the magnitude of the recoil velocity for PDIS and the direction of recoil changed

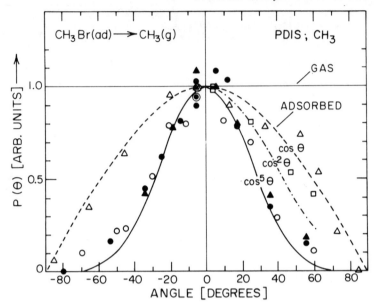

Figure 3 *Angular distributions for the methyl photofragment from the 222 nm PDIS of* CH_3Br. *The horizontal line would result for gas-phase photolysis with the unpolarized laser. The solid triangles and circles* ($\sim \cos^5 \theta$) *are results from separate experiments on* CH_3Br *ice. The open circles and squares* ($\sim \cos^2 \theta$) *are from PDIS of* CH_3Br *on an annealed LiF(001) surface at high (0.18 L/pulse dosing) and low (0.01 L/pulse dosing) coverage, respectively. The open triangles* ($\sim \cos \theta$) *are from the PDIS of* CH_3Br *on an unannealed LiF substrate at low coverage (0.01 L/pulse dosing).* ($1L \equiv 10^{-6}$ torr s) (Reproduced with permission from *J. Chem. Phys.*, 1988, **89**, 1475.)

markedly on adsorption can be seen in Figures 2 and 3 (later we show examples in which the preferred recoil direction peaks away from the surface normal). The comparison of gas-phase and adsorbed state data for the translational energy of photofragment, CH_3, from the photolysis of $CH_3Br/LiF(001)$ at 222 nm in Figure 2, reveals a significant shift in the most probable energy to lower translational energy, and a broadening of the distribution up to the limit of the available energy. Dependence of the form of the distribution on the character of the substrate and phase of the adsorbate is also apparent. Unannealed substrates broadened and shifted the translational energy distribution. Multilayer 'ice' adsorbate distributions were again different, but in this case independent of the substrate.

These early observations, and others relating to the bimodal angular distribution of atomic H from PDIS of $H_2S(ad)$, as well as the sensitivity of the HS vibrational excitation in $H_2S(ad) \rightarrow HS + S(g)$ to surface coverage,[49,51] showed that adsorbate–substrate interaction affected the molecular dynamics of PDIS and also (given the observed angular distributions) that the substrate was capable of *aligning* the adsorbate molecules giving rise to a preferred angle between the bond axis and the surface plane. The existence of

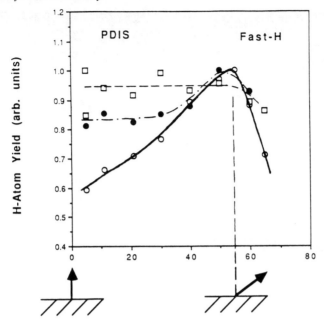

Figure 4 *Angular distribution, P(θ), of energetic H atoms leaving the surface from PDIS of HBr on LiF(001) at coverages of 0.2 ML (open circles), 0.5 ML (closed circles), and 1.0 ML (open squares). At low surface coverage the distribution is peaked at 50 ± 5° to the surface normal; the anisotropy is degraded at higher coverages due to increased probability of collisions with adsorbate*
(Reproduced with permission from reference 77)

orientation as well as alignment was indicated in the early work by the observation of narrow photofragment energy distributions peaked at a high energy, only moderately displaced from that for gas-phase photolysis. These narrow energy distributions suggested that the departing atoms or radicals were ejected so that they moved *away* from the surface without having their energy substantially degraded by surface collisions.[49-51]

Characterization of adsorbate structure in these systems by means independent of PDIS is, however, vital, and has begun. The methods used to date are He diffraction and FTIR. The system $CH_3Br/LiF(001)$ has been the subject of a He diffraction study by Scoles and co-workers,[83] establishing the existence of *order* in the substrate, as well as confirming the *alignment*, and *orientation* previously inferred, namely a C—Br axis normal to the surface with CH_3 up, alternating with Br up.[49,50]

Measurements of FTIR spectra for HBr/LiF(001) using polarized infrared radiation showed that HBr on LiF(001) was oriented with the molecular axis tilted at 21 ± 5° to the surface plane.[84] A red-shift of 300 cm^{-1} from the gas-phase absorption indicated that the HBr was

Figure 5 *Van der Waals representation of* HBr *adsorbate on* LiF(001) *(small spheres in the substrate are* Li^+, *large spheres are* F^-) *at 100 K, based on Monte Carlo calculations. A sample pair of adsorbate molecules in a configuration representative of high coverages is shown (with lightly shaded atoms as* H, *heavily shaded as* Br). *The method used in the calculations was that discussed in detail in reference 85*
(Reproduced with permission from reference 77)

hydrogen-bonded to the surface in the configuration Br—H $\cdots$ F^-. Hence, the H was oriented downwards. This alignment and orientation is consistent with the results of a dynamical study of HBr adsorbed on LiF(001), in which fast H atoms were observed to leave the surface at close to the complementary angle to the surface normal.[77] The angular distribution of H from the PDIS of HBr on LiF(001) as a function of coverage is shown in Figure 4. Photodissociation of HBr at low coverages (sub-monolayers) resulted in energetic H atoms peaked at 50 ± 5° to the surface normal. The H atom angular distribution became more isotropic with increasing coverage due, it was thought, to the increased probability of H + HBr collisions (this is currently the subject of detailed computer modelling).

The orientation of HBr on LiF, which can be seen to be consistent with the above, is shown in Figure 5. The figure is, however, based on an independent computation of the geometry of HBr adsorbed on LiF(001).[85] The minimum energy of a sample of 200 HBr molecules adsorbed on LiF(001) was computed including ionic, dispersion, and repulsive forces; this led to a theoretical value of 23° for the tilt to the surface plane, with the H ends of the HBr molecules pointing downward towards adjacent F^-, in agreement with the FTIR study. A further prediction from this calculation was that the nearest neighbour HBr–HBr geometry at coverages approaching a monolayer consisted largely of 90° angles between successive HBrs, with the H of one HBr directed at the Br of its neighbour, suggestive of hydrogen bonding in the adsorbate.

The extent to which photodissociation dynamics in the adsorbed state have been probed through measurements of the product state internal energy distribution of newly-formed fragments has been limited. There can be no doubt, however, that such studies will be feasible and informative. Product state-resolution has been achieved using the technique of resonance enhanced multiphoton ionization time of flight (REMPI-TOF) spectroscopy in the case of dissociative desorption from multilayers of CH_3I[73] and NO_2.[72,86] In these studies species in the multilayer were photolysed by irradiating with a UV laser. Fragments detached from the surface were probed a short time later by REMPI-TOF, using a second tunable laser source. Similar studies at sub-monolayer coverages can be expected in the near future.

Semiconductors

Investigations of photoprocesses at semiconductor surfaces have centred largely on Si, and to a smaller extent on GaAs. The most commonly studied adsorbates to date are the transition metal carbonyls, $Fe(CO)_5$,[19,87-89] $Mo(CO)_6$,[19,20,90,91] and $W(CO)_6$,[19,92] for which there is much interest in exploring and understanding the photochemical deposition of films of molecular thickness. These systems offer the prospect of applications in microelectronics, namely writing, etching, and doping.

Early work by Creighton[90] on the photodecomposition of $Mo(CO)_6$ adsorbed on Si(100) used pulsed excimer radiation at 248 nm in conjunction with pre- and post-desorption TDS (thermal desorption spectroscopy) and AES (Auger electron spectroscopy). It was established that $Mo(CO)_6$ physisorbed on Si(100) partially dissociated to yield CO upon irradiation with UV, with a measured cross-section similar to that observed in the gas-phase. The small magnitude of the temperature rise due to laser heating indicated that the decomposition was initiated by PDIS of the adsorbate rather than thermal decomposition. It was clear that PDIS was a major pathway despite the possibility of quenching by the semi-conductor substrate. The conclusion was substantiated in independent experiments on $Fe(CO)_5/Si(100)$ by Celii et al.,[87] using a pulsed N_2 laser at 337 nm to initiate photodissociation.

A clear separation of photolytic and pyrolytic processes in $Mo(CO)_6/Si(111)$ was demonstrated by Ho and co-workers,[20] as illustrated in Figure 6. Using CW laser radiation at 514 and 257 nm (Ar^+ fundamental and second harmonic wavelengths) and mass spectrometric detection of CO, in conjunction with TPD and high resolution electron energy loss spectroscopy (HREELS), the photoinduced decomposition of $Mo(CO)_6$ was monitored. High power radiation at 514 nm produced an appreciable temperature rise at the surface, as shown in Figure 6a. The mass spectrometer trace, giving the amount of CO reaching the detector as a function of time, also in Figure 6a, is identical to that obtained in TDS for the same temperature profile. This CO^+ signal resulted from the cracking in the MS (mass spectrometer) ionizer of thermally desorbed $Mo(CO)_6$. In contrast, the temporal profile for CO^+ under conditions of 257 nm irradiation, Figure 6b, showed an immediate onset and exponential decay over several minutes. Spectra

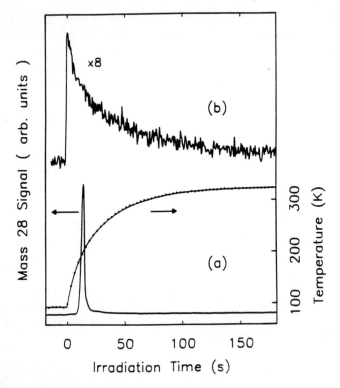

Figure 6 *Desorption signal of mass* 28 *from* $Mo(CO)_6$ *adsorbed on* Si(111)–(7 × 7) *induced by* (a) 1.77 mW cm^{-2} *of* 514 nm *radiation, and* (b) 2.2 mW cm^{-2} *of* 257 nm *radiation. In both cases irradiation started at* $t \equiv 0$. *Also shown in* (a) *is the measured sample temperature (circles) with a theoretical fit (solid line) based on a thermal diffusion equation for high power laser heating. The temporal profile of the desorption signal in* (a) *is precisely that for TDS with the same temperature profile*
(Reproduced with permission from *J. Vac. Sci. Technol.*, 1987, **A5**, 1608.)

obtained by HREELS following UV irradiation confirmed a distinct change in the adsorbate arising due to fragmentation.

Other more recent work from the same laboratory, using an arc lamp source, demonstrated that the photoyield of CO tracked the absorption spectrum of $Mo(CO)_6$.[93,94] The low adsorption energies in these systems (~ 0.5 eV[19]) either do not perturb the electronic potentials of the adsorbed species, or perturb both ground and excited states similarly. Charge transfer processes were not observed on the silicon substrate. However, upon pre-dosing the Si(111) surface with K, the work function of the substrate was lowered to the extent that electrons of sufficient energy were photogenerated to initiate dissociation of $Mo(CO)_6$. Significantly, this dissociation channel was then observed for wavelengths extending from the UV to at least 800 nm.[93,94]

The effect of photogenerated electrons on photochemistry in the adsorbed

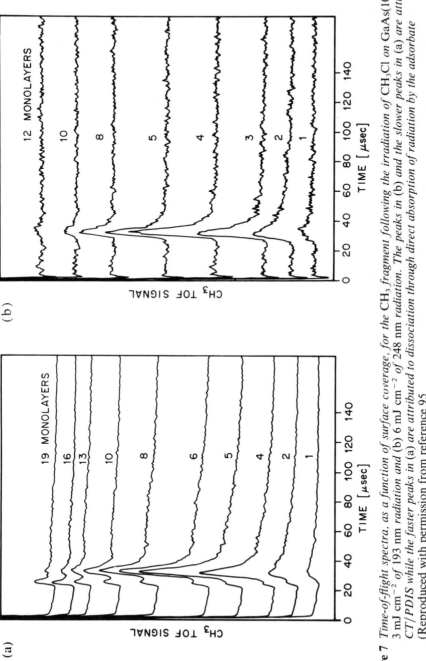

Figure 7 Time-of-flight spectra, as a function of surface coverage, for the CH_3 fragment following the irradiation of CH_3Cl on GaAs(100) by (a) 3 mJ cm^{-2} of 193 nm radiation and (b) 6 mJ cm^{-2} of 248 nm radiation. The peaks in (b) and the slower peaks in (a) are attributed to CT/PDIS while the faster peaks in (a) are attributed to dissociation through direct absorption of radiation by the adsorbate (Reproduced with permission from reference 95

state in undoped semiconductor systems is evident in recent work by Cowin and co-workers.[95] Figure 7 gives data for the photoinduced decomposition of CH_3Cl adsorbed on GaAs(100). Figure 7a shows TOF spectra of CH_3 detached following pulsed irradiation at 193 nm with increasing coverage of CH_3Cl, while the data of Figure 7b were recorded under similar conditions but at 248 nm. Two photolytic channels were resolved in the case of 193 nm irradiation. The first, of higher energy (early time of arrival) was ascribed to dissociation through direct absorption by the adsorbate since it persisted at all coverages and had a wavelength onset comparable to that which would be expected for unperturbed photofragmentation in the gas-phase. By contrast, at 248 nm, where the gas-phase cross-section for dissociation was negligible, only the second (lower energy) channel was apparent. This second channel was attributed to photoelectron-induced fragmentation of $CH_3Cl(ad)$; the $CH_3Cl(ad)$ is thought to capture a photoelectron from the GaAs substrate to form $CH_3Cl^-(ad)$ which then dissociates to yield Cl^- and CH_3, of which only the latter escapes the surface. The dependence of the yield of the observed photofragment, CH_3, on coverage was thought to reflect the mean free path of the photoelectrons in the adsorbate. The process is analogous to the dissociative attachment of electrons observed in the gas-phase species.[96] The same process, termed charge transfer photodissociation (CT/PDIS), has been studied at metallic surfaces (see below).

Metals

Photolytic bond fission in sub-monolayers of adsorbates on metals is again observed, in many cases, to compete effectively with energy transfer to the (metal) surface. This is to be expected if PDIS takes place on a sub-picosecond (10^{-13}–10^{-14} s) time scale, by photoexcitation to a repulsive state. Investigations to date have had as their objective the establishment of mechanisms for PDIS. The two limiting cases which have emerged are: fission of a bond in the adsorbate directly through photoabsorption, or the excitation of charge carriers in the substrate which interact with the adsorbate molecules leading indirectly to photodissociation. The degree to which adsorbate or substrate absorption contribute to PDIS is system-dependent; various experimental techniques have been applied to elucidate the mechanism.

The example discussed above of PDIS in CH_3Cl/GaAs(100) demonstrated a separation of PDIS channels resulting from adsorbate or substrate absorption of UV radiation. An analogous study on CH_3Cl/Ni(111) in the same laboratory showed a similar separation in the TOF spectra of detached Cl (Cl^- was not observed).[67] This separation depended on a difference in the kinetic energy imparted to the photofragments in the two pathways, PDIS and CT/PDIS.

Increasingly compelling evidence for CT/PDIS has come from a variety of sources. (a) From observation of a hundred-fold enhancement in photodissociation cross-section in the case of HCl/Ag(111) which decreased progress-

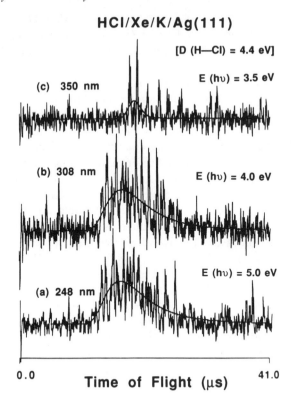

Figure 8 *TOF spectra of atomic H resulting from PDIS of HCl on Xe/K/Ag(111) at* (a) 248 nm, (b) 308 nm, *and* (c) 350 nm. *The signal was averaged over 50 laser pulses and normalized to a photon intensity of* 1.2×10^{17} *photons* cm^{-2} *per pulse. The bond dissociation energy of HCl in the gas phase corresponds to* ~ 280 nm *and consequently the detection of H atoms in* (b) *and* (c), *where the incident radiation was of lower energy than this, was attributed to CT/PDIS at the gas–solid interface*
(Reproduced with permission from *J. Phys. Chem.*, 1990, **94**, 5664.)

ively as the work function of the substrate was increased by chlorination.[58] (b) From the observation of photodissociation of HCl on K (with an Xe layer over the K to protect it) at wavelengths out to 350 nm, corresponding to a photon energy 0.9 eV less than the H—Cl bond dissociation energy.[59] [D_o(H—Cl) = 4.431 ± 0.002 eV[97] which is equivalent to a wavelength of ~ 280 nm.] This latter result excluded PDIS to yield H + Cl, but was understandable if the photoproduct was the more stable H + Cl$^-$ anion resulting from CT/PDIS. Figure 8 displays the experimental data in the form of TOF spectra of photodetached H atoms from a study of PDIS from HCl adsorbed on Xe/K/Ag(111). A significant yield of H atoms was measured for 248 and 308 nm irradiation, and a lower, but detectable, yield was apparent for 350 nm irradiation.[59] (c) From other work in which adsorbate photodissociation has been shown to occur throughout the visible region out to the red (800 nm). The examples investigated were $Mo(CO)_6$ on potassium-covered Cu or Si,[94] and also HBr and HI on potassium-covered Ag.[98] (d) From the observation of the anion, Cl^-, product of CT/PDIS in CCl_4 /Ag(111) which gives direct evidence of electron transfer to the adsorbate.[98]

The findings of another recent study[99] led the authors to stress the importance of the 'local' work function in determining the rate of photoinduced charge transfer dissociation. For CH_3Cl on Pt(111) irradiated with UV from a Hg arc the decreasing work function in the coverage range 0.25– ~ 1.0 ML correlated with increasing total photoelectron yield, but not with an increase of rate coefficient for photodissociation. The authors postulated that this was due to the fact that it was the *local* work function at each adsorbate molecule which governed local electron flux, and hence CT/PDIS rate per molecule, and this quantity would not be a function of coverage. Further research will be required to establish the role of the local work function in contrast to the overall work function.

The wavelength dependence of cross-sections for PDIS has provided a further indicator of the underlying mechanism. This was first achieved in experiments employing CW radiation from Hg-lamp sources (work using tunable pulsed dye lasers is referred to below). An extensive series of studies by White and co workers[78] on the methyl halides (CH_3X; X = Cl, Br, I) and ethyl chloride (C_2H_5Cl) on Pt[31–34,36–39, 100,101] and on Ag(111)[40,102–104] exemplifies the value of tunable arc lamps in adsorbate photochemistry. In all cases, the cross-sections for PDIS were markedly red-shifted relative to that of the gas-phase species, consistent with a CT/PDIS mechanism which can occur at lower photon energy. In some cases, PDIS was observed for wavelengths corresponding to an energy less than the work function of the metal substrate, suggesting electron tunnelling. As would be expected the process was of particular importance for the first layer of adsorbate adjacent to the metal. A possible alternative mechanism, at the present time, would be PDIS of an adsorbate–substrate complex.

Further evidence for tunnelling of electrons in CT/PDIS comes from the observation of H atoms formed in substantial yield when HI was adsorbed on Ag(111) (protected by Xe) and irradiated with 308 nm pulsed laser light; the photon energy was 0.75 eV less than the work function for clean silver, indicative of tunnelling.[59] No CT/PDIS was observed in the same work at 350 nm; 1.25 eV below the work function. The photolysis of $Mo(CO)_6$ on Cu(111) or Si(111) at long wavelengths was also suggestive of tunnelling.[94]

Additional evidence comes from the observation for photolysis of HBr or HI on potassium using pulsed tunable dye lasers, of two markedly different slopes for the CT/PDIS cross-section, σ_{CT}, namely a shallow slope at low frequency (long wavelength) and a steep slope at high frequency, could most readily be explained by the existence of a region of low frequencies that gave CT/PDIS by tunnelling followed by a region of higher frequencies in which CT/PDIS resulted from electrons that had been promoted to, or beyond, the vacuum level of the metallic substrate, *i.e.* free electrons.[98]

The study of PDIS using tunable arc lamps for cross-section measure-

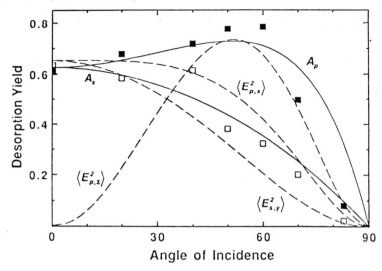

Figure 9 *Desorption yield of NO from Pd(111) as a function of the angle of incidence for excitation with light being polarized parallel (p:■) and perpendicular (s:□) to the plane of incidence. The curves give the absorption in the metal substrate (———) and the electric field intensities at the interface (— — —). The absorption differs for both cases of polarization. In the case of s-polarization the field intensity has only a component in the plane of the surface $\langle E_{s,y}^2 \rangle$. In the case of p-polarization there are two components, one in the surface plane $\langle E_{p,x}^2 \rangle$ and one normal to it $\langle E_{p,z}^2 \rangle$, which have to be considered (Reproduced with permission from J. Chem. Phys., 1990, **92**, 3154.)*

ments, or tunable pulsed dye lasers for dynamical studies, will yield still more dividends in the coming period.

Another valuable tool is the use of polarized light to initiate photochemistry.[72,77] This technique is best illustrated by a recent study by Hasselbrink *et al.*[72] As part of a study of the photodissociation of NO_2 on NO-covered Pd(111) the angular dependence of the NO photofragment yield for *s*- and *p*-polarized light was measured. These results are shown in Figure 9, where NO yields are compared with calculated curves giving the angular variations of the absorption, $A_i (i = s, p)$, in the metal substrate and the square of the surface electric field strength, $\langle E_{i,k}^2 \rangle$; ($k = x, y, z$), at the interface for both *s*- and *p*-polarized incident light (*z* corresponds to the surface normal). The predicted curves arise through the Fresnel equations, using the known bulk optical constants of the Pd substrate. The fact that the angular dependences of the *s*- and *p*-induced yields tracked the calculated curves for absorption by the surface, indicated that substrate excitation was responsible for initiating the adsorbate dissociation.

The polarization dependence of PDIS has been used to good effect in other parallel studies to discriminate between substrate-mediated PDIS and PDIS as a result of direct absorption by the adsorbate, as well as yielding information on alignment of the adsorbate with respect to the surface.[43,77,105–109]

4 Photoreaction; PRXN

An important objective in adsorbate photochemistry is the study of *surface aligned photoreaction*, PRXN.[49,51,53,56,77,110,111] As noted in Section 2, PRXN denotes, in general, reaction between a photofragment coming from one adsorbed molecule (no. 1), reacting with a co-adsorbed species (no. 2), which is a second molecule of the same substance or of some other substance that has not undergone photodissociation. The interest in PRXN stems from the fact that reaction can involve restricted angles of approach, θ_i, and impact parameters, b_i. The first of these parameters, θ_i, will be restricted if the molecule undergoing PDIS (no. 1) is *aligned* relative to the second reagent (no. 2). A stricter restriction applies if molecule no. 1 is *oriented*, so that a preferred reactive group is photoejected in the direction of molecule no. 2. If, in addition, the co-adsorbed species' are held at a preferred separation due to *ordering* (they need not be in register with the substrate for ordering to be present) then there will be a corresponding restriction on the distance by which the reagent coming from no. 1 misses the centre-of-mass of the molecule under attack, no. 2, *i.e.* a restriction on the impact parameter in PRXN.

By varying the substrate, or the coverage, the geometry of the adsorbates can be altered. This offers the prospect of controlling PRXN. Since the outcome of reaction is known to depend sensitively in many cases on θ_i and b_i, this approach should ultimately lead to an intelligent and intelligible catalysis of photoreaction.

This possibility has not yet been fully realized, though PRXN has been observed under conditions where alignment and orientation have been shown to be present, *e.g.* for HBr/LiF(001).[77,84] Enough progress has been made that one can cite evidence for photoreaction in the adsorbed state on all three of our broad categories of substrate, as indicated below.

Insulators

The evidence for photoreaction in the adsorbed state comes from (*a*) the detection of reaction products leaving the surface (at coverages in the range ~ 0.1–1.0 ML plus (*b*) demonstration, under these conditions, that photofragments leaving the surface suffer no gas collisions, hence reaction of photofragments in a desorbed gas cloud above the surface is ruled out. In some cases (*b*) may be supplemented by (*c*) demonstration that the reaction dynamics are sensitive to coverage [$\vec{S}$ + OCS(ad) → S_2(g) + CS(ad/g)[56]] and (*d*) that the reaction product energy-distribution differs from that expected for gas-phase reaction [$\vec{H}$ + HBr(ad) → H_2(g) + Br(ad/g)[77]].

In early studies of H_2S adsorbed on LiF(001),[49,51] and in more recent work on OCS/LiF(001),[56] PRXN was observed for surface coverages down to ~ 0.1 ML. The photoinduced yield of H_2 from H_2S/LiF(001):

$$\vec{H} + H_2S(ad) \rightarrow H_2(g) + HS(ad/g) \qquad (1)$$

and S_2 from OCS/LiF(001):

$$\vec{S} + OCS(ad) \rightarrow S_2(g) + CO(ad/g) \qquad (2)$$

was linear in incident laser fluence, indicative of a single photon process, and the dynamics of PRXN were found to vary with adsorbate coverage. The symbols $\vec{H}$ and $\vec{S}$ in equations (1) and (2) are used to denote photofragments in transition between the adsorbed and the gaseous state; these fragments would travel a distance of the order of angstroms in the course of equations (1) or (2). In other contexts this might be described as a 'transition state'. The entrance valley of the reactive potential energy surface has been truncated (see Figure 2 in reference 111 and its accompanying text) so that only part of the transition state region is traversed. In addition, in view of the aligned nature of the collision and the single collision energy (governed by hν), there exists the possibility of forming long-lived transition states which oscillate for extended periods perpendicularly to the reaction co-ordinate.

In both equations (1) and (2) there was evidence to suggest the existence of two channels for PRXN, termed 'direct' and 'indirect'. Direct PRXN occurred at all coverages yielding products of higher translational energy as a result of what may be pictured as a simple 'pick-up' exchange interaction.

In the case of H_2S/LiF(001), for example, it was speculated that PDIS followed by pick-up abstraction was responsible for the 'direct' PRXN. Indirect PRXN product H_2(g) of lower translational energy was likely to be a consequence of secondary encounters between PRXN product H_2 and H_2S(ad); the relative yield for this route increased with coverage in the sub-monolayer regime. These secondary encounters also resulted in an angular distribution for the H_2 product which was broader than that for the direct PRXN pathway.

The dual dynamics may be linked to different geometries in the adsorbate; H_2 formed due to a reactive encounter between H_2S(ad) and an $\vec{H}$ atom that was observed to be photoejected at $\sim 45°$ to the surface normal was unlikely to suffer further collisions, whereas H_2 formed from photolytic $\vec{H}$ photoejected in the plane of the surface (a necessary complement to Hs observed to be ejected normal to the surface, since H_2S is bent at $\sim 90°$) has a high probability of colliding with neighbouring adsorbate molecules before leaving the surface. This speculation (as yet unconfirmed) has the merit of being consistent with the observed angular distribution of the unreacted $\vec{H}$. Figure 10 shows the angular distribution for PDIS of H_2S. It is a bimodal distribution suggesting the existence of two preferred geometries; one in which the S—H bond is oriented chiefly along the surface normal (the complement of the $\vec{H}$ leaving at 90°, and a second in which the S—H bond undergoing PDIS points at $\sim 45°$ to the normal. The two orientations could correlate, respectively, with direct and indirect PRXN. This speculation serves to illustrate the type of dynamical insights presently being sought in PRXN.

For OCS(ad)/LiF(001) the indirect pathway gives evidence of a higher

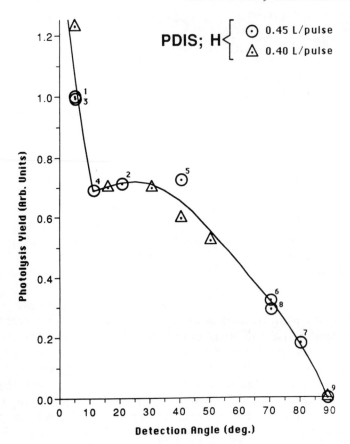

Figure 10 *High coverage angular distribution of the H photofragment yield (normalized to the laser power) from the 222 nm PDIS of H_2S adsorbed on LiF. The circled data were taken at a dosing of 0.45 L laser pulse^{-1} and a laser power of 8.2 ± 0.3 mJ pulse^{-1}. The triangle data were taken on a separate LiF crystal at a dose of 0.4 L pulse^{-1} and a laser power of 4.1 ± 0.2 mJ pulse^{-1}. The sequence of data taking is indicated for the circles. The solid curve was drawn to emphasize the bimodal nature of the angular distribution*
(Reproduced with permission from *J. Chem. Phys.*, 1988, **89**, 1498.)

degree of complexity leading to adsorption–desorption of the S_2 PRXN product. The relative probability of the indirect dynamics increases with coverage in the sub-monolayer regime. The observation of the adsorption–desorption type of 'indirect' dynamics (in contrast to the previous example) is understandable in light of the fact that S_2 is formed with much lower translational energy, and the heat of adsorption for S_2/LiF(001) exceeds that for H_2/LiF(001).

In the recent studies of HBr/LiF(001) it was possible to compare the results of an FTIR study of the geometry of HBr(ad),[84] a theoretical study of

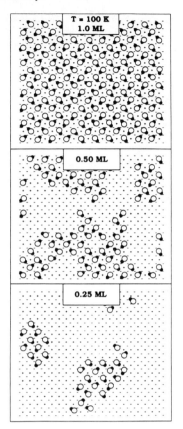

Figure 11 *Pattern of organization in an* HBr *adlayer on* LiF(001) *at* T = 100 K *as a function of coverage. At 1.0 ML 50% of the sites are occupied. The order observed is that of a predominance of 90° configurations, with the H-end of one* HBr *pointed towards the* Br-*end of an adjacent* HBr
(Reproduced with permission from *J. Chem. Phys.*, 1991, **94**, 978.)

the adsorbate structure,[85] and a dynamical study of the detached products from PRXN.[77] As discussed above, the domination of electrostatic forces evident in the formation of hydrogen bonding was primarily responsible for the downward tilted Br—H geometry. Both FTIR and theory indicated a geometry in which the H-end of HBr pointed toward the surface with the molecular axis at an angle of ~ 20° from the surface plane. The results from the dynamical study of PDIS were consistent with this finding if it was supposed that a significant fraction of $\vec{H}$ bounced off the rough LiF surface at a slightly less than specular angle. Recent dynamical studies lend credence to this view of the effect of surface roughness (see Figure 23 of reference 85, and reference 112).

Figure 11 shows the patterns of organization in an HBr adlayer over LiF(001) computed in the theoretical study. These patterns indicated order in the adsorbate, with the majority of HBr molecules aligned such that the

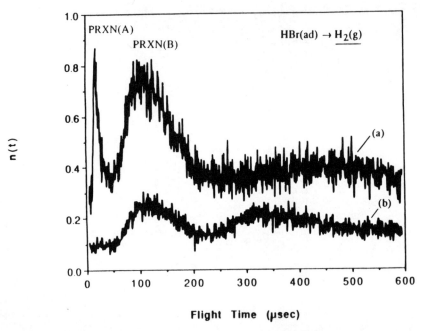

Figure 12 *Time-of-flight mass spectra of mass 2 (H_2) following the irradiation of 1 ML of HBr adsorbed on LiF(001) with* (a) 193 nm *and* (b) 248 nm *radiation. Two reaction channels are distinguished: the first,* PRXN(A), *requires the presence of H atoms through PDIS of HBr monomer and, consequently appears only in* (a); *while the second,* PRXN(B), *is attributed to a bimolecular reaction initiated by electronic excitation in the dimer,* $(HBr)_2$, *and occurs over a wider range of excitation wavelength spanned by* (a) *and* (b) *above* (Reproduced with permission from reference 77)

H-end of one HBr pointed towards the Br-end of an adjacent HBr; *i.e.* the hydrogen bonding already noted for adsorbate–substrate interaction, extends to adsorbate–adsorbate interaction. The H_2 product of the PRXN abstraction reaction

$$\vec{H} + HBr(ad) \rightarrow H_2(g) + Br(ad/g) \tag{3}$$

was observed by TOF mass spectrometry in the experimental study of this system. The $\vec{H}$ represents, as before, a photorecoiling atom in transition from the adsorbed to the gaseous state.

Figure 12 shows the TOF spectra of H_2 recorded following the irradiation, at 193 and 248 nm, of a monolayer of HBr adsorbed on LiF(001). Two channels producing H_2 were apparent: the higher energy channel, PRXN(A), depended on the photolysis of HBr(ad) to yield atomic $\vec{H}$, and its occurrence required short wavelengths of incident radiation capable of giving PDIS. The channel giving lower-energy H_2, PRXN(B), was present at both incident wavelengths employed, *i.e.* even in the absence of atomic $\vec{H}$, and was attributed to the bimolecular reaction

$$2\text{HBr}(\text{ad}) + h\nu \rightarrow \text{H}_2(\text{g}) + \text{Br}_2(\text{g}) \tag{4}$$

Other evidence, such as the detection of equal yields of H_2 and Br_2[53,77] and also equal yields of H_2 and Cl_2 in an analogous study of HCl/LiF(001),[53] indicated that the likely mechanism for PRXN(B) was electronic excitation of the dimer, $(HX)_2$ (X = Cl, Br), leading to bimolecular reaction, denoted (B), to give H_2 and X_2.

The photolytic channel PRXN(A) is of particular interest in studies of surface aligned photoreaction. The translational energy distribution of the H_2 produced in the surface reaction was found, experimentally, to be shifted to lower energy relative to that expected for the gas-phase process. The angular distribution of the H_2 product was peaked at the surface normal, showing an approximate $\cos^2\theta$ dependence. As previously stated, and displayed in Figure 4, high energy H atom photoproducts from PDIS of HBr had an angular distribution which was peaked at ~ 50° from the surface normal at low coverages, but became increasingly isotropic in the monolayer regime and above. This is likely to be due to inelastic scattering off the surface and neighbouring adsorbates.

Given the calculated geometry ($\vec{H}$ approaches the neighbouring HBr at 90° to the H—Br axis, with impact parameter b ~ 0) and measured orientation it is not surprising that a proportion of these collisions were reactive and resulted in the generation of H_2 product. The case is, however, less than ideal for the study of PRXN since the dynamics are complicated by the initial direction of $\vec{H}$ trajectories, which are tilted downwards and therefore involve substantial interaction with the surface, as well as with neighbouring adsorbates.[85] Trajectory calculations which incorporate these interactions are providing a theoretical link between adsorbate geometry and the resultant dynamics.[112] Measurement of the vibrational energy distribution of the H_2 product from the surface reaction would be of interest since vibration is less liable to modification in secondary encounters with the surface.

Although studies of surface aligned photoreaction are few in number, the use of surface alignment as a means of obtaining (or controlling) the details of reaction dynamics shows promise. The combination of structure determinations by FTIR and He diffraction, dynamical studies to probe all energy and angular distributions, and more detailed theoretical calculations to tie the experimental findings together, are required and will soon be available.

Semiconductors

There are relatively few examples of photoinduced reaction of adsorbates on semi-conductor substrates, However, a related investigation, in which the reaction was with the surface rather than between adsorbates, was performed by Houle[14,15,113] who found that band-gap radiation enhanced fluorine insertion into Si—Si bonds in the etching of Si by XeF_2. The silicon fluoride reaction products were monitored as a function of time, laser power, and dopant. Illumination of the Si surface with photons of energy greater than the band gap was postulated to result in an increase in minority carriers (holes for n-type and electrons for p-type) at the surface. As fluorine is an efficient electron acceptor, and an ionized Si—Si bond constitutes a hole, this charge transfer was viewed as a rate-determining step in promoting fluorine attack and insertion.

Another reactive process is the synthesis of N_2O from pre-adsorbed NO at semiconductor surfaces. Both thermal and photoinduced synthesis have been observed by Ho and co-workers for NO adsorbed on Si(111)–(7 × 7)[20,21,25,26] and GaAs(110).[23,24] The isolation of photoreaction and identification of its mechanism is complicated by the additional formation of N_2O both thermally (by laser heating) and due to reactive adsorption of NO at the clean surface. For NO/Si(111)–(7 × 7), through a combination of the techniques of HREELS to identify chemisorbed species, TDS to measure surface coverages, and photon induced desorption (termed PID in this work) to monitor species detached upon irradiation, progress was made in identifying the various contributions to N_2O production.[25]

The temporal profile of photodesorbed N_2O was distinctly non-thermal, rising directly after irradiation and falling off thereafter. The observed time-dependence was suggestive of bimolecular photoreaction to form N_2O [$2NO(ad) + h\nu \rightarrow N_2O(g) + O(ad)$] or photodissociation of NO(ad) to yield N(ad) which combined with further NO(ad) in a Langmuir–Hinshelwood type process to liberate $N_2O(g)$.

Metals

It is to be expected that translationally and internally excited photofragments will readily be deactivated and adsorbed at metal surfaces, leading to radical combination reactions or exchange reactions during subsequent warming of the substrates. This does not, however, constitute photoreaction as we define it, since the absorption of light does not lead *directly* to reaction. Nonetheless it is an important process. If it took place within, for example, a picosecond of light-absorption, it would be properly classified as photoreaction since the initial photofragment direction of motion and impact parameter might affect the reactive collision. If, on the other hand, it took place after (on the order of) a picosecond it would be of interest as a secondary source of chemically similar reaction products to those formed by PRXN with restricted transition state geometry.

The irradiation of ethyl chloride adsorbed on Pt(111) provided an example of secondary reaction following PDIS of the parent molecule.[37,38] No reaction products were detached from the Pt surface during irradiation. Subsequent warming of the substrate in the course of TPD did, however, result in desorption of ethylene (C_2H_4) and ethane (C_2H_6). Although PDIS was thought to be insignificantly quenched, the kinetic energy of the ethyl fragment was sufficiently low that the probability of escaping as a radical from the surface was negligible. The reaction products may have been formed in the course of TPD.

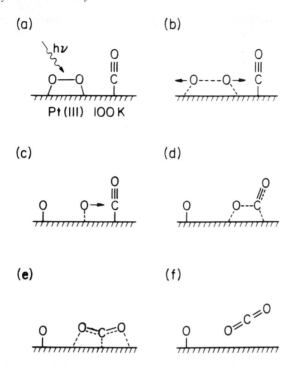

Figure 13 *Schematic of the photoreaction between co-adsorbed O_2 and CO on Pt(111) at 100 K. (a) preferential absorption of radiation by 'peroxy'-bonded O_2, (b) dissociation of O_2, (c) translationally hot O(ad) undergoes a bimolecular encounter with a co-adsorbed CO molecule, (d) and (e) formation and stabilization of CO_2(ad) transition state and (f) desorption of energetic CO_2(g)*
(Reproduced with permission from 'Desorption Induced by Electronic Transitions', DIET IV, Springer–Verlag, Berlin, 1990, p. 48)

The studies of photoreaction at a metal surface which have the closest bearing on PRXN have involved the photoinduced reaction of O_2 with a second, but different, type of adsorbate on the surface of Pt(111). Preliminary work with co-adsorbates O_2(ad) + CO(ad) on Pt(111) gave rise to the evolution of CO_2(g) reaction product.[28] When CO was co-adsorbed with atomic O no reaction was observed, suggesting that 'hot' O atoms were required from PDIS of molecular O_2 for reaction to occur. The following binary PRXN was postulated to follow photolysis of O_2(ad),

$$\vec{O} + CO(ad) \rightarrow CO_2(g) \quad (5)$$

The $\vec{O}$ denotes an O atom which photorecoils a distance on the order of angstroms before encountering an adjacent CO(ad). This $\vec{O}$, in common with the $\vec{H}$ and $\vec{S}$ which took part in PRXN on LiF(001) described above [equations (1)–(3)], may be in transition between the adsorbed and the gaseous states, *i.e.* it may react *en route* from (ad) to (g) as suggested by equation (5).

A schematic of this process is shown in Figure 13. That the primary step was PDIS of O_2 on Pt(111) was supported by the fact that the wavelength dependence of the CO_2 yield followed that for PDIS of O_2. The latter was red-shifted with respect to the gas-phase photodissociation.[97] This red-shift was ascribed to alteration in electronic structure on adsorption.[93] The observed wavelength dependence resembled the absorption spectrum of surface peroxides, suggesting that the incident radiation preferentially dissociated O_2 molecules adsorbed parallel to the surface.[27]

In the above case, O_2/Pt(111), it appeared that the mechanism of PDIS which initiated PRXN involved direct adsorbate electronic excitation. Evidence for a different, indirect, PDIS pathway has been presented for O_2/Ag(110). Irradiation was with radiation linearly polarized in the surface plane or, alternatively, along the surface normal.[43] Since O_2 is bound solely in a 'peroxo' configuration, with the O—O bond lying parallel to the surface,[114] the insensitivity of PDIS to the angle between the electric vector of the light and the molecular transition moment was taken as evidence of a substrate-mediated excitation mechanism, namely CT/PDIS (charge transfer photodissociation).

5 Photodesorption; PDES

Insulators

The term photodesorption, PDES, is used here (see Section 2 above) to denote a surface-mediated process. It frequently results in detached species with translational energies an order of magnitude lower than that of fragments arising from photodissociation, PDIS. In this section we consider PDES from insulator substrates.

The magnitude of the band-gap in insulators precludes a charge transfer mechanism for PDES in the mid-UV; the band-gap for LiF, for example, is about 14.2 eV (equivalent to a wavelength of ~ 87 nm).[115] For some insulators photodesorption is thermal in origin.[62,65] For a material that is largely transparent to UV, such as LiF, this need not be the case. Taking CH_3Br/LiF(001) as an example,[50] PDES was characterized by the TOF detection of intact CH_3Br as a function of detection angle, coverage, laser fluence, and wavelength. The translational energy of the CH_3Br was nonthermal. The Boltzmann distribution with the same peak energy would have had a temperature $T \approx 1100$ K, yet the temperature increase of the substrate was calculated to be in the region of milliKelvin, *i.e.* orders of magnitude lower.

The most probable translational energies of CH_3Br were low (< 0.1 eV) relative to that of the CH_3 fragments (~ 1.7 eV), so that any CH_3^+ species from cracking of CH_3Br in the ionizer of the mass-spectrometer arrived significantly later in time at the detector. The efficiency of PDES was relatively high, up to 0.1 monolayer desorbed per laser pulse. It occurred with comparable efficiency at a longer wavelength where PDIS (photodissociation) did not take place. Irradiating the crystal bulk from the side, as opposed to the surface directly, did not diminish PDES.

A mechanism for PDES that seemed plausible, based on these observations, was that of a photoacoustic disturbance.[50] On this model, impurity centres in LiF would absorb UV radiation and their de-excitation would excite phonons. Impurity centres in LiF and other alkali halide crystals are (Farben or) F-centres which consist of an electron in a halide-ion vacancy.

The absorption maxima of the F-band in LiF occurs around 243 nm[116] with a width sufficiently broad to encompass the experimental wavelengths employed. A further source of impurity absorption over part of this spectral range is likely to be Ti which remains in the Harshaw equipment from the manufacture of scintillator crystals, and which evidences itself in the UV/vis absorption spectrum of LiF samples.[117]

Semiconductors and Metals

The nature of PDES in semiconductor and metal systems can differ from that described above for insulator systems, since the material is no longer transparent to UV, and band-gap excitation is possible. In the majority of adsorbate and conductor systems studied to date PDES appears to be driven by some type of charge transfer mechanism.[21,23,24,44,45,47,65,71,73,87-89,105-107,109,118-123] A number of studies exist for photodesorption of NO; Ho[93] has collected PDES cross-section data for many of these systems involving a variety of semiconductors and metal substrates.

Ho and co-workers have investigated the details of PDES of NO from Si(111)[20,21,25,26] and GaAs(110)[23,24] surfaces using a combination of photon induced desorption and HREELS. Measurements of the wavelength dependence of the NO PDES yield (which approximately followed that of the substrate absorption cross-section), the power dependence of the NO yield at fixed photon wavelength, and the temporal profile of the desorbed NO were most satisfactorily explained by a simple charge-carrier model, originally used to explain the photodesorption of CO_2 from ZnO.[123]

The model, which started by calculating the rate equations for the time-dependent concentration of the four relevant species, NO, NO^-, free electrons, and free holes, predicted a square root dependence of PDES yield on power at fixed wavelength, on wavelength at fixed power, and on reciprocal time. It was assumed in this model that only those holes generated within a distance δ from the surface were effective, and desorption of NO occurred when a hole was captured by NO^-.

A straightforward extension to the basic model[124] to include the high power regime and to allow for systems in which the extent of the various photoprocesses was substantial, was able to fit the experimental data adequately for both NO/GaAs(110) and NO/Si(111)–(7 × 7). It was thought that PDES proceeded through the interaction of photogenerated holes with the $2\pi^*$ antibonding orbital of the NO–Si complex.

In addition to the hot-hole mechanism for PDES from semiconductor substrates put forward for NO/Si(111) and NO/GaAs(110), there was substantial evidence to suggest that PDES of NO from metals was driven by the photogeneration of hot electrons.[119,122,125]

Dynamical studies of PDES from NO/Pt(111)[119,122,125] due to Cavanagh, King and co-workers, have probed the dependence of the translational and internal energy distributions of NO desorbing from the surface following pulsed laser irradiation as a function of incident wavelength. The detached NO was detected using LIF (laser induced fluorescence) to measure the vibrational and rotational state distributions. It was also possible, through variation of the time delay between pulse and probe lasers, to measure the translational energy distribution for selected quantum states of the desorbed NO. The results showed that the mean translational energy of the various internal energy states was independent of laser fluence (ruling out a thermal mechanism) but was sensitive to laser wavelength. The translational energy was 35% lower at 1064 nm than at 532 or 355 nm, and no desorption was observed at 1907 nm. The extent of vibrational excitation was also seen to be wavelength dependent. The $v'' = 1$ species, when produced, had the same translational energy and rotational distribution as $v'' = 0$ species.

Further experiments explored the effect of laser angle of incidence and polarization on PDES and confirmed that the NO yield scaled with the number of photons absorbed by the substrate.[119] The results pointed to a desorption mechanism driven by photogenerated charge carriers produced in the near-surface region of the substrate. Similar PDES features were observed for NO adsorbed on Ni(100)–O,[118] and in the system NO_2/Pd(111)[72] (discussed above with respect to PDIS).

A mechanism which seems to be consistent with the experimental observations, supported by recent theoretical modelling for NO/Pt(111),[120] is one in which hot electrons drive PDES. In this model, optically excited electrons scatter into an unoccupied valence electron resonance of the adsorbate, i.e. a Franck–Condon transition to an excited state ionic potential occurs due to electron capture. The temporary negative ion, which experiences an enhanced attraction towards the substrate, propagates on the excited potential for $\sim 10^{-4}$–10^{-5} s and subsequently decays back to the ground state potential energy surface. Species which project onto continuum states in the region of the repulsive inner wall of the ground state potential, where the total energy is greater than the neutral adsorbate binding energy, desorb as neutrals.

A noteworthy advance in methodology has recently been reported by Chuang and co-workers which concerns the two-dimensional imaging of photodesorbed species from surfaces.[126] The technique is based on the principles employed by Chandler *et al.* in probing the dynamics of photofragmentation in the gas-phase.[127,128] Through the coupling of a microchannel plate, phosphor screen and CCD (charge-coupled device) camera it is possible to measure the angular distributions of detached species, ionized by REMPI, in a single shot experiment, with an important multiplexing advantage. The experimental arrangement could readily yield the angular depend-

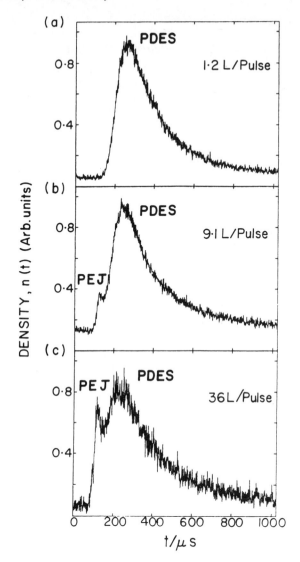

Figure 14 *Time-of-arrival spectra of intact H_2S photodetached from LiF(001) at 222 nm pulsed-laser irradiation, as a function of dosage (under the experimental conditions employed 1 L ~ 1 ML).* (a) *At a dosing of* 1.2 L pulse^{-1} *only the slow (low translational energy) PDES was present.* (b) *At* 9.1 L pulse^{-1} *fast (higher translational energy) species due to PEJ were clearly visible, as a shoulder on the slow PDES, at early times.* (c) *At* 36 L pulse^{-1} *PEJ increased further in comparison to the PDES (primarily due to a diminution in the PDES channel)*
(Reproduced with permission from *J. Chem. Phys.*, 1988, **89**, 1498.)

ence of translational or internal energy in species photoformed at a surface in any of the photoprocesses reviewed in this article.

6 Photoejection; PEJ

At coverages ≳ 1 monolayer a second channel for photoemission of intact molecules has been observed in a number of systems; $H_2S/LiF(001)$,[51] $OCS/LiF(001)$,[57] vinyl chloride/LiF(001)[110] and $HBr/LiF(001)$.[54] This desorption channel has been termed 'photoejection', PEJ. Approximate studies of the wavelength dependence of PEJ performed using a variety of excimer laser wavelengths indicated that PEJ, in contrast to PDES, involved absorption of radiation by the adsorbate. The 'ejected' material could have energies in the region of 0.5 eV, *i.e.* up to 10 times that for PDES. The translational energy distribution was markedly non-Boltzmann (see, for example, Figure 2 of reference 57). The occurrence of PEJ and its dependence on coverage is evident in Figure 14 for the 222 nm irradiation of H_2S adsorbed on LiF(001). At sub-monolayer and monolayer coverages intact H_2S molecules arriving at the ionizer of the MS result from PDES. At increased coverages an increasing proportion of higher energy species were observed due to PEJ.

The observations were consistent with a mechanism in which PEJ occurred from the uppermost layer in a quenching encounter with an electronically-excited molecule in the layer beneath. The process was thought to be comparable to the well studied event of $E \to V, R, T$ (electronic to vibrational, rotational, and translational energy transfer) in gas-phase quenching collisions. The narrow angular distributions of photoejected molecules, *e.g.* $P(\theta) \sim \cos^{18}\theta$ for OCS,[57] was ascribed to 'channelling' of the ejected species through repulsion by adjacent molecules of adsorbate which act as a molecular collimator for the departing molecule.

When a thick layer of HBr ice was irradiated with 193 nm radiation, desorbed monomers and clusters could be detected by angularly resolved TOF–MS.[54] From the high energy of the $(HBr)_n$, $n \leq 4$, it appeared that this was a further example of PEJ.

The clusters were ejected from the ice with distinctive angular distributions. Figure 15 shows polar plots of the angular distributions of HBr monomer, dimer, trimer, and tetramer ejected from HBr ice prepared by a 900 L dose (1L = 1 Langmuir $\equiv 10^{-6}$ torr s). The HBr monomer was found to peak normal to the surface, whereas dimers, trimers, and tetramers exhibited a peak at *ca.* 40° to the normal.

The angular distributions are likely to be related to the structure of a crystalline HBr layer. Inspection of the structure of an orthorhombic HBr lattice indicates that linear collision cascades along the [011] direction would result in a peak directed at about 45° to the normal, as observed. Thus, monomer ejection (the major pathway) could be envisaged as described above for PEJ of OCS, while energy transfer from deeper inside the crystal layer, directed along close-packed directions in the HBr lattice, may well be responsible for the ejection of the higher-order clusters. Adiabatic cooling of the expanding gas as it leaves the surface cannot be responsible for the observed cluster formation, since this mechanism would require that the angular distribution of clusters resemble that of the monomers that consti-

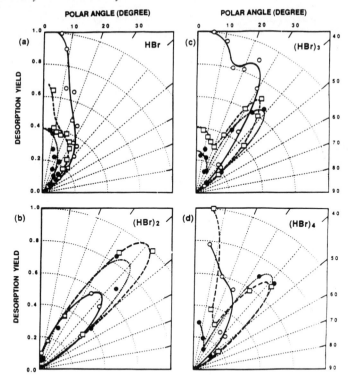

Figure 15 *Angular distributions of* HBr *monomer, dimer, trimer, and tetramer, from the first (solid line), second (dashed line), and third (dotted line) batches of five laser shots, in an experiment using a* 900 L *dose of* HBr *and a laser fluence of* 3 MW cm^{-2}. *The distributions have been normalized to unity at the peak. Clusters ejected at* ~ 40° *to the surface normal are attributed to linear collision cascades along the* [011] *direction in the* HBr *crystalline solid* (Reproduced with permission from *J. Phys. Chem.*, 1988, **92**, 6859.)

tute the major part of the gas-cloud. Once again ordering in the adsorbed state is responsible for profoundly modifying the dynamics of a process as compared with its gas-phase analogue.

Acknowledgements. Work performed at the University of Toronto was funded by the Natural Sciences and Engineering Research Council of Canada, the Venture Research Unit (BP Canadian Holdings, Limited), and the Ontario Laser and Lightwave Research Centre. HR thanks the Science and Engineering Research Council of Britain for the award of a NATO Fellowship held at the University of Toronto (1987–89), and Mrs P. J. Rieley for her assistance with this manuscript.

References

1. *Faraday Discuss. Chem. Soc.*, 1974, **58**.
2. H. Gerischer, *Faraday Discuss. Chem. Soc.*, 1974, **58**, 219.

3. T. B. Grimley, *Faraday Discuss. Chem. Soc.*, 1974, **58**, 7.
4. T. F. Deutsch, D. J. Ehrlich, and R. M. Osgood, Jr., *Appl. Phys. Lett.*, 1979, **35**, 175.
5. D. J. Ehrlich, R. M. Osgood, Jr., and T. F. Deutsch, *Appl. Phys. Lett.*, 1981, **38**, 946.
6. J. Y. Tsao and D. J. Ehrlich, *J. Chem. Phys.*, 1984, **81**, 4620.
7. J. Y. Tsao and D. J. Ehrlich, *J. Cryst. Growth*, 1984, **68**, 176.
8. R. M. Osgood, *Annu. Rev. Phys. Chem.*, 1983, **34**, 77.
9. *Chem. Rev.*, 1989, **89(6)**.
10. T. J. Chuang, *Surf. Sci. Rep.*, 1983, **3**, 1.
11. J. S. Foord and R. B. Jackman, *Chem. Phys. Lett.*, 1984, **112**, 190.
12. G. M. Goncher and C. B. Harris, *J. Chem. Phys.*, 1982, **77**, 3767.
13. G. M. Goncher, C. A. Parsons, and C. B. Harris, *J. Phys. Chem.*, 1984, **88**, 4200.
14. F. A. Houle, *J. Chem. Phys.*, 1983, **79**, 4237.
15. F. A. Houle, *J. Chem. Phys.*, 1984, **80**, 4851.
16. N. Nishi, H. Shinohara, and T. Okuyama, *J. Chem. Phys.*, 1984, **80**, 3898.
17. E. B. D. Bourdon, J. P. Cowin, I. Harrison, J. C. Polanyi, J. Segner, C. D. Stanners, and P. A. Young, *J. Phys. Chem.*, 1984, **88**, 6100.
18. C. E. Bartosch, N. S. Gluck, W. Ho, and Z. Ying, *Phys. Rev. Lett.*, 1986, **57**, 1425.
19. N. S. Gluck, Z. Ying, C. E. Bartosch, and W. Ho, *J. Chem. Phys.*, 1987, **86**, 4957.
20. Z. Ying, C. E. Bartosch, N. S. Gluck, and W. Ho, *J. Vac. Sci. Technol.*, 1987, **A5**, 1608.
21. Z. Ying and W. Ho, *Phys. Rev. Lett.*, 1987, **60**, 57.
22. Z. Ying and W. Ho, *Surf. Sci.*, 1988, **198**, 473.
23. S. K. So, F. J. Kao, and W. Ho, *J. Vac. Sci. Technol.*, 1988, **A6**, 1435.
24. S. K. So and W. Ho, *Appl. Phys. A*, 1988, **47**, 213.
25. Z. C. Ying and W. Ho, *J. Chem. Phys.*, 1989, **91**, 2689.
26. Z. C. Ying and W. Ho, *J. Chem. Phys.*, 1989, **91**, 5050.
27. W. D. Meiher and W. Ho, *J. Chem. Phys.*, 1989, **91**, 2755.
28. W. D. Meiher and W. Ho, *J. Chem. Phys.*, 1990, **92**, 5162.
29. T. A. Germer and W. Ho, *J. Chem. Phys.*, 1990, **93**, 1474.
30. B. Roop, S. A. Costello, C. M. Greenlief, and J. M. White, *Chem. Phys. Lett.*, 1988, **143**, 38.
31. Y. Zhou, W. M. Feng, M. A. Henderson, B. Roop, and J. M. White, *J. Am. Chem. Soc.*, 1988, **110**, 4447.
32. Z.-M. Liu, S. Akhter, B. Roop, and J. M. White, *J. Am. Chem. Soc.*, 1988, **110**, 8708.
33. S. A. Costello, B. Roop, Z.-M. Liu, and J. M. White, *J. Phys. Chem.*, 1988, **92**, 1019.
34. Z.-M. Liu, S. A. Costello, B. Roop, S. R. Coon, S. Akhter, and J. M. White, *J. Phys. Chem.*, 1989, **93**, 7681.
35. X.-Y. Zhu, S. R. Hatch, A. Campion, and J. M. White, *J. Chem. Phys.*, 1989, **91**, 5011.
36. B. Roop, K. G. Lloyd, S. A. Costello, A. Campion, and J. M. White, *J. Chem. Phys.*, 1989, **91**, 5103.
37. K. G. Lloyd, B. Roop, A. Campion, and J. M. White, *Surf. Sci.*, 1989, **214**, 227.
38. K. G. Lloyd, A. Campion, and J. M. White, *Catal. Lett.*, 1989, **2**, 105.

39. B. Roop, Y. Zhou, Z.-M. Liu, M. A. Henderson, K. G. Lloyd, A. Campion, and J. M. White, *J. Vac. Sci. Technol.*, 1989, **A7**, 2121.
40. X.-L. Zhou and J. M. White, *Chem. Phys. Lett.*, 1990, **167**, 205.
41. X.-L. Zhou and J. M. White, *J. Chem. Phys.*, 1990, **92**, 1504.
42. X.-L. Zhou and J. M. White, *J. Phys. Chem.*, 1990, **94**, 2643.
43. S. R. Hatch, X.-Y. Zhu, J. M. White, and A. Campion, *J. Chem. Phys.*, 1990, **92**, 2681.
44. X. Guo, L. Hanley, and J. T. Yates, Jr., *J. Chem. Phys.*, 1989, **90**, 5200.
45. L. Hanley, X. Guo, and J. T. Yates, Jr., *J. Chem. Phys.*, 1989, **91**, 7220.
46. X. Guo, A. Hoffman, and J. T. Yates, Jr., *J. Chem. Phys.*, 1989, **90**, 5787.
47. X. Guo, J. Yoshinobu, and J. T. Yates, Jr., *J. Chem. Phys.*, 1990, **92**, 4320.
48. V. H. Grassian and G. C. Pimentel, *J. Chem. Phys.*, 1988, **88**, 4484.
49. E. B. D. Bourdon, P. Das, I. Harrison, J. C. Polanyi, J. Segner, C. D. Stanners, R. J. Williams, and P. A. Young, *Faraday Discuss. Chem. Soc.*, 1986, **82**, 343.
50. I. Harrison, J. C. Polanyi, and P. A. Young, *J. Chem. Phys.*, 1988, **89**, 1475.
51. I. Harrison, J. C. Polanyi, and P. A. Young, *J. Chem. Phys.*, 1988, **89**, 1498.
52. St. J. Dixon-Warren, I. Harrison, K. Leggett, M. S. Matyjaszczyk, J. C. Polanyi, and P. A. Young, *J. Chem. Phys.*, 1988, **88**, 4092.
53. C.-C. Cho, J. C. Polanyi, and C. D. Stanners, *J. Chem. Phys.*, 1989, **90**, 598.
54. C.-C. Cho, J. C. Polanyi, and C. D. Stanners, *J. Phys. Chem.*, 1989, **92**, 6859.
55. K. Leggett, J. C. Polanyi, and P. A. Young, *J. Chem. Phys.*, 1990, **93**, 3645.
56. St. J. Dixon-Warren, K. Leggett, M. S. Matyjaszczyk, J. C. Polanyi, and P. A. Young, *J. Chem. Phys.*, 1990, **93**, 3659.
57. J. C. Polanyi and P. A. Young, *J. Chem. Phys.*, 1990, **93**, 3673.
58. C.-C. Cho, B. A. Collings, R. E. Hammer, J. C. Polanyi, C. D. Stanners, J. H. Wang, and G.-Q. Xu, *J. Phys. Chem.*, 1989, **93**, 7761.
59. St. J. Dixon-Warren, J. C. Polanyi, C. D. Stanners, and G.-Q. Xu, *J. Phys. Chem.*, 1990, **94**, 5664.
60. K. Domen and T. J. Chuang, *Phys. Rev. Lett.*, 1987, **59**, 1484.
61. T. J. Chuang and K. Domen, *J. Vac. Sci. Technol.*, 1989, **B7**, 1200.
62. K. Domen and T. J. Chuang, *J. Chem. Phys.*, 1989, **90**, 3318.
63. K. Domen and T. J. Chuang, *J. Chem. Phys.*, 1989, **90**, 3332.
64. A. Modl, K. Domen, and T. J. Chuang, *Chem. Phys. Lett.*, 1989, **154**, 187.
65. F. L. Tabares, E. P. Marsh, G. A. Bach, and J. P. Cowin, *J. Chem. Phys.*, 1987, **86**, 738.
66. E. P. Marsh, F. L. Tabares, M. R. Schneider, and J. P. Cowin, *J. Vac. Sci. Technol.*, 1987, **A5**, 519.
67. E. P. Marsh, T. L. Gilton, W. Meier, M. R. Schneider, and J. P. Cowin, *Phys. Rev. Lett.*, 1988, **61**, 2725.
68. E. P. Marsh, M. R. Schneider, T. L. Gilton, F. L. Tabares, W. Meier, and J. P. Cowin, *Phys. Rev. Lett.*, 1988, **60**, 2551.
69. T. L. Gilton, C. P. Dehnbostel, and J. P. Cowin, *J. Chem. Phys.*, 1989, **91**, 1937
70. E. P. Marsh, F. L. Tabares, M. R. Schneider, T. L. Gilton, W. Meier, and J. P. Cowin, *J. Chem. Phys.*, 1990, **92**, 2004.
71. M. Wolf, S. Nettesheim, J. M. White, E. Hasselbrink, and G. Ertl, *J. Chem. Phys.*, 1990, **92**, 1509.
72. E. Hasselbrink, S. Jakubith, S. Nettesheim, M. Wolf, A. Cassuto, and G. Ertl, *J. Chem. Phys.*, 1990, **92**, 3154.
73. J. Kutzner, G. Lindeke, K. H. Welge and D. Feldmann, *J. Chem. Phys.*, 1989, **90**, 548.

74. L. M. Cousins and S. R. Leone, *J. Mater. Res.*, 1988, **3**, 1158.
75. L. M. Cousins, R. J. Levis, and S. R. Leone, *J. Phys. Chem.*, 1989, **93**, 5325.
76. M. Kawasaki, H. Sato, and N. Nishi, *J. Appl. Phys.*, 1989, **65**, 792.
77. E. B. D. Bourdon, C.-C. Cho, P. Das, J. C. Polanyi, C. D. Stanners, and G.-Q. Xu, *J. Chem. Phys.*, in press.
78. J. M. White, in 'Springer Series in Surface Science', Vol. 22, Springer–Verlag, Berlin, 1990, p. 29.
79. 'Desorption Induced by Electronic Transitions', DIET I, ed. N. Tolk, M. Traum, J. Tully, and T. Madey, Springer–Verlag, Berlin, 1983.
80. 'Desorption Induced by Electronic Transitions', DIET II, ed. W. Brenig and D. Menzel, Springer–Verlag, Berlin, 1985.
81. 'Desorption Induced by Electronic Transitions', DIET III, ed. R. H. Stulen and M. L. Knotek, Springer–Verlag, Berlin, 1988.
82. 'Desorption Induced by Electronic Transitions', DIET IV, ed. G. Betz and P. Varga, Springer–Verlag, Berlin, 1990.
83. P. Rowntree, G. Scoles and J. Xu, (personal communication).
84. P. M. Blass, R. C. Jackson, J. C. Polanyi, and H. Weiss, *J. Chem. Phys.*, 1991, **94**, 7003.
85. J. C. Polanyi, R. J. Williams, and S. F. O'Shea, *J. Chem. Phys.*, 1991, **94**, 978.
86. St. J. Dixon-Warren, J. C. Polanyi, H. Rieley, and J. G. Shapter, in preparation.
87. F. G. Celii, P. M. Whitmore, and K. C. Janda, *J. Phys. Chem.*, 1988, **92**, 1604.
88. J. R. Swanson, C. M. Friend, and Y. J. Chabal, *J. Chem. Phys.*, 1987, **87**, 5028.
89. J. R. Swanson, C. M. Friend, and Y. J. Chabal, *J. Chem. Phys.*, 1988, **89**, 2593.
90. J. R. Creighton, *J. Appl. Phys.*, 1986, **59**, 410.
91. N. S. Gluck, G. J. Wolga, C. E. Bartosch, W. Ho, and Z. Ying, *J. Appl. Phys.*, 1987, **61**, 998.
92. J. R. Swanson, F. A. Flitsch, and C. M. Friend, *Surf. Sci.*, 1990, **226**, 147.
93. W. Ho, in 'Desorption Induced by Electronic Transitions, DIET IV', ed. G. Betz and P. Varga, Springer–Verlag, Berlin, 1990, p. 48; D. V. Chakarov and W. Ho, *J. Chem. Phys.* 1991, **94**, 4075.
94. Z. C. Ying and W. Ho, *Phys. Rev. Let.*, 1990, **65**, 741.
95. M. R. Schneider, C. P. Dehnbostel, T. L. Gilton, and J. P. Cowin, submitted for publication.
96. L. G. Christophorou, D. L. McCorkle, and A. A. Christodoulides, in 'Electron Molecule Interactions and Their Applications', ed. L. G. Christophorou, Academic, New York, 1984.
97. H. Okabe, 'Photochemistry of Small Molecules', Wiley-Interscience, New York, 1978.
98. St. J. Dixon-Warren, E. T. Jensen, J. C. Polanyi, G.-Q. Xu, G. H. Yang, and H. C. Yeng, *Faraday Discuss. Chem. Soc.*, in press.
99. S. K. Jo and J. M. White, *J. Chem. Phys.*, 1990, **94**, 6852.
100. B. Roop, S. A. Costello, Z.-M. Liu, and J. M. White, in 'Springer Series in Surface Science', Vol. 14, ed. F. W. de. Witte, Springer–Verlag, Berlin, 1988, p. 343.
101. J. M. White, *Proc. SPIE-Int. Soc. Opt. Eng.*, 1989, **1056** (Photochem. Thin Films), 129.
102. X.-L. Zhou and J. M. White, *Surf. Sci.*, 1991, **241**, 244.

103. X.-L. Zhou and J. M. White, *Surf. Sci.*, 1991, **241**, 259.
104. X.-L. Zhou and J. M. White, *Surf. Sci.*, 1991, **241**, 270.
105. M. Wolf, E. Hasselbrink, J. M. White, and G. Ertl, *J. Chem. Phys.*, 1990, **93**, 5327.
106. J. Yoshinobu, X. Guo, and J. T. Yates, Jr., *Chem. Phys. Lett.*, 1990, **169**, 209.
107. J. M. White, S. Hatch, X.-Y. Zhu, and A. Campion, *J. Chem. Phys.*, 1989, **91**, 5011.
108. X.-Y. Zhu and J. M. White, *J. Chem. Phys.*, 1991, **94**, 1555.
109. M. Wolf, S. Nettesheim, J. M. White, E. Hasselbrink, and G. Ertl, *J. Chem. Phys.*, 1991, **94**, 4609.
110. St. J. Dixon-Warren, M. S. Matyjaszczyk, J. C. Polanyi, H. Rieley, and J. G. Shapter, *J. Phys. Chem.*, 1991, **95**, 1333.
111. J. C. Polanyi and R. J. Williams, *J. Chem. Phys.*, 1988, **88**, 3363.
112. V. J. Barclay, D. Jack, J. C. Polanyi, and Y. Zieri, in preparation.
113. F. A. Houle, *Chem. Phys. Lett.*, 1983, **95**, 5.
114. D. A. Outka, J. Stohr, W. Jark, P. Stevens, J. Soloman, and R. J. Madix, *Phys. Rev. B*, 1987, **35**, 4119.
115. F. Luty, in 'Physics of Color Centers', ed. W. B. Fowler, Academic, New York, 1968.
116. H. Rabin and M. Reich, *Phys. Rev. A*, 1964, **101**, 135.
117. H. Weiss, unpublished work, University of Toronto, 1989.
118. F. Budde, A. V. Hamza, P. M. Ferm, G. Ertl, D. Weide, P. Andresen, and H.-J. Freund, *Phys. Rev. Lett.*, 1988, **60**, 1518.
119. S. A. Buntin, L. J. Richter, D. S. King, and R. R. Cavanagh, *J. Chem. Phys.*, 1989, **91**, 6429.
120. J. W. Gadzuk, S. A. Buntin, L. J. Richter, R. R. Cavanagh, and D. S. King, *Surf. Sci.*, 1990, **235**, 317.
121. D. S. King and R. R. Cavanagh, *Adv. Chem. Phys.*, 1989, **76**, 45.
122. L. J. Richter, S. A. Buntin, R. R. Cavanagh, and D. S. King, *J. Chem. Phys.*, 1988, **89**, 5344.
123. Y. Shapira, R. B. McQuistran, and D. Lichtman, *Phys. Rev. B*, 1977, **15**, 2163.
124. W. Ho, *Proc. SPIE-Int. Soc. Opt. Eng.* 1989, **1056** (Photochem. Thin Films), 157.
125. S. A. Buntin, L. J. Richter, R. R. Cavanagh, and D. S. King, *Phys. Rev. Lett.*, 1988, **61**, 1321.
126. R. Schwartzwald, A. Modl, and T. J. Chuang, *Surf. Sci.*, 1991, **242**, 437.
127. D. W. Chandler, J. W. Thoman Jr., M. H. M. Janssen, and D. H. Parker, *Chem. Phys. Lett.*, 1989, **156**, 151.
128. D. H. Parker, Z. W. Wang, M. H. M. Janssen, and D. W. Chandler, *J. Chem. Phys.*, 1989, **90**, 60.

Subject Index

Accommodation, 192, 196
accommodation coefficient, 1, 196
adsorbate photochemistry, 329
alignment, 336
alkane activation, 205
angular distribution, 18, 30, 81, 143, 151, 279, 337, 348, 359
anti-correlation between rotational and phonon excitation, 102
Arrhenius expression, 58, 223
Auger process, 263, 271

Baule formula, 10, 14, 16
binary collision models, 16
Born approximation,
  distorted wave, 42

Chemisorption,
  barrier, 51, 117, 118, 142, 145
  direct, 49, 68, 172
  dissociative, 47, 139, 172
    influence of initial rotational excitation, 82
    influence of initial vibrational excitation, 82
  precursor-mediated, 49, 58, 67, 141, 172, 187, 205

Debye–Waller factor, 38, 40, 44, 102
desorption, 49, 144, 204, 220, 227
  Auger stimulated, 271
  electron stimulated (see electron stimulated desorption)
  isothermal, 222
  light induced (see photodesorption)
  temperature jump, 222
  temperature programmed, 222
    high resolution, 250
desorption energy,
  differential, 230, 232
detailed balance, 149, 151, 164
diffraction, 48
diffusion, 236, 240

Elastic scattering, 35, 38
  non-specular, 48
  specular, 48
electron stimulated desorption, 257
  charge transfer mechanism, 274
  dissociative attachment mechanisms, 269
  experimental observations, 276
  ion angular distributions (ESDIAD), 282
  models of:
    Antoniewicz model, 260
    Knotek–Feibelman model, 262
    Menzel–Gomer–Redhead (MGR) model, 259
    quantum mechanical, 264
  products of:
    electronically excited neutrals, 279

Subject Index

ions, 276
Eley–Rideal reaction mechanism, 50, 179
  comparison with Langmuir–Hinshelwood reaction mechanism, 184
energy scaling, 198, 213
  normal, 50, 78, 147
  total, 50, 80
energy transfer,
  translation–rotation, 22, 101
  translation–vibration, 107, 112

Force constant, 7
forced oscillator model, 35, 42
Fourier transform, 36
  fast, 60, 96
Franck–Condon transition, 260, 356

Gas–surface interactions – specific systems,
  Ag/Mo(110), 231
  Ar/Pt(111), 19, 194
  Ar/Ru(001), 250
  Au/Mo(110), 225
  $CH_4(CD_4)$/W, 131
  C_2H_6/Ir(110) – (1 × 2), 205
  cyclo–C_5H_8/Ag(221), 300
  CO/Cr(110), 296
  CO/Ni(110), 294
  CO/Pt, 26, 32
  CO/Pt(111), 285
  CO/Se/Pt(111), 288
  CO/Pt(112), 289
  CO/Ru(001), 284
  F/Si(100), 302
  H/Ni(111) in the presence of alkali metal co-adsorbates, 303
  H_2/Cu, 137
  $H_2(D_2)$/Cu(100), 42, 100, 122, 147
  H_2/Cu(110), 145, 147
  H_2/Cu(310), 147
  H_2/Ni(100), 68
  H_2O/Ni(110), 304
  Na^+/Cu(110), 20
  N_2/Ni(430), 235
  N_2/Re, 130
  N_2/W(110), 78
  NH_3/Ni(110), 312
  NH_3/Si(100), 309
  NO/Ag(111), 23, 27, 34, 101, 112
  NO/Pt(112), 308
  O_2/Ag(110), 316
  PF_3/Ni(111), 314
  pyridine/Ir(111), 320
  Xe/Ni(111), 241
  Xe/Pt(111), 18, 202

Harmonic lattice,
  one-dimensional, 6, 35
  three-dimensional, 37
heat of adsorption,
  isosteric, 229
heterogeneous catalysis, 47

Image potential, 2
impact parameter, 3
inelastic scattering, 1, 49, 100
infrared catastrophe, 44
islands, 239, 243
isotope effect, 131, 149, 157

Jellium, 56, 142

Kinetic lattice gas model, 245
Kirkwood approximation, 247

Langevin equation
  generalized, 62
  local, 66
Langmuir–Hinshelwood reaction mechanism, 50, 147, 175, 333
  comparison with Eley–Rideal reaction mechanism, 184
Langmuir kinetics, 246
laser micro-fabrication, 330

Markovian master equation, 245
modes of vibration,
  bulk, 8
  normal, 7
  surface, 8

Onsager coefficients, 238
orientation, 34, 103, 188, 300, 337

Pauli repulsion, 23, 142, 261
permeation, 144
phonon
  band, 35

density of states, 37, 64
  surface, 42
photodesorption
  CH$_3$Br from LiF(001), 354
  NO from Pt(111), 356
  NO from Si(111), 355
    charge carrier model, 355
photodissociation, 334
  charge transfer mechanism, 332, 342
  CH$_3$Br on LiF(001), 335
  CH$_3$Cl on GaAs(100), 342
  CH$_3$Cl on Ni(111), 342
  HBr on LiF(001), 337
  HCl on Ag(111), 342
  Mo(CO)$_6$ on Si(111), 339
  polarization dependence, 345
photoejection, 358
  (HBr)$_n$ clusters from HBr ice, 359
photoreaction,
  surface aligned, 346
  HBr on LiF(001), 348
  H$_2$S on LiF(001), 346
  NO on Si(111) – (7×7), 352
  O$_2$ + CO on Pt(111), 353
  OCS on LiF(001), 347
physisorption, 23
  kinetics, 249
polarization, 23
potential energy functions,
  H$_2$/Cu one-dimensional, 146
  Morse, 9, 56, 99, 117
    modified, 25
  van der Waal's, 2, 9, 25, 51, 100
potential energy surfaces, 23, 50, 91
  adiabatic, 92
  diabatic, 92
  H$_2$/Cu, 114, 160

H$_2$/Mg(100), 51
H$_2$/Ni(100), 68, 73
LEPS, 55
N$_2$/Fe, 129
N$_2$/W(110), 80
NO/Ag(111), 23, 103, 112
probability density distribution, 266

Quasi-chemical approximation, 230, 244

Redhead formula, 224

Schrödinger equation,
  time-dependent, 60, 96
spherical harmonics, 99
sticking probability, 42, 149, 152, 240
surface recoil effects, 114

Transition state theory, 175, 237
  canonical, 67
  canonical variational, 67
trapping, 14, 42, 100, 192
  resonant, 43
trapping-desorption, 49, 194
trapping threshold, 14
tunnelling, 131, 344
two-hole states, 271

Vibrational adiabaticity, 122
vibrationally enhanced dissociation, 156

Wavefront, 13
wavepacket, 61, 104
Wigner-Polanyi equation, 208, 223